Gas-Expanded Liquids and Near-Critical Media

ACS SYMPOSIUM SERIES **1006**

Gas-Expanded Liquids and Near-Critical Media

Green Chemistry and Engineering

Keith W. Hutchenson, Editor
DuPont Central Research and Development

Aaron M. Scurto, Editor
The University of Kansas

Bala Subramaniam, Editor
The University of Kansas

**Sponsored by the
Division of Industrial and Engineering Chemistry, Inc.**

American Chemical Society, Washington, DC

The paper used in this publication meets the minimum requirements of American National Standard for Information Sciences—Permanence of Paper for Printed Library Materials, ANSI Z39.48–1984.

ISBN: 978–0–8412–6971–2

Copyright © 2009 American Chemical Society

Distributed by Oxford University Press

PRINTED IN THE UNITED STATES OF AMERICA

Foreword

The ACS Symposium Series was first published in 1974 to provide a mechanism for publishing symposia quickly in book form. The purpose of the series is to publish timely, comprehensive books developed from ACS sponsored symposia based on current scientific research. Occasionally, books are developed from symposia sponsored by other organizations when the topic is of keen interest to the chemistry audience.

Before agreeing to publish a book, the proposed table of contents is reviewed for appropriate and comprehensive coverage and for interest to the audience. Some papers may be excluded to better focus the book; others may be added to provide comprehensiveness. When appropriate, overview or introductory chapters are added. Drafts of chapters are peer-reviewed prior to final acceptance or rejection, and manuscripts are prepared in camera-ready format.

As a rule, only original research papers and original review papers are included in the volumes. Verbatim reproductions of previously published papers are not accepted.

ACS Books Department

Contents

Preface xi

Introduction and Overview

1. Gas-Expanded Liquids: Fundamentals and Applications 3
   Aaron M. Scurto, Keith Hutchenson, and Bala Subramaniam

Thermodynamics and Transport Properties: Experiment and Modeling

2. Phase Equilibrium, Structure, and Transport Properties
   of Carbon-Dioxide Expanded Liquids: A Molecular
   Simulation Study 41
   Brian B. Laird, Yao A. Houndonougbo, and Krzysztof Kuczera

3. On the Molecular-Based Modeling of Dilute Ternary Systems
   in Compressible Media: Formal Results and Thermodynamic
   Pitfalls 66
   Ariel A. Chialvo, Sebastian Chialvo, and J. Michael Simonson

4. Viewing the Cybotactic Structure of Gas-Expanded Liquids 81
   John L. Gohres, Rigoberto Hernandez, Charles L. Liotta,
   and Charles A. Eckert

5. Solvatochromism and Solvation Dynamics in CO_2-Expanded
   Liquids 95
   Chet Swalina, Sergei Arzhantzev, Hongping Li, and
   Mark Maroncelli

6. Phase Behavior and Equilibria of Ionic Liquids and
   Refrigerants: 1-Ethyl-3-methyl-imidazolium
   Bis(trifluoromethylsulfonyl)imide ([EMIm][Tf$_2$N]) and R-134a 112
   Wei Ren, Aaron M. Scurto, Mark B. Shiflett, and
   Akimichi Yokozeki

Reactions

7. **In Situ Alkylcarbonic Acid Catalysts Formed in CO$_2$-Expanded Alcohols** 131
   Jason P. Hallett, Charles A. Eckert, and Charles L. Liotta

8. **Catalytic Oxidation Reactions in Carbon Dioxide-Expanded Liquids Using the Green Oxidants Oxygen and Hydrogen Peroxide** 145
   Daryle H. Busch and Bala Subramaniam

9. **Hydrogenation of CO$_2$-Expanded Liquid Terpenes: Phase Equilibrium-Controlled Kinetics** 191
   Ewa Bogel-Łukasik, Ana Serbanovic, Rafal Bogel-Łukasik,
   Anna Banet-Osuna, Vesna Najdanovic-Visak, and
   Manuel Nunes da Ponte

10. **Hydroformylation in CO$_2$-Expanded Media** 202
    Ruihu Wang, Hu Cai, Hong Jin, ZhuanZhuan Xie,
    Bala Subramaniam, and Jon A. Tunge

11. **Hydrogenation in Biphasic Ionic Liquid–Carbon Dioxide Systems** 218
    Azita Ahosseini, Wei Ren, and Aaron M. Scurto

12. **Sulfur Trioxide Containing Caprolactamium Hydrogen Sulfate: An Expanded Ionic Liquid for Large-Scale Production of ε-Caprolactam** 235
    I. T. Horváth, V. Fábos, D. Lantos, A. Bodor, A.-M. Bálint,
    L. T. Mika, O. E. Sielcken, and A. D. Cuiper

Materials Processing

13. **Emulsion-Templated Porous Materials Using Concentrated Carbon Dioxide-in-Water Emulsions and Inexpensive Hydrocarbon Surfactants** 243
    Colin D. Wood, Bien Tan, Jun-Young Lee, and Andrew I. Cooper

14. **Fluoropolymer Synthesis in Carbon Dioxide-Expanded Liquids: A Practical Approach to Avoid the Use of Perfluorooctanoic Acid** 259
    Libin Du, Joseph M. DeSimone, and George W. Roberts

15. **Green Methods for Processing and Utilizing Metal Complexes** 274
    Joseph G. Nguyen, Chad A. Johnson, Sarika Sharma,
    Bala Subramaniam, and A. S. Borovik

16. **Application of Gas Expanded Liquids for Nanoparticle Processing: Experiment and Theory** 290
    Christopher L. Kitchens, Christopher B. Roberts, Juncheng Liu,
    W. Robert Ashurst, Madhu Anand, Gregory Von White II,
    Kendall M. Hurst, and Steven R. Saunders

17. **Development of a Novel Precipitation Technique for the Production of Highly Respirable Powders: The Atomized Rapid Injection for Solvent Extraction Process** 309
    Neil R. Foster and Roderick Sih

Indexes

Author Index 351

Subject Index 353

Preface

Gas-expanded liquids (GXLs) and near-critical fluids are tunable solvent media for conducting chemical reactions, separations, and materials processing. In recent years, there has been an expanding body of fundamental and applied research in the use of these media for chemical processing, particularly for applications to sustainable processes based on the principles of green chemistry and engineering.

This symposium series volume was developed from a selection of papers presented in a three-session symposium entitled *Green Chemistry and Engineering with Gas-Expanded Liquids and Near-Critical Media* at the 234th American Chemical Society (ACS) National Meeting that was held from August 19–23, 2007, in Boston, Massachusetts. We believe this to be the first compilation of research in this general area. The chapters were selected to represent the breadth of applications under investigation in this technical field and to reflect the diversity of papers presented at the symposium.

This volume spans research and development activities ranging from theoretical and experimental investigations of the underlying fundamental science to developmental aspects of the technology for commercial applications. The first chapter provides an introduction and overview of gas-expanded liquids including discussion on applications such as chemical reactions, particle formation, and materials processing. This chapter is followed by chapters that are organized into three general topical areas: thermodynamics and transport properties (including both experimental and modeling studies), reactions, and materials processing. The chapters in these sections include both new developments in the field as well as reviews of specific topical areas. We hope that this compilation will be a useful reference source for those interested in learning about gas-expanded liquids and potential applications.

We thank the authors and speakers for their valuable contributions in presenting papers at the symposium and in preparing chapters for this volume. We thank the reviewers for their careful evaluations and valuable suggestions for improving the quality of these chapters. We are grateful to Dr. Philip Jessop, Chair, Green Chemistry and Engineering Subdivision, ACS Division of Industrial and Engineering Chemistry for the invitation to organize the aforementioned symposium at the 2007 Boston ACS meeting. Finally, we thank Mr. James Busse of the University of Kansas for his invaluable help with the cover design.

Keith W. Hutchenson
DuPont Central Research and Development
Experimental Station
P.O. Box 80304
Wilmington, DE 19880–0304
(302) 695–1389 (telephone)
keith.w.hutchenson@usa.dupont.com (email)

Aaron M. Scurto
Department of Chemical and Petroleum Engineering
Center for Environmentally Beneficial Catalysis
The University of Kansas
4132 Learned Hall
1530 West 15th Street
Lawrence, KS 66045–7609
(785) 864–4947 (telephone)
ascurto@ku.edu (email)

Bala Subramaniam
Department of Chemical and Petroleum Engineering
Center for Environmentally Beneficial Catalysis
The University of Kansas
1501 Wakarusa Drive, Building A, Suite 110
Lawrence, KS 66047–1803
(785) 864–2903 (telephone)
bsubramaniam@ku.edu (email)

Gas-Expanded Liquids and Near-Critical Media

Introduction and Overview

Chapter 1

Gas-Expanded Liquids: Fundamentals and Applications

Aaron M. Scurto[1], Keith Hutchenson[2], and Bala Subramaniam[1,*]

[1]Department of Chemistry and Petroleum Engineering and NSF–ERC Center for Environmentally Beneficial Catalysis, University of Kansas, Lawrence, KS 66045 (fax: +1 785–864–4967; email: ascurto@ku.edu and bsubramaniam@ku.edu)
[2]DuPont Central Research and Development, Experimental Station, Wilmington, DE 19880–0304
(email: Keith.W.Hutchenson@usa.dupont.com)

The physical and thermodynamic properties within the continuum of gas-expanded liquids (GXLs), CO_2-expanded liquids in particular, are reviewed along with the many applications of GXLs in chemical reactions, particle formation and materials processing. The potential and outlook for practically viable applications of CXLs are also discussed.

Introduction

In recent years, alternative media for performing chemical reactions have been the subject of many investigations (1,2,3,4). Among them, supercritical CO_2 ($scCO_2$) (5,6,7,8,9,10), water (11,12), CO_2-expanded liquids (CXLs) (13), ionic liquids (ILs) (14, 15) and switchable solvents (16) have received much attention as benign media. Supercritical CO_2 as a benign medium satisfies several green chemistry and engineering principles such as pollution prevention, lower toxicity and use of an "abundantly available" resource with no increase in environmental burden. In many cases however, $scCO_2$-based reactions are limited by inadequate solubilities of reagents or catalysts. Further, the use of CO_2 results in relatively low reaction rates for reactions that have polar transition states and requires high process pressures, typically exceeding 100 bar. These drawbacks increase reactor volumes and energy consumption, partly nullifying the environmental advantages. In contrast, conventional organic solvents are chosen with dielectric properties that help solubilize the reagents and improve the rate of the desired reaction. However, the drawbacks with conventional organic solvents are toxicity and vapor emissions that may also form explosive mixtures with air.

In recent years, gas-expanded liquids (GXLs) have emerged as promising media for performing separations, extractions, reactions, and other applications. A GXL is a mixed solvent composed of a compressible gas (such as CO_2 or ethane) dissolved in an organic solvent. CO_2-expanded liquids (CXLs) are the most commonly used class of GXLs, representing a continuum of liquid media ranging from the neat organic solvent to $scCO_2$. CXLs have been exploited in a variety of applications including separations, fine particle precipitation, polymer processing, and as reaction media for catalytic reactions. The chapters that follow present several examples of applications in these areas.

A CXL provides the ability to combine the beneficial properties of dense CO_2 (improved permanent gas solubilities and mass transfer properties) and organic solvents (enhanced solubility of solid and liquid solutes including homogeneous catalysts) in an optimal manner for a given application. In some cases, the gas solubility enhancement in CXLs is up to two orders of magnitude relative to organic solvents at ambient conditions. In CXLs, a commensurate enhancement is seen in reaction rates limited by gas solubility limitations in conventional solvents. Process advantages of CXLs include enhanced safety due to fire suppression capability and mild process pressures (tens of bars) compared to the use of $scCO_2$. Environmental advantages accrue from replacement of organic solvents with CO_2. It should be noted that in the case of light gas conversions (such as ethylene or propylene epoxidation) in liquid phase media, the availability of the light gas is enhanced by expanding the liquid phase with the light gas substrate (17). It should also be noted that CXLs do not completely eliminate the organic solvent, unless CO_2-expanded neat reactions are employed. Despite the lower pressures of GXLs/CXLs than typical supercritical systems, the compression of gases (especially CO_2) still requires energy that should be factored in any life-cycle analysis for proposed "green"/ sustainable processes.

A detailed review of gas-expanded liquids has appeared in 2007 (13). This overview includes developments since that review and places emphasis on the subjects covered in this symposium book. The specific subjects are highlighted in the sections that follow.

GXL Classification

Jessop and Subramaniam (13) classified CXLs based on the ability of the liquids to dissolve CO_2. Class I liquids such as water have insufficient ability to dissolve CO_2 (Table I) and therefore do not expand significantly. Class II liquids such as methanol, hexane, and most other traditional organic solvents, dissolve large amounts of CO_2, expand appreciably, and undergo significant changes in physical properties. The volumetric expansion of class II liquids is strongly dependent on the mole fraction of CO_2 in the liquid phase (18). In contrast, Class III liquids, such as ionic liquids, liquid polymers, and crude oil, dissolve relatively smaller amounts of CO_2 and hence expand only moderately. Consequently, some properties such as viscosity change significantly while others, such as polarity, do not.

It is worth noting that when expansion is plotted as a function of CO_2 mass fraction in the liquid phase (Figure 1), the ionic liquids and liquid polymers expand as much as class II solvents, at equal CO_2 content.

Gases such as ethane, fluoroform and other compressible gases are also capable of expanding liquids. Noncompressible gases (i.e., those having critical temperatures far below the temperature of the experiment) are generally incapable of expanding solvents due to low solubility.

Physical-Chemical Properties of Gas-Expanded Liquids

Phase Equilibrium and Volume Expansion

The underlying principle of GXL/CXL technology is expansion of an organic liquid phase by the dissolution of the compressible gas, most commonly CO_2. At certain temperatures and pressures of the system, equilibrium of the CO_2 and the liquid will exist, whereby both the liquid component and the CO_2 will partition between the two phases: some of the liquid will volatilize or solubilize into the vapor/fluid phase; and the CO_2 will dissolve into the liquid phase. Figure 2 illustrates the vapor-liquid equilibrium (VLE) envelope of a typical GXL system: methanol and CO_2 at 40°C from the data of Yoon *et al.* (19). The bubble point curve represents the solubility of CO_2 in the liquid phase; this is often the most pertinent data for GXL systems. As shown, the solubility of CO_2 in methanol is large over the pressure range shown. The solubility increase with pressure is nearly linear until approximately 30 bar, indicating the possibility to use Henry's Law constants to correlate or predict the solubility. After this region, however, the increase is non-linear and terminates at the mixture critical point (see below). The dew point curve is the solubility of the liquid component in the CO_2-rich phase. Methanol has a low solubility in the vapor phase (<6% mole methanol) except at very low pressures approaching the methanol vapor pressure and above the mixture critical point. Also of note is the solubility minimum of methanol in CO_2 at approximately 50 bar, above which the solubility of methanol in CO_2 increases with pressure. Here, the CO_2 phase

Table I. A comparison of different classes of solvents and their expansion behavior at 40 °C under CO₂ pressure

Class	Solvent	P, bar	Volumetric expansion, %	wt% CO_2	mol% CO_2
I	H_2O	70	na	4.8	2.0
II	MeCN	69	387	83	82
	1,4-dioxane	69	954	79	89
	DMF	69	281	52	65
III	[bmim]BF₄	70	17	15	47
	PEG-400	80	25	16	63
	PPG-2700[a]	60	25	12	89

[a] 35°C

SOURCE: Adapted from reference 13. Copyright 2007 American Chemical Society.

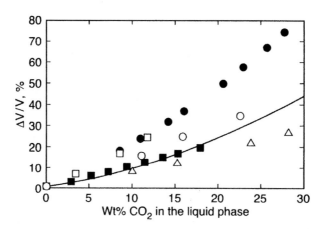

Figure 1. Expansion of liquids at 40 °C, as a function of the mass fraction of CO_2 dissolved in MeCN (△), ethyl acetate (●), [BMIm][BF₄] (filled squares), crude oil (line), PPG (□), and PEG (○). All data for 40 °C except PPG at 35 °C and crude oil at 43 °C. (Reproduced from reference 13. Copyright 2007 American Chemical Society.)

becomes denser at these pressures, which increases the solvation and solubility of methanol. This phenomenon indicates that a small composition region exists where, at high pressures, the mixture is one phase, but with decreasing pressure partially condenses into a liquid and vapor phase, followed by a single vapor phase at low pressures; this phenomenon is called retrograde condensation. Compositions below the bubble point indicate a single liquid phase, and concentrations greater than the dew point indicate a single vapor phase.

The mixture critical point is of special interest to GXLs and represents the highest pressure and composition at a given pressure in which both a liquid and vapor/fluid phase can coexist. In other words, the mixture critical point is the highest point that one can operate in a GXL before changing to the supercritical fluid state. Figure 2 illustrates that the mixture critical point for CO_2 and methanol is the termination of both the bubble and dew point curves: above the mixture critical point, the two components are infinitely miscible (critical). The pressure and composition of the mixture critical point changes with temperature, often to higher pressures and more concentrated in the liquid component. Table II illustrates the mixture critical points of several common organic liquids with CO_2 at different temperatures. The mixture critical point is different than the pure gas critical point; for instance, a pure gas above its critical point (supercritical fluid) can form a vapor phase with a second liquid component. Only beyond the mixture critical point will both components technically reside in a supercritical state. With the addition of a third component (e.g., liquid reactant, product, etc.), the mixture critical point may shift to higher or lower pressures and compositions. Thus, the critical point for the multi-component GXL systems should be measured to determine the exact limits of operation.

As the CO_2 dissolves into the liquid component, the volume of the liquid phase expands. The expansion for methanol and CO_2 from Subramaniam and coworkers (20) is shown in Figure 3a. The fractional expansion above the volume at ambient pressure increases with increasing pressure. As the liquid approaches the mixture critical point, the volume expansion significantly increases and technically becomes infinite at the mixture critical point of 82.4 bar. While the volume increases, the density for most organic solvents actually increases (decrease in molar volume) due to the higher increase in the number of moles/mass of CO_2 in the liquid phase as shown in Figure 3b from the data of Sih *et al.* (21).

Behavior of Ionic Liquids, Water and Polymers with CO_2

The vapor-liquid equilibrium and volume expansion properties of many different organic solvents are often quite similar. However, several other types of solvents can have markedly different behavior. For instance, the VLE of a typical ionic liquid (IL), 1-hexyl-3-methyl-imidazolium bis(trifluoromethyl-

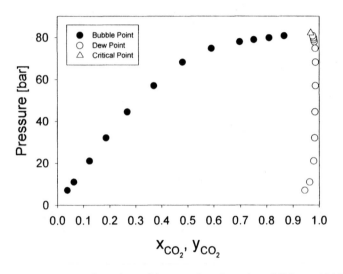

Figure 2. Vapor-liquid equilibrium of methanol and CO_2 at 40 °C from the data of Yoon et al. (19).

Table II. Mixture critical points of some common organic liquids with CO_2.

| Solvent | Mixture Critical Point | | | |
	T [°C]	P [bar]	x_{CO2}	Ref.
Methanol	40	82.1	0.968	(19)
	50	111.4	-	(22)
Propan-2-ol	35.5	77.1	0.986	(23)
	48.9	90.9	0.943	(23)
Acetone	38.7	80.4	0.977	(24)
	59.1	95.8	0.930	(24)
Acetonitrile	50	93.0	-	(25)
	60	104	-	(25)

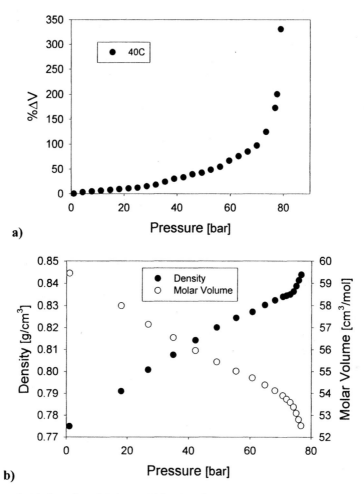

*Figure 3. Methanol and CO_2 at 40 °C: a) Volume expansion with pressure (20);
b) Liquid density and molar volume with pressure from the data of Sih et al. (21).*

sulfonyl)imide ([HMIm][Tf$_2$N]) and CO$_2$ are illustrated in Figure 4 from Ahosseini *et al.* (Chapter 11). Here, the bubble point curve is similar to organic solvents at lower pressures, but, at higher pressures, it abruptly becomes steeper indicating only a marginal increase in solubility with pressure. Importantly, the VLE behavior of ILs and CO$_2$ does not have a known dew point. The solubility of the IL in CO$_2$ even at higher pressures is immeasurably small. Thus, for any biphasic system with ILs and CO$_2$, the CO$_2$ phase will be free from any contamination of the IL. The mixture critical point for IL and CO$_2$ systems is not seen even at high pressures. Brennecke and coworkers (26) have demonstrated that a mixture critical point does not exist for 1-butyl-3-methyl-imidazolium hexafluorophosphate and CO$_2$ even to 3100 bar. In addition, the volume expansion of ionic liquids is much less than with organic solvents below their mixture critical points. For example, the volume expansion of [HMIm][Tf$_2$N] with CO$_2$ at 70°C and 120 bar is only 25% despite the fact that the CO$_2$ solubility at these conditions is approximately 70 mol%; see Chapter 11 for further discussion. Polymers often have a similar behavior to ILs in both solubility and volume expansion. Water has a low solubility for CO$_2$ compared to organic solvents and its volume expansion is minimal.

Transport Properties

The presence of CO$_2$ in the liquid phase can significantly affect the transport properties. CO$_2$ in the vapor, liquid and supercritical state has lower viscosity

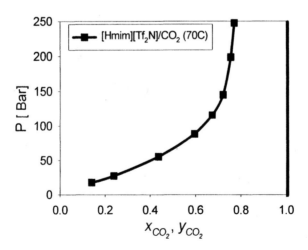

Figure 4. Vapor-liquid equilibrium data of [HMIm][Tf$_2$N] and CO$_2$ at 70 °C from the data of Ahosseini et al. (Chapter 11) . Line is smoothed data.

and higher diffusivity than most organic liquids. As more CO_2 dissolves in the liquid state, the viscosity decreases accordingly. Figure 5 illustrates the viscosity of methanol with CO_2 pressure at 40°C from the data of Sih *et al.* (21). The viscosity decreases nearly 80% from pure methanol to CXL methanol at 77 bar. This viscosity reduction is especially striking for ionic liquids, which often have higher viscosity than organic solvents (27).

The Stokes-Einstein relationship establishes an inverse proportionality between diffusivity and viscosity. As the viscosity decreases, the diffusivity correspondingly increases. The diffusivity of various solutes in CO_2-expanded methanol has been measured by Eckert and coworkers (28). The diffusivity of benzene in methanol and CO_2 mixtures at 40°C and 150 bar is shown in Figure 6. The diffusivity of benzene increases by over 200% on replacing pure methanol with 75% CO_2. These enhanced transport properties allow more efficient processes such as reactions and particle formation, as will be discussed below.

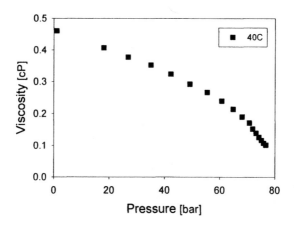

Figure 5. Viscosity of methanol and CO_2 with pressure at 40 °C from the data of Sih et al. (21).

Polarity

Many different measures of polarity can be used to judge solvents (29). Often dipole moment, dielectric constant, etc. have been used to quantify polarity. For GXLs, Roškar *et al.*(30) have measured the dielectric constant of methanol with CO_2 at 35°C, and the data are shown in Figure 7. From ambient conditions to approximately 40 bar of CO_2, the dielectric constant changes very little. However with higher pressure, the dielectric constant significantly decreases by about an order of magnitude at 76 bar of CO_2 pressure. This

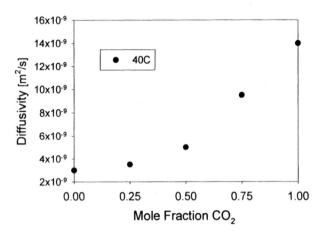

Figure 6. Diffusivity of benzene in methanol and CO_2 solutions at 40 ℃ and 150 bar (28)

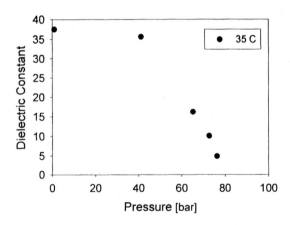

Figure 7. Dielectric constant of methanol and CO_2 solutions at 35 ℃ from the data of Roškar et al. (30).

behavior has ramifications for solutes dissolved in the solution as the media would be inefficient at stabilization of charged and polar compounds.

While many different physical and chemical phenomena correlate with dipole moment or dielectric constant, they often do not quantitatively represent the different aspects of solvent polarity. Kamlet-Taft (KT) parameters, however, differentiate "polarity" into acidity (α), basicity (β), and dipolarity/polarizability (π^*) (31). Acidity, α, is a measure of the ability to donate a proton in a solvent-to-solute hydrogen bond or to accept electrons (32). β is the measure of the ability to accept a proton in a solvent-to-solute hydrogen bond or to donate electrons (33), and π^* is a measure of the solvent's ability to stabilize a charge or dipole (31,34). Table III illustrates the KT parameters of several common solvents. The KT parameters of pure CO_2 are very low and change little with pressure/density. Sigma *et. al.* (35) have published KT values for CO_2 demonstrating that π^* ranges from approximately 0 to 0.1 at pressures of 88.8-222.6 bar at ~40°C, and β between -0.1 to 0.1 under the same conditions. Also, shown in the literature are α values ranging from 0 to 0.2 from a pressure range of 81.1-96.2 bar at 45°C (36).

As CO_2 dissolves into the liquid solvents, the polarity of the mixed solvent changes to different degrees in different composition/pressure regimes. Figure 8 illustrates the effect of CO_2 on the α and β parameter of methanol at 35°C from the data of Wyatt *et al.* (37). As shown, the acidity or α parameter decreases very little up to approximately 70 bar. The basicity or β-parameter also decreases slightly to approximately 55 bar after which it dramatically decreases as the mixture pressure approaches the critical point. The polarizability/dipolarity (π^*), as shown in Figure 9, decreases moderately to ~60 bar, and then substantially drops in the near critical area.

From the KT parameters of the pure solvents from Table III and the VLE data of Figure 2, one could possibly estimate the mixture parameters based upon the mole fraction average. As the KT of CO_2 are much smaller that those of

Table III. Kamlet-Taft parameters for common organic solvents and CO_2.

Solvent	KT Parameters			
	α	β	π^*	Ref.
Methanol	0.91	0.63	0.70	(38)
Acetone	0.11	0.52	0.72	(38)
Acetonitrile	0.23	0.38	0.79	(38)
n-Hexane	0.00	0.00	-0.11	(29)
Low P Gas Phase	0.00	0.00	-1.23	(29)
CO_2 [a]	0 to 0.195	-0.14 to 0.09	-0.01 to 0.1	(35,36)

[a] various densities

methanol, one would expect a continual decrease with pressure. The mixture KT parameters are measured from solvatochromic dyes dissolved in the solution. The probes actually report the local environment (cybotactic area) of the probe and may differ significantly from the bulk values (29). However, the local molecular environment of the probe may also be more similar to that of different solutes, reactants, etc. dissolved in the GXL and would thus experience the local, not bulk, polarity.

Multi-Component and Multi-Phase Equilibrium

Solid Components

The ability of CO_2 or a dissolved gas to modify the solubility of dissolved solids in a liquid forms the basis for the gas or supercritical anti-solvent processes that will be discussed below. Under certain conditions, the solubility of a solid in the liquid dramatically decreases as more of the gas dissolves in the liquid phase. Thus, the dissolved gas acts as an anti-solvent for the dissolved solid solute. The enhanced mass transport properties of GXLs allow precision crystallization and particle formation of a wide-variety of solutes, from organics to polymers and proteins (39).

Particle formation processes are governed by the thermodynamic condition of ternary (or higher) solid-liquid-vapor equilibrium (SLV). The SLV conditions are determined by the pressure, temperature, initial concentration of the solid solute, and the concentration of the CO_2 in the liquid phase. At a constant temperature and initial concentration of solid in the liquid, there will be one pressure (concentration of CO_2 in the liquid phase) that will begin to induce precipitation of the solid solute. Above this pressure/concentration-CO_2, the solubility of the solid will decrease and more of the solid will precipitate. While this process is central to crystallization/particle formation, other processes are also affected. For instance, solid metal complexes or precursors are used in homogeneous catalysis in GXLs. The conditions of SLV equilibrium are important to prevent precipitation of the catalyst. However, controlled precipitation of the catalyst may be useful for post-reaction separations; efficient separations are often the limiting process in practical implementation of homogeneous catalysis.

Liquids

In a similar manner, the dissolved gas may affect the solubility/miscibility of liquid mixtures. Here, the dissolved gas can induce a phase-separation of two

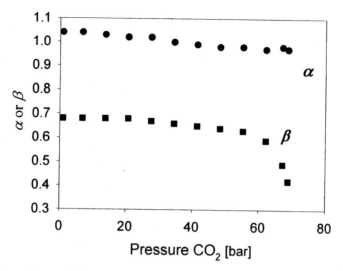

Figure 8. α and β parameters in methanol and CO₂ solutions at 35 °C from the data of Wyatt et al. (37).

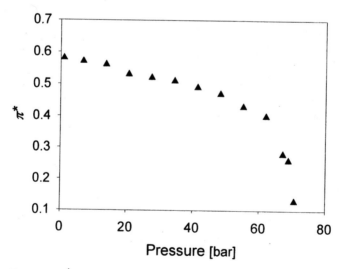

Figure 9. π parameter in methanol and CO₂ solutions at 35 °C from the data of Wyatt et al. (37).*

(or more) liquids in solution to form vapor-liquid-liquid equilibrium (VLLE) at the lower-critical end point (LCEP). Surprisingly, this behavior is often seen between water and polar solvents that are completely miscible at ambient conditions; for a review see Adrian *et al.* (40). Maurer and coworkers (41) have used this phenomena for inducing separation of biomolecules from aqueous solutions with compressed CO_2. While CO_2 is the most common compressed gas, this behavior has been observed with ethylene, ethane and N_2O (41).

Once two liquid phases form, each component and any dissolved solute(s) will partition between the two liquid phases. The ratio of each compound in each liquid phase is a function of the temperature, pressure CO_2 and initial loading of the liquids (42,43). For some systems with VLLE, a further increase in pressure may pass above the mixture critical points of one of the liquid phases (K-point) and return the system to vapor-liquid equilibrium with one liquid phase. This usually occurs in systems with large differences between the binary mixture critical points with CO_2. This is the case with ionic liquids and organic solvents or water (42,44,45); and with water/polar solvents (40). In contrast, the presence of such gases of CO_2 can actually induce two immiscible liquids to become miscible. Eckert and coworkers (46) have shown that CO_2 may induce miscibility in liquid-liquid systems consisting of fluorinated or perfluorinated solvents and organics solvents (e.g., tetrahydrofuran, etc.).

Gases

The solubility of permanent or reaction gases (H_2, CO, O_2, etc.) in liquids is important for many processes, especially catalysis. The presence of CO_2 often can increase the solubility of the gases beyond that achievable with the same pressure of the pure gas. For instance, Subramaniam and coworkers (47) found enhanced solubility of H_2 and CO in organic solvents and olefins with mixtures of 50% to 75% CO_2. Foster and coworkers (48) measured the ternary phase equilibrium of CO_2 + H_2 + methanol at 25°C and found that below 60 bar the hydrogen solubility was similar or slightly decreased with CO_2 pressure over the same pressure of pure H_2. However, above 60 bar the H_2 solubility increased with increased pressure. Brennecke and coworkers (49) found enhanced solubility of O_2 and CH_4 in ionic liquids with increased CO_2 pressure. Leitner and coworkers (50) found enhanced H_2 solubility in ionic liquids using an NMR technique which they used to explain the enhanced selectivity observed in the hydrogenation of aryl-imines.

Applications

Non-Catalytic Reactions

Selective precipitation of a product by using CO_2 as an antisolvent can be used to overcome equilibrium limitations as shown for the synthesis of copper indomethacin in CO_2-expanded DMF (51). During the liquid phase esterification of acetic acid with ethanol, increased conversion was observed when the reaction mixture was expanded with CO_2 (60 °C, 59 bar), possibly because of preferential partitioning of the product ester into the CO_2-rich phase, which would drive the equilibrium further to the right (52). CO_2 has also been exploited as a reactant to promote a Pictet-Spengler reaction. Dunetz *et al.* (53) used the high concentration of CO_2 in expanded neat substrate to convert an arylethylamine to a carbamate salt which was subsequently reacted with a carbonate ester and an aldehyde to form an acyliminium that is particularly active towards the Pictet-Spengler reaction. The yield of tetrahydroisoquinoline increased with CO_2 expansion up to 120-130 bar, but decreased at higher pressures, presumably due to increased extraction of the dialkylcarbonate from the liquid phase by the denser CO_2. Xie *et al.* (54) showed that during the homogeneous hydrogenation of nitriles to primary amines in CO_2-expanded THF, the CO_2 converts the amine product into the insoluble carbamate salt which can be filtered and reconverted to the amine by heating. Improved yields to the primary amines are reported.

GXLs have also been exploited to mitigate the gel effect seen during polymerizations. The gel effect causes the polymer molecular weights to increase due to the increasing viscosity of the polymer solution during reaction. The free radical polymerization of styrene in CO_2-expanded THF yielded lower molecular weights and narrower molecular weight distributions than in unexpanded THF (55). The polymerization of methyl methacrylate (MMA) in CO_2-expanded solvents also yielded lower molecular weights with increasing pressures (55,56). The lowering of viscosity has also been exploited in ultrasound-activated free radical polymerizations. The ultrasound initiates radical formation due to the high temperatures and shear induced during cavitation. However, higher viscosities hinder cavitation. Even low pressures (1-7 bar) of CO_2 are shown to increase the conversion and the molecular weight of poly(methyl methacrylate) prepared in CO_2-expanded monomer (57,58).

Catalytic Reactions

Homogeneous Catalysis

Hydrogenations: Jessop and coworkers (59) showed that the choice of the expansion gas significantly affects the rate of hydrogenation of CO_2 in liquid

MeOH/NEt$_3$ mixtures. The turnover frequency was 770 h^{-1} with no expansion gas, 160 h^{-1} with ethane (40 bar) and 910 h^{-1} with CHF$_3$ as the expansion gas. The decreased rate in the ethane-expanded solvent was attributed to the low polarity of ethane.

As demonstrated during the hydrogenations of α,β-unsaturated carboxylic acids catalyzed by Ru BINAP complexes such as Ru(OAc)$_2$(BINAP), the enantioselectivities are strongly dependent on the availability of H$_2$ in solution (60). For substrates such as atropic acids (2-arylacrylic acids), higher H$_2$ availability improves the selectivity, while for others such as tiglic acid (2-methyl-2-butenoic acid), increased H$_2$ availability results in lower selectivity. These contrasting effects are also demonstrated during asymmetric hydrogenations in ionic liquids (ILs). The enantioselectivity during hydrogenation of tiglic acid in [BMIm][PF$_6$] (where "BMIm" is 1-n-butyl-3-methylimidazolium) is superior [H$_2$ = 5 bar; ee = 93%] to that in CO$_2$-expanded [BMIm][PF$_6$] (H$_2$ = 5 bar; CO$_2$ = 70 bar; ee = 79%) wherein the H$_2$ availability is improved with CO$_2$ (50). In contrast, the hydrogenation of atropic acid is greatly improved in selectivity (ee increases from 32 to 57%) when the IL is expanded with 50 bar CO$_2$ (61). Leitner and coworkers (50) demonstrated that hydrogenation of N-(1-phenylethylidene)aniline proceeded to only 3 % conversion in 22 h in [EMIm][Tf$_2$N] (1-ethyl-3-methyl-imidazolium bis(trifluoromethylsulfonyl)imide) and to >99% in CO$_2$-expanded [EMIm][Tf$_2$N].

During the asymmetric hydrogenation of 2-(6'-methoxy-2'-naphthyl)acrylic acid, an atropic acid, with [RuCl$_2$(BINAP)(cymene)]$_2$ in CO$_2$-expanded methanol, the reaction was faster but less selective than in normal methanol (62). In contrast, using RuCl$_2$(BINAP) as catalyst decreased the reaction rate in CO$_2$-expanded methanol compared to neat methanol (63).

Facile hydrogenation of compounds such as vinylnaphthalene with a RhCl(PPh$_3$)$_3$ catalyst which is a solid at ambient conditions was demonstrated by melting the solid with compressed CO$_2$ and performing the reaction in the melt phase at 33°C, which is well below the normal melting point of the solid (64). This reaction technique is particularly useful in the case of thermally labile substrates. Scurto and Leitner (65) discovered that CO$_2$ induced a melting point depression of an ionic solid (tetrabutylammonium tetrafluoroborate) greater than 100°C below its normal melting point of 156°C. They used this induced melt for the hydrogenation, hydroformylation and hydroboration of vinyl-naphthalene using rhodium complexes.

Hydroformylations: Jin and Subramaniam (66) demonstrated homogeneous hydroformylation of 1-octene using an unmodified rhodium catalyst (Rh(acac)(CO)$_2$) employing CXLs as reaction media. At 60 °C, the turnover numbers (TONs) for aldehydes formation in CO$_2$-expanded acetone were significantly higher than those obtained in either neat acetone or scCO$_2$.

In subsequent studies, the performance of several rhodium catalysts, $Rh(acac)(CO)_2$, $Rh(acac)[P(OPh)_3]_2$, $Rh(acac)(CO)[P(OAr)_3]$ and two phosphorous ligands, PPh_3 and biphephos, was compared in neat organic solvents and in CXLs (47). For all catalysts, enhanced turnover frequencies (TOFs) were observed in CXLs. For the most active catalyst, $Rh(acac)(CO)_2$ modified by biphephos ligand, the selectivity to aldehyde products was improved from approximately 70% in neat solvent to nearly 95% in CXL media. The enhanced rates and selectivity are attributed to increased syngas availability in the CXL phase. In contrast, the Rh complex-catalyzed hydroformylation of 1-hexene in CO_2-expanded toluene was more rapid than in $scCO_2$ but slower than in normal toluene (67). The high CO_2 pressure was believed to dissolve some of the 1-hexene out of the liquid phase into the CO_2 phase, thereby lowering the concentration of hexene available to the catalyst.

Recently the use of CO_2-IL biphasic media for continuous-flow hydroformylation has been reported (68,69). Constant activity for up to three days was demonstrated in 1-octene hydroformylation catalyzed by a Rh-based catalyst in $[BMIm][PF_6]/scCO_2$ biphasic media. Compared to the industrial cobalt-catalyzed processes, higher TOFs were observed; the selectivity to linear aldehyde (70%) is comparable to those attained in the industrial processes (70-80%). However, the authors point out that the air/moisture sensitivity of the ILs and the ligands may lead to the deactivation and leaching of the rhodium catalyst. More recently, Cole-Hamilton and coworkers reported a solventless homogeneous hydroformylation process using dense CO_2 to transport the reactants into and transfer the products out of the reactor leaving behind the insoluble Rh-based catalyst complex in the reactor solution (70).

By melting vinylnaphthalene with compressed CO_2, enantioselective hydroformylation with a Rh catalyst was demonstrated in the CO_2-based melt phase (71) similar to the hydrogenation described earlier.

O_2 Oxidations: Carbon dioxide-expanded liquids provide both rate and safety advantages for oxidations. The homogeneous catalytic O_2-oxidation of 2,6-di-*tert*-butylphenol, DTBP, by Co(salen*) was studied in $scCO_2$, in CO_2-expanded acetonitrile ($V/V_0 = 2$), and in the neat organic solvent (72,73). The TOF in the CO_2-expanded MeCN (at 60-90 bar) is between one and two orders of magnitude greater than in $scCO_2$ (at 207 bar). The TOF and DTBQ selectivity are lower in neat MeCN (O_2 bubbling, 28 °C, 1 bar). The oxidation of cyclohexene by O_2 was investigated with a non-fluorinated iron porphyrin catalyst, (5,10,15,20-tetraphenyl-21H,23H-porphyrinato)iron(III) chloride, Fe(TPP)Cl, in addition to a fluorinated catalyst (5,10,15,20-tetrakis-(pentafluorophenyl)-21H,23H-porphyrinato)iron(III) chloride, Fe(PFTPP)Cl (72). While Fe(TPP)Cl is insoluble and displays little activity in $scCO_2$, it displays high activity in CO_2-expanded MeCN. The cyclohexene conversion

obtained with the fluorinated catalyst Fe(PFTPP)Cl in the CO_2-expanded solvent at 90 bar and 80°C is 41%, which is nearly 7-fold greater than that reported for $scCO_2$ with the identical catalyst at the same temperature (at 345 bar). Also, an approximately 1.5-fold higher epoxidation selectivity over $scCO_2$ is obtained in CO_2-expanded CH_3CN.

The MC (Mid-century) process for the oxidation of methylbenzenes to carboxylic acids represents an important industrial process for the synthesis of polymer intermediates for fabrication into fibers, resins and films. For example, terephthalic acid (TPA) is prepared in greater than 90% yield by liquid phase oxidation of p-xylene at $ca.$ 200°C in the presence of Co/Mn/Br catalysts. Among the drawbacks is the low O_2 solubility of O_2 in the liquid phase reaction medium at the high operating temperature, which also increases waste production by partial oxidation of both acetic acid solvent and the carboxylic acid product. Recently, the liquid phase oxidation of toluene and p-toluic acid with Co/Zr(acac)$_4$ dissolved in acetic acid medium was investigated to demonstrate the advantages of CO_2-expanded media (74). In these experiments, the Zr/Co molar ratio was kept at 1/6 and, accordingly, the acac/Co molar ratio was 4/6, and the CO_2 pressure was applied at 26 bar. Experiments involving the oxidation of toluene, with either N_2 or CO_2 as the inert component, were run at 80°, 60° and 50°C, and the benzoic acid yields are as follows.

80°C: N_2/O_2, 93.3% (5 hrs); CO_2/O_2, 91.3% (5 hrs).
60°C: N_2/O_2, 87.3% (28 hrs); CO_2/O_2, 89.0% (24 hrs)
50°C: N_2/O_2, 41.1% (48 hrs); CO_2/O_2, 80.5% (48 hrs)

Clearly, the CO_2 effect is far more evident at lower reaction temperatures where the acetic acid expansion is more pronounced. It is plausible that the burning of acetic acid solvent will also be significantly curtailed at the lower temperatures.

H_2O_2-based Oxidation: The reaction between H_2O_2 and CO_2 yields a peroxycarbonic acid species, an oxidant that facilitates olefin epoxidation (75). Using such a system with NaOH as a base, relatively low conversion ($\sim$ 3%) of propylene to propylene oxide (PO) is reported (76). The low conversions are indicative of mass transfer limitations in biphasic systems.

By employing CO_2-expanded $CH_3CN/H_2O_2/H_2O$ homogeneous mixtures, it was shown that olefin epoxidation reactions may be performed in a homogeneous CO_2-expanded phase containing the olefin, CO_2 and H_2O_2 (in aqueous solution), thereby eliminating the interphase mass transfer limitations associated with biphasic systems (77). By employing pyridine as the base to stabilize the peroxy acids, one to two orders of magnitude enhancement in epoxidation rates (compared to the biphasic system) was achieved with >85% epoxidation selectivity. Recently, this approach was applied to propylene oxidation by creating a homogeneous mixture of dense CO_2, an organic solvent such as acetonitrile or methanol, and water (78). When using pyridine as a base, the PO yield at 40°C is on the order of 10% after 12 h at roughly 48 bar. In the presence of an added catalyst (methyltrioxorhenium or MTO), the PO yield is significantly enhanced to 80% in 3 h with PyNO as the base. More recently, it was found that when N_2/C_3H_6 (instead of CO_2/C_3H_6) was used as the pressuring medium over the homogeneous $H_2O_2/H_2O/MeOH/PyNO$ mixture containing dissolved catalyst, remarkably high activity (92% PO yield in 1 h) was achieved at approximately 14 bar (17). Because lower olefins exhibit high solubility in methanol, it seems plausible that the propylene by itself acts as an effective expansion gas while the CO_2 swelling of the reaction phase lowers the reactant concentration. Thus, N_2/C_3H_6 is the preferred expansion gas in this case. Furthermore, the PO may be easily separated from the aqueous phase by distillation, thereby recycling the catalyst.

Polymerizations: Catalytic chain transfer polymerizations are free radical polymerizations that use a homogeneous catalyst to terminate one chain and start a new one. During the polymerization of methyl methacrylate, the chain transfer is suspected of being a diffusion-controlled reaction in which the Co(II) catalyst abstracts a hydrogen atom from the polymer radical (R•) and transfers it to a monomer to start the growth of a new chain. Zwolak *et al.* (79) reported that the rate of chain transfer was 4-times greater in CO_2-expanded methyl methacrylate (60 bar, 50 °C) than in neat monomer. The improvement was attributed to the lower viscosity of the CO_2-expanded solution.

Multiphase Catalysis

The properties of GXLs may also be exploited in heterogeneous catalysis. For example, adding CO_2 to an organic liquid phase in a fluid-solid catalytic system should enhance gas solubilities and improve the mass transfer properties of the expanded phase. Further, gas expanded ternary phases may be employed to reduce immiscible liquid phases to a single phase.

Hydrogenations: Employing dense CO_2 as the solvent medium and a Pd/Al_2O_3 catalyst in a stirred reactor, it was reported that the hydrogenation rate of a CO_2-expanded unsaturated ketone was higher than that of the unswollen ketone (80). A similar observation was reported during the Pd/C hydrogenation of pinene in $scCO_2$, wherein the reaction rate was higher at lower pressures (where a condensed phase exists) compared to single-phase operation at supercritical conditions (81). More recently, the rate constant for the hydrogenation of the aromatic rings in polystyrene (PS) was found to be higher in CO_2-expanded decahydronaphthalene (DHN) than in neat DHN (82). Chan and Tan (83) reported enhanced rates in the presence of a CO_2-expanded toluene phase during the Pt/γ-Al_2O_3 catalyzed hydrogenation of tetralin to decalin at 280 °C in a trickle bed reactor.

The Pt-catalyzed hydrogenation of oleic acid (a solid substrate) at 35 °C is limited by solid-solid mass transfer limitations and reaches a maximum of 90% conversion after 25 h. In contrast, in the presence of 55 bar CO_2 the oleic acid melts at 35°C and the reaction proceeds to 97% conversion after only 1 h (84).

During the $NiCl_2$-catalyzed reduction of benzonitrile to benzylamine by $NaBH_4$, Xie *et al.* (85) showed that CO_2 expansion of the reaction mixture traps the primary amine as a carbamate salt, thereby preventing its further reaction with the imine intermediate. Thus, the benzylamine yield was 98% in CO_2-expanded ethanol but <0.01% in normal ethanol.

Selective Oxidations: During a study of Pd/Al_2O_3 catalyzed partial oxidation of octanol, Baiker's group report significantly enhanced oxidation rates at intermediate pressures (where two phases exist) (86). It would appear that in these examples, the reactions in condensed phases benefit from increased concentrations of the substrate relative to the single supercritical phase while also enjoying adequate O_2 availability in the liquid phase.

During the O_2-based oxidation of cyclohexene on a MCM-41 encapsulated iron porphyrin chloride complex (P = 1–127 bar, T = 25–50 °C, t = 4–12 h), conversion and product yields in CO_2-expanded acetonitrile (~30% CO_2) almost doubled compared to the neat organic solvent (87). For the oxidation of 2,6-di-*tert*-butylphenol (DTBP) to 2,6-di-*tert*-butyl-1,4-benzoquinone (DTBQ) and 3,5.3',5'-tetra-*tert*-butyl-4,4'-diphenoquinone (TTBDQ), a series of porous materials with immobilized Co(II) complexes were screened as catalysts in neat acetonitrile, supercritical carbon dioxide (*sc*CO_2), and CO_2-expanded acetonitrile (88). In this case, the highest conversions were found in *sc*CO_2 suggesting that *sc*CO_2, rather than CXL or liquid reaction media, provides the best mass transfer of O_2 and of substrates through the porous catalysts.

Hydroformylation: Abraham and coworkers (89) investigated 1-hexene hydroformylation over a rhodium–phosphine catalyst immobilized on a silica support and found that the rates in CO_2-expanded toluene and in *sc*CO_2 were comparable but faster than in normal toluene. However, the activity declined with time due to possible catalyst leaching.

Acid Catalysis: The acylation of anisole with acetic anhydride was investigated in a continuous slurry reactor over mesoporous supported solid acid catalysts based on Nafion® (SAC-13) and heteropolyacids (90). The CXL media gave lower conversion and, surprisingly, faster deactivation compared to a liquid phase reaction despite the use of polar cosolvents like nitromethane. GC/FID analysis of the soxhlet extract and IR analysis of the spent catalyst indicated that the deactivation is due to heavy molecules (possibly di- and tri-acylated products) formed by the interaction of acetic anhydride with para-methoxyacetophenone (*p*-MOAP) in the micropores of Nafion® aggregates due to the slow desorption rate of the primary product.

Added CF_3CO_2H has been shown to catalyze a Friedel-Crafts alkylation of anisole in CO_2-expanded anisole (95 °C, 42 bar), but the reaction was no faster than that in normal anisole (91). *In-situ* generation of acids by the reaction of CO_2 with alcohols or water is desirable because depressurization leads to self-neutralization (92,93). The formation of the dimethyl acetal of cyclohexanone is up to 130 times faster in CO_2-expanded methanol than in normal methanol without any added acid. In-situ acids, generated with alcohol and CO_2, also catalyze the hydrolysis of β-pinene to terpineol and other alcohols with good selectivity for alcohols rather than hydrocarbons (93).

Addition of CO_2 to pressurized hot water accelerates reactions that can proceed by acid catalysis. CO_2 dissolved in water at 250 °C promotes the decarboxylation of benzoic acid (94), the dehydration of cyclohexanol to cyclohexene and the alkylation of p-cresol to 2-*tert*-butyl-4-methylphenol (95). The hydration of cyclohexene to cyclohexanol at 300 °C showed a 5-fold rate increase as the CO_2 pressure was increased from 0 to 55 bar.

Particle Formation

Several techniques based on CXLs have been reported in the literature for preparing fine particles for such applications as pharmaceutical compounds (96,97) and explosives (98). The main techniques are briefly described below (schemes shown in Figure 10).

Gas Antisolvent Solvent (GAS) Technique: The GAS method was first described by Krukonis and Gallagher (99). In this process, CO_2 is introduced into a vessel containing a solution of the solute in an organic solvent (Figure 10a). Expansion of the solvent by CO_2 lowers the solubility of the solute, eventually causing it to precipitate. Thus, the CO_2 acts as an antisolvent for the solute. The precipitated solute is recovered after flushing the organic solvent from the vessel with pressurized CO_2 (100). GAS has also been used to grow large crystals that are useful for crystallographic studies. Jessop's group dissolved a fluorous compound in a CO_2-expanded solvent and then slowly released the CO_2 pressure resulting in a supersaturated solution from which the compound slowly crystallized (101).

DELOS: A variation of the GAS process is the DELOS (Depressurization of an Expanded Liquid Organic Solution) technique (102,103) in which a solute is dissolved in an organic solvent and the solution is expanded with a compressed gas such as CO_2. The concentration of the solute in solution is such that the expansion does not trigger precipitation. However, when the pressure is then suddenly lowered, the dissolved CO_2 rapidly elutes out of solution and causes a dramatic temperature drop that triggers precipitation of the solute as a fine powder.

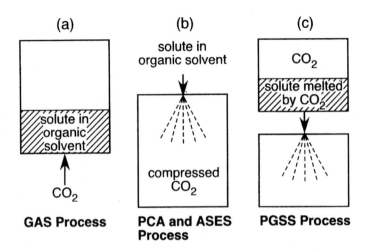

Figure 10. Processes for particle formation with compressed gases.

PCA and ASES: The PCA (Precipitation with Compressed Antisolvents) process (Figure 10b) complements the GAS process. The solution of solute in organic solvent is sprayed into a vessel containing pressurized CO_2. The solvent droplets expand rapidly causing nucleation and precipitation of the solute as fine particles. Following spraying, the organic solvent is dried from the particles with fresh CO_2 and the particles are recovered. A variation of PCA is ASES (Aerosol Solvent Extraction System) in which the solvent is completely dissolved into the $scCO_2$. ASES has also been performed as a continuous operation with the CO_2 flowing either concurrently or countercurrently with respect to the organic solution, thereby extracting and continuously removing the organic solvent from the chamber (104). Hanna and York (105) showed that the use of a co-axial nozzle to feed the organic solution in the inner nozzle and supercritical CO_2 in the outer nozzle produced finer particles than in the conventional PCA or ASES processes. This variation was called SEDS (Solution Enhanced Dispersion by Supercritical fluids). Subramaniam and coworkers (106,107,108) employed an ultrasonic nozzle and $scCO_2$ as the energizing medium to form fine droplets of the drug solution. The $scCO_2$ also selectively extracts the solvent from the droplets, precipitating the drug. Submicron particles of hydrocortisone, ibuprofen, insulin and paclitaxel were formed using this technique.

While a majority of PCA studies have involved the precipitation of organic compounds, a few recent studies have investigated different types of compounds. For example, Hutchings *et al.* (109) precipitated vanadium phosphate (an active catalyst for the oxidation of butane to maleic anhydride) from isopropanol by the

PCA technique. Johnson *et al.* (110) prepared rod-shaped nanoparticles (~60 nm in diameter and ~700 nm in length) of metal complexes containing salen ligands by PCA from CH_2Cl_2-solutions. The Co(salen) version of these nanoparticles show near-stoichiometric O_2 binding properties and room temperature NO decomposition activity, properties not seen with the commercial Co(salen) starting material from which the nanoparticles were synthesized.

PGSS: The PGSS (Particles from Gas-Saturated Solution) process uses dense CO_2 to melt a solid that is close to its melting point, T_m. The molten compound, actually swollen by CO_2, is then expanded rapidly across a small nozzle (Figure 10c). The rapid depressurization and cooling of the melt results in a fine powder. Applications include Nifedipine, a calcium-channel blocker (T_m 172-174 °C), which after processing by PGSS at 185 °C yielded smaller particle sizes that displayed a more rapid rate of dissolution in water than the unprocessed material (111). The process temperature, however, is too high for handling most pharmaceuticals. To practice PGSS at lower temperatures, the micronization of a mixture of poly(ethylene glycol) (PEG, MW 4,000) with Nifedipine was attempted at 50 °C. A finely powdered co-precipitate was obtained without observable degradation. Using a similar approach, Shine and Gelb (112) encapsulated a vaccine into particles of polycaprolactone.

Particle Precipitation from Reverse Emulsions: By expanding reverse emulsions with CO_2, compounds such as proteins (113) or enzymes (114) dissolved in the aqueous core of such emulsions can be precipitated as particles. For example, trypsin was precipitated as 10 nm particles from reverse emulsions consisting of a decane-based continuous phase containing 100 mM surfactant, water (water/surfactant mole ratio $w_0 = 20$) and trypsin (0.5 to 0.8 mg/ml). By expansion with CO_2 (50 bar, 20 °C), the protein was precipitated.

Inorganic particles may also be prepared by this technique. By mixing two reverse emulsions, one containing $ZnSO_4$ dissolved in the aqueous core and the other Na_2S (both in AOT-stabilized water in iso-octane emulsions), inorganic ZnS particles form in the combined emulsion. The ZnS particles are collected by expansion of the emulsion with CO_2 (115).

Particle Fractionation and Coating: Roberts and coworkers (116,117) showed how nanoparticles dispersed in a solvent may be fractionated by gradual addition of CO_2 to the solvent. Gold or silver nanoparticles protected by alkanethiols were fractionated in this manner. After each fraction precipitated at various CO_2 pressures, an Archimedes screw was used to move the remaining suspension to prevent later fractions from precipitating onto fractions that had already precipitated. Roberts' group has also exploited this technique to form an even monolayer of nanoparticles from a suspension of the particles in solution. Expansion of the solvent by CO_2, followed by raising the temperature to 40 °C and flushing with fresh CO_2 allows the precipitation of the particles and complete removal of the solvent without disruption of the even monolayer (118).

Subramaniam's group demonstrated a fluidized bed particle coating process based on the PCA technique (119,120). The process is similar to a Wurster coater except that the air or other permanent gas is replaced by $scCO_2$. Here, the $scCO_2$ serves to not only fluidize the core substrate particles but also to dry the substrates sprayed with the coating solution, thereby forming the coating. The process was demonstrated by coating glass and non-pareil sugar beads (1-2 mm in diameter) with a thin layer of RG503H polymer, either as a deposit of fine particles or as a continuous film.

Polymer Processing

The field of polymer processing using CO_2-induced lowering of T_m and T_g has been extensively reviewed (121,122,123,124). For a variety of polymers, CO_2 dissolution lowers the melting temperature, the crystallization temperature and the glass transition temperature, reduces the viscosity of the CO_2/polymer mixture, and raises the rate and degree of crystallization. These effects have been exploited to manipulate particle size (125) and morphology (126,127) of polymers and to facilitate extrusion (128), foaming (129), impregnation (130), and mixing or co-molding of polymers (131).The CO_2-induced plasticization combined with the foaming effect of CO_2 when it is released from the polymer melt results in a high-surface area porous extruded polymer (132). Polymers swollen by CO_2 at mild temperatures have been mixed with heat-sensitive species such as an enzyme and co-precipitated using PGSS (123,133). Impregnation of active ingredients into polymers avoiding thermal degradation is particularly attractive for pharmaceutical applications.

Separations

GXLs have been exploited to purify compounds and to separate mixed solutes in solution. Expansion with CO_2 was used to separate citric acid from oxalic acid (134) and to isolate and purify β-carotene from a mixture of carotene oxidation products (135). Recently, crude terephthalic acid was purified by dissolving in dimethyl formamide and recrystallization using CO_2-expansion to cause supersaturation and precipitation (136).

Post-reaction catalyst separation schemes often involve the use of solvents to switch the polarity of the medium to precipitate homogenous catalysts such as transition metal complexes. The use of CO_2 expansion has been exploited to effect such polarity switches for separating oxidation catalysts post-reaction (137). Isothermal CO_2 addition to a solution of CH_3CN containing the dissolved catalyst, [{N,N'-bis(3,5-di-*tert*-butylsalicylidene)1,2-cyclohexanediiminato(2-)}

cobalt(II)] or Co(salen*) caused the catalyst to precipitate after a certain level of CO_2 expansion. The CO_2-expansion technique was also used post-reaction to precipitate Rh complexes at relatively mild pressures and temperatures (47).

CO_2 can make two-liquids miscible and split miscible phases into two liquids. The addition of compressed CO_2 to fluorous biphasic chemical systems induces a phase split, with the catalyst remaining in the fluorous phase and the product in the organic phase. This CO_2 "miscibility switch" was demonstrated to effect post-reaction separation (138). In another application, compressed CO_2 was added to mixed organic-aqueous tunable solvents (OATS) to facilitate phase separation and recycle enzymes (139). For example, the dimethyl ether (DME)-water system has been employed for the alcohol dehydrogenase-catalyzed reduction of hydrophobic ketones. Following reaction, the DME-rich phase, containing the product, is separated from the aqueous phase by depressurization and evaporation to recover DME. The enzyme is retained in the aqueous phase, which is recycled.

The Jessop and Eckert/Liotta groups together reported (140) a system for using CO_2 expansion of a solvent to shuttle a catalyst from a solid phase to a solution phase and back again. The method was demonstrated using hydrogenation of styrene with a fluorous version of Wilkinson's catalyst as a test reaction. Catalyst precursor, styrene, cyclohexane, and fluorous silica were placed in a vessel at 40 °C. H_2 and CO_2 (60 bar) were added, at which point the solvent expanded and the catalyst became soluble. After the reaction, the gases were vented, causing the catalyst to leave the solution and become entrapped by the fluorous silica. The liquid product could be decanted, and the fluorous silica/catalyst solid phase was re-used repeatedly. Rh losses to the liquid product were too low to be detectable by ICP/AA. These results suggest opportunities to design catalyst ligands to permit facile catalyst separation with GXLs.

Process engineering issues

Economics

There are relatively few known examples of commercialization in the areas of chemical manufacturing and particle formation processes employing dense CO_2, despite nearly two decades of research in these areas and the promise of many potential applications. This is in contrast to the many industrial scale CO_2-based plants worldwide in the food and natural products industries (such as the decaffeination of coffee beans, extraction of hops and spices, etc.) and other CO_2 applications including dry cleaning and precision cleaning (141,142), paints and coatings (Union Carbide process), polymer processing (DuPont Fluoropolymers Plant) and hydrogenations. For GXL-based technologies to gain wider acceptance in chemical manufacturing, process economic analysis based on plant scale simulations and demonstration of satisfactory pilot plant scale performance are necessary. Reliable knowledge of the phase behavior, intrinsic kinetics, fluid dynamics and transport properties involving GXLs is essential for rational process design and simulation.

In homogeneous catalysis, the preferred operating pressures for GXL processes (wherein both reaction and environmental benefits are optimized) are on the order of tens of bars. Process intensification at lower pressures favors process economics. However, this must be supported by quantitative comparisons of the relative economics of the GXL and competing processes. Quantitative economic analysis, even at an early research stage of process development, typically provides valuable research guidance by identifying performance targets (operating pressure range, temperature range, extent of catalyst recovery, product purity, etc.) for economic viability. An example of such an analysis of a CXL-based hydroformylation process is provided by Fang et al. (143). As explained in a previous section, the hydroformylation turnover frequencies (TOFs) were up to four-fold higher in CXLs than those in neat organic solvent. The enhanced rates were achieved at milder conditions (30-60 °C and 4-12 MPa) compared to industrial processes (140-200 °C and 5-30 MPa). Preliminary economic and environmental analyses of the CXL process were benchmarked against a simulated conventional hydroformylation process for which non-proprietary data were obtained mostly from the Exxon process. The simulation results indicate that the CXL process has clear potential to be economically viable and environmentally favorable subject to nearly quantitative recovery and recycle of the rhodium-based catalysts.

For the simulated conventional process, acetic acid discharged during the catalyst recovery steps is the dominant source of adverse environmental impact. These analyses have provided guidance in catalyst design and in choosing materials and operating conditions that favor process economics while lessening the environmental footprint.

Safety

When using compressed CO_2 to expand less volatile liquid phases, the vapor-liquid equilibrium is such that the vapor phase is dominated by CO_2. The presence of CO_2 in the vapor phase provides safety benefits in the case of O_2-based oxidation reactions by reducing the propensity to form explosive hydrocarbon/O_2 mixtures in the vapor phase. "Calculated Adiabatic Reaction Temperature" (CART) is the most widely used method to estimate vapor phase hazardous behavior. The temperature rise of a combustion reaction, based on vapor phase composition, is calculated under adiabatic conditions. The Gibbs free energy minimization method is typically employed for predicting the flammability of fuels and chemicals such as lower alkanes, esters and carboxylic acids (144). It is well known that inert gases arrest flames when used in substantial quantities (>50% by volume in an O_2/inert/fuel mixture). In general, triatomic inert gases (CO_2, water) reduce the flammable region to a greater extent than diatomic or monoatomic inert gases (145). In the case of each inert, the flammable region vanishes completely beyond a certain concentration. As shown in Figure 11, for methane+O_2 mixtures, this critical concentration is around 20 mole% for CO_2 (146). In other words, an adequate presence of CO_2 (instead of N_2) in an O_2-containing vapor phase reduces the flammability of various hydrocarbon+O_2 mixtures (147). This implies the possibility of operating with O_2-enriched atmospheres (where O_2 concentration exceeds 21 mole%) in the presence of CO_2 with reduced or even no flammability hazards.

Rajagopalan (136) estimated CART values for vapor phase acetic acid vapor/O_2/inert mixtures in the composition ranges encountered in the Mid-Century (MC) process for terephthalic acid manufacture using both CO_2 and nitrogen as inerts. It was found that CO_2 significantly reduces the CART (by tens of degrees) as an inert over N_2, even under the high MC process operating temperatures that are far removed from the critical temperature. It is thus possible to perform reliable calculations to guide the choice of operating

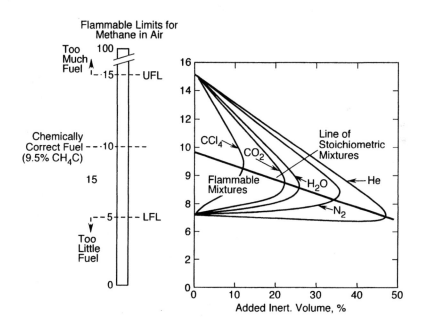

Figure 11. Effect of inerts on the flammability envelope of methane [Taken from reference 136]

conditions (pressure, temperature and composition) that would ensure safe reactor operation in the non-flammable region and to provide adequate pressure-relief in the case of runaway reactions. It is a good practice to perform such *a priori* safety calculations, especially in the case of O_2-based oxidations.

It is well known that the isobaric heat capacity of CO_2 goes through a maximum when the temperature is tuned around the critical temperature of CO_2. The magnitude of the maximum is greater at pressures closer to the critical pressure. Jin and Subramaniam (148) have shown that the high heat capacities of dense CO_2 may be used advantageously to avoid thermal runaway regions during exothermic reactions.

Summary and Outlook

GXLs have potential applications in a wide range of applications such as facilitated separations, reactions, materials processing, and particle formation as described in this chapter. In many of these applications, the use of GXLs provides significant improvements in one or more process performance measures such as reaction rate, transport rate, product selectivity, product quality, and environmental friendliness. CXLs are already being exploited in applications requiring viscosity reduction (such as enhanced oil recovery). Future commercial applications that exploit viscosity reduction will be most likely in polymer processing, particle formation and ionic liquid/CO_2 systems. Other promising applications include multifunctional CO_2-aided continuous catalytic reaction systems that integrate reaction and separation. Several of these applications are highlighted in the chapters that follow.

References

1. Tundo, P.; Anastas, P.; Black, D. S.; Breen, J.; Collins, T.; Memoli, S.; Miyamoto, J.; Poliakoff, M.; Tumas, W. *Pure Appl. Chem.* **2000**, *72*, 1207.
2. DeSimone, J. M. *Science* **2002**, *297*, 799.
3. Adams, D. J.; Dyson, P. J.; Tavener, S. J. *Chemistry In Alternative Reaction Media*; Wiley: Chichester, England, 2004.
4. Eckert, C. A.; Liotta, C. L.; Bush, D.; Brown, J. S.; Hallett, J. P. *J. Phys. Chem. B* **2004**, *108*, 18108.
5. Morgenstern, D. A.; LeLacheur, R. M.; Morita, D. K.; Borkowsky, S. L.; Feng, S.; Brown, G. H.; Luan, L.; Gross, M. F.; Burk, M. J.; Tumas, W. In *Green Chemistry : Designing Chemistry for the Environment*; Anastas, P. T., Williamson, T. C., Eds.; American Chemical Society: Washington, DC, 1996; Vol. 626.

6. Jessop, P. G.; Leitner, W., Eds.*Chemical Synthesis using Supercritical Fluids*;; VCH/Wiley: Weinheim, 1999.

7. DeSimone, J. M.; Tumas, W., Eds.; *Green Chemistry Using Liquid and Supercritical Carbon Dioxide*; New York: Oxford University Press, 2003.

8. Gordon, C. M.; Leitner, W. *Chimica Oggi/Chem. Today* **2004**, *22*, 39.

9. Beckman, E. J. *Environ. Sci. Technol.* **2002**, *36*, 347A.

10. Licence, P.; Poliakoff, M. In *Multiphase Homogeneous Catalysis*; Cornils, B., Herrmann, W. A., Horváth, I. T., Leitner, W., Mecking, S., Olivier-Bourbigou, H., Vogt, D., Eds.; Wiley-VCH: Weinheim, Germany, 2005.

11. Li, C.-J.; Chan, T.-H. *Organic Reactions in Aqueous Media*; Wiley: NY, 1997.

12. Cornils, B.; Herrmann, W. A., Eds. *Aqueous-Phase Organometallic Catalysis*; Wiley-VCH: Weinheim, 1998.

13. Jessop, P.G.; Subramaniam, B. *Chem. Rev.* **2007**, *107*, 2666.

14. Wasserscheid, P.; Welton, T. *Ionic Liquids in Synthesis*; VCH-Wiley: Weinheim, 2002.

15. Rogers, R.D.; Seddon, K.R.; Volkov, S., Eds.; *Green Industrial Applications of Ionic Liquids*; Springer, 2003.

16. Jessop, P. G.; Heldebrant, D. J.; Xiaowang, L.; Eckert, C. A.; Liotta, C. L. *Nature* **2005**, *436*, 1102.

17. Lee, H-J., Shi, T-P.; Busch, D. H.; Subramaniam, B. *Chem. Eng. Sci.,* **2007**, *62*, 7282.

18. Kordikowski, A.; Schenk, A. P.; Van Nielen, R. M.; Peters, C. J. *J. Supercrit. Fluids* **1995**, *8*, 205.

19. Yoon, J.-H.; Lee, H.-S.; Lee, H., *J. Chem. Eng. Data* **1993**, *38*, 53.

20. Jin, H. PhD Dissertation, University of Kansas, Lawrence, KS, 2006.

21. Sih, R.; Dehghani, F.; Foster, N. R., *J. Supercrit. Fluids* **2007**, *41*, 148.

22. Brunner, E. *J. Chem. Thermodyn.* **1985**, *17*, 671.

23. Gurdial, G. S.; Foster, N. R.; Yun, S. L. J.; Tilly, K. D., In *Supercritical Fluid Engineering Science: Fundamentals and Applications*, Kiran, E.; Brennecke, J. F., Eds. ACS Symposium Series 514: Washington, D.C., 1992; pp 34-45.

24. Reaves, J. T.; Griffith, A. T.; Roberts, C. B. *J. Chem. Eng. Data* **1998**, *43*, 683.

25. Ziegler, J. W.; Dorsey, J. G.; Chester, T. L.; Innis, D. P. *Anal. Chem.* **1995**, *67*, 456.

26. Blanchard, L.; Gu, Z.; Brennecke, J., *J. Phys. Chem. B* **2001**, *105*, 2437.

27. Kelkar, M. S.; Maginn, E. J., *J. Phys. Chem. B* **2007**, *111*, 4867.

28. Maxey, N. B. Ph.D. Dissertation, Georgia Institute of Technology, Atlanta, 2006.

29. Reichardt, C., *Solvent and Solvent Effects in Organic Chemistry* 3ed.; Wiley-VCH: Weinheim, Germany, 2003; p 629.

30. Roškar, V.; Dombro, R. A.; Prentice, G. A.; Westgate, C. R.; McHugh, M. A., *Fluid Phase Equil.* **1992**, *77*, 241.
31. Kamlet, M. J.; Abboud, J. L.; Taft, R. W., *J. Amer. Chem. Soc.* **1977**, *99*, 6027.
32. Kamlet, M. J.; Taft, R. W., *J. Amer. Chem. Soc.* **1976**, *98*, 2886.
33. Kamlet, M. J.; Taft, R. W., *J. Amer. Chem. Soc.* **1976**, *98*, 377.
34. Laurence, C.; Nicolet, P.; Dalati, M. T.; Abboud, J. M.; Notario, R. *J. Phys. Chem.* **1994**, *98*, 5807.
35. Sigma, M. E.; Linley, S. M.; Leffler, J. E., *J. Amer. Chem. Soc.* **1985**, *107*, 1471.
36. Ikushima, Y.; Saito, N.; Arai, M.; Arai, K., *Bull. Chem. Soc. Jpn.* **1991**, *64*, 2224.
37. Wyatt, V. T.; Bush, D.; Lu, J.; Hallett, J. P.; Liotta, C. L.; Eckert, C. A., *J. Supercrit. Fluids* **2005**, *36*, 16-22.
38. Schleicher, J. MS Thesis, University of Kansas, Lawrence, 2007.
39. York, P., *Pharm. Sci. Techn. Today* **1999**, *2*, 430.
40. Adrian, T.; Wendland, M.; Hasse, H.; Maurer, G., *J. Supercrit. Fluids* **1998**, *12*, 185.
41. Adrian, T.; Freitag, J.; Maurer, G., *Biotech. Bioeng.* **2000**, *69*, 559.
42. Aki, S. N. V. K.; Scurto, A. M.; Brennecke, J. F., *Ind. Eng. Chem. Res.* **2006**, *45*, 5574.
43. Pfohl, O.; Petersen, J.; Dohrn, R.; Brunner, G., *J. Supercrit. Fluids* **1997**, *10*, 95.
44. Scurto, A. M.; Aki, S. N. V. K.; Brennecke, J. F., *J. Am. Chem. Soc.* **2002**, *124*, 10276.
45. Scurto, A. M.; Aki, S. N. V. K.; Brennecke, J. F., *Chem. Commun.* **2003**, 572.
46. West, K. N.; Hallett, J. P.; Jones, R. S.; Bush, D.; Liotta, C. L.; Eckert, C. A., *Ind. Eng. Chem. Res.* **2004**, *43*, 4827.
47. Jin, H.; Subramaniam, B.; Ghosh, A.; Tunge, J., *AIChE J.* **2006**, *52*, 2575.
48. Bezanehtak, K.; Dehghani, F.; Foster, N. R., *J. Chem. Eng. Data* **2004**, *49*, 430.
49. Hert, D. G.; Anderson, J. L.; Aki, S. N. V. K.; Brennecke, J. F., *Chem. Commun.* **2005**, 2603.
50. Solinas, M.; Pfaltz, A.; Cozzi, P.; Leitner, W., *J. Am. Chem. Soc.* **2004**, *126*, 16142.
51. Warwick, B.; Dehghani, F.; Foster, N. R.; Biffin, J. R.; Regtop, H. L. *Ind. Eng. Chem. Res.* **2000**, *39*, 4571.
52. Blanchard, L. A.; Brennecke, J. F. *Green Chem.* **2001**, *3*, 17.
53. Dunetz, J. R.; Ciccolini, R. P.; Fröling, M.; Paap, S. M.; Allen, A. J.; Holmes, A. B.; Tester, J. W.; Danheiser, R. L. *Chem. Commun.* **2005**, 4465.
54. Xie, X.; Liotta, C. L.; Eckert, C. A. *Ind. Eng. Chem. Res.* **2004**, *43*, 7907.

55. Liu, J.; Han, B.; Liu, Z.; Wang, J.; Huo, Q. *J. Supercrit. Fluids* **2001**, *20*, 171.
56. Xu, Q.; Han, B.; Yan, H. *J. Appl. Polym. Sci.* **2003**, *88*, 2427.
57. Kemmere, M. F.; Kuijpers, M. W. A.; Prickaerts, R. M. H.; Keurentjes, J. T. F. *Macromol. Mater. Eng.* **2005**, *290*, 302.
58. Kuijpers, M. W. A.; Jacobs, L. J. M.; Kemmere, M. F.; Keurentjes, J. T. F. *AIChE J.* **2005**, *51*, 1726.
59. Thomas, C. A.; Bonilla, R. J.; Huang, Y.; Jessop, P. G. *Can. J. Chem.* **2001**, *79*, 719.
60. Sun, Y.; Landau, R. N.; Wang, J.; LeBlond, C.; Blackmond, D. G. *J. Am. Chem. Soc.* **1996**, *118*, 1348.
61. Jessop, P. G.; Stanley, R.; Brown, R. A.; Eckert, C. A.; Liotta, C. L.; Ngo, T. T.; Pollet, P. *Green Chem.* **2003**, *5*, 123.
62. Combes, G. B.; Dehghani, F.; Lucien, F. P.; Dillow, A. K.; Foster, N. R. In *Reaction Engineering for Pollution Prevention*; Abraham, M. A., Hesketh, R. P., Eds.; Elsevier: Amsterdam, 2000.
63. Combes, G.; Coen, E.; Dehghani, F.; Foster, N. *J. Supercrit. Fluids* **2005**, *36*, 127.
64. Jessop, P. G.; DeHaai, S.; Wynne, D. C.; Nakawatase, D. *Chem. Commun.* **2000**, 693.
65. Scurto, A.M.; Leitner, W. *Chem. Commun.* **2006**, 3681.
66. Jin, H.; Subramaniam, B. *Chem. Eng. Sci.* **2004**, *59*, 4887.
67. Hemminger, O.; Marteel, A.; Mason, M. R.; Davies, J. A.; Tadd, A. R.; Abraham, M. A. *Green Chem.* **2002**, *4*, 507.
68. Webb, P. B.; Kunene, T. E.; Cole-Hamilton, D. J. *Green Chem.* **2005**, *7*, 373.
69. Webb, P. B.; Sellin, M. F.; Kunene, T. E.; Williamson, S.; Slawin, A. M. Z.; Cole-Hamilton, D. J. *J. Am. Chem. Soc.* **2003**, *125*, 15577.
70. Frisch, A.C.; Webb, P. B.; Zhao, G.; Muldoon, M. J.; Pogorzelec, P. J.; Cole-Hamilton, D.J. *Dalton Trans.*, **2007**, 5531.
71. Francio, G.; Wittmann, K.; Leitner, W. *J. Organomet. Chem.* **2001**, *621*, 130.
72. Wei, M.; Musie, G. T.; Busch, D. H.; Subramaniam, B. *J. Am. Chem. Soc.* **2002**, *124*, 2513.
73. Wei, M.; Musie, G. T.; Busch, D. H.; Subramaniam, B. *Green Chem.* **2004**, *6*, 387.
74. X. Zuo, B. Subramaniam and D. H. Busch, *Ind. Eng. Chem. Res.*, **2008**, *47*, 546.
75. Nolen, S. A.; Lu, J.; Brown, J. S.; Pollet, P.; Eason, B. C.; Griffith, K. N.; Glaser, R.; Bush, D.; Lamb, D. R.; Liotta, C. L.; Eckert, C. A.; Thiele, G. F.; Bartels, K. A. *Ind. Eng. Chem. Res.* **2002**, *41*, 316.
76. Hancu, D.; Green, H.; Beckman, E. J. *Ind. Eng. Chem. Res.* **2002**, *41*, 4466.

34

77. Rajagopalan, B.; Wei, M.; Musie, G. T.; Subramaniam, B.; Busch, D. H. *Ind. Eng. Chem. Res.* **2003**, *42*, 6505.
78. H-J Lee, T-P Shi, B. Subramaniam and D. H. Busch, in S. R. Schmidt (Ed.), *Catalysis of Organic Reactions,* Chemical Industries Series, Vol. 115, CRC Press, Taylor & Francis Group LLC, Boca Raton, FL., pgs. 447-451 (2006).
79. Zwolak, G.; Jayasinghe, N. S.; Lucien, F. P. *J. Supercrit. Fluids* **2006**, *38*, 420.
80. Devetta, L.; Giovanzana, A.; Canu, P.; Bertucco, A.; Minder, B. J. *Catal. Today* **1999**, *48*, 337.
81. Chouchi, D.; Gourgouillon, D.; Courel, M.; Vital, J.; Nunes da Ponte, M. *Ind. Eng. Chem. Res.* **2001**, *40*, 2551.
82. Xu, D.; Carbonell, R. G.; Kiserow, D. J.; Roberts, G. W. *Ind. Eng. Chem. Res.* **2005**, *44*, 6164.
83. Chan, J. C.; Tan, C. S. *Energy Fuels* **2006**, *20*, 771.
84. Jessop, P. G.; DeHaai, S.; Wynne, D. C.; Nakawatase, D. *Chem. Commun.* **2000**, 693.
85. Xie, X.; Liotta, C. L.; Eckert, C. A. *Ind. Eng. Chem. Res.* **2004**, *43*, 7907.
86. Jenzer, G.; Schneider, M. S.; Wandeler, R.; Mallat, T.; Baiker, A. *J. Catal.* **2001**, *199*, 141.
87. Kerler, B.; Robinson, R. E.; Borovik, A. S.; Subramaniam, B. *Appl. Catal., B* **2004**, *49*, 91.
88. Sharma, S.; Kerler, B.; Subramaniam, B.; Borovik, A. S. *Green Chem.* **2006**, *8*, 972.
89. Hemminger, O.; Marteel, A.; Mason, M. R.; Davies, J. A.; Tadd, A. R.; Abraham, M. A. *Green Chem.* **2002**, *4*, 507.
90. Sarsani, V. S. R.; Lyon, C. J.; Hutchenson, K. W.; Harmer, M. A.; Subramaniam, B. *J. Catal.* **2007**, *245*, 184.
91. Chateauneuf, J. E.; Nie, K. *Adv. Environ. Res.* **2000**, *4*, 307.
92. West, K. N.; Wheeler, C.; McCarney, J. P.; Griffith, K. N.; Bush, D.; Liotta, C. L.; Eckert, C. A. *J. Phys. Chem. A* **2001**, *105*, 3947.
93. Chamblee, T. S.; Weikel, R. R.; Nolen, S. A.; Liotta, C. L.; Eckert, C. A. *Green Chem.* **2004**, *6*, 382.
94. Alemán, P. A.; Boix, C.; Poliakoff, M. *Green Chem.* **1999**, *1*, 65.
95. Hunter, S. E.; Savage, P. E. *Chem. Eng. Sci.* **2004**, *59*, 4903.
96. Fages, J.; Lochard, H.; Letourneau, J.-J.; Sauceau, M.; Rodier, E. *Powder Technol.* **2004**, *141*, 219.
97. B. Subramaniam, R. A. Rajewski and W. K. Snavely, *J. Pharm. Sci.* **1997**, *86*, 885.
98. Pourmortazavi, S. M.; Hajimirsadeghi, S. S. *Ind. Eng. Chem. Res.* **2005**, *44*, 6523.
99. Krukonis, V. J.; Gallagher, P. M.; Coffey, M. P. U.S. 5,360,478, 1994 (application 1991, CAN 122:34594).

100. Jung, J.; Perrut, M. *J. Supercrit. Fluids* **2001**, *20*, 179.
101. Jessop, P. G.; Olmstead, M. M.; Ablan, C. D.; Grabenauer, M.; Sheppard, D.; Eckert, C. A.; Liotta, C. L. *Inorg. Chem.* **2002**, *41*, 3463.
102. Ventosa, N.; Sala, S.; Veciana, J. *J. Supercrit. Fluids* **2003**, *26*, 33.
103. Gimeno, M.; Ventosa, N.; Sala, S.; Veciana, J. *Cryst. Growth & Design* **2006**, *6*, 23.
104. Bleich, J.; Müller, B. W.; Wassmus, W. *Int. J. Pharm.* **1993**, *97*, 111.
105. Hanna, M.; York, P. WO 95/01221 A1, 1994 (AN 2005:769844).
106. Subramaniam, B.; Saim, S.; Rajewski, R. A.; Stella, V. J. In *Green Engineering*; Anastas, P. T., Heine, L. G., Williamson, T. C., Eds.; Oxford University Press: Washington, D.C., 2000.
107. Snavely, W. K.; Subramaniam, B.; Rajewski, R. A.; DeFelippis, M. R. *J. Pharm. Sci.* **2002**, *91*, 2026.
108. Niu, F.; Roby, K. F.; Rajewski, R. A.; Decedue, C.; Subramaniam, B. In *Polymeric Drug Delivery Volume II - Polymeric Matrices and Drug Particle Engineering*; Svenson, S., Ed.; Oxford University Press: Washington, DC, 2006.
109. Hutchings, G. J.; Bartley, J. K.; Webster, J. M.; Lopez-Sanchez, J. A.; Gilbert, D. J.; Kiely, C. J.; Carley, A. F.; Howdle, S. M.; Sajip, S.; Caldarelli, S.; Rhodes, C.; Volta, J. C.; Poliakoff, M. *J. Catal.* **2001**, *197*, 232.
110. Johnson, C. A.; Sharma, S.; Subramaniam, B.; Borovik, A. S. *J. Am. Chem. Soc.* **2005**, *127*, 9698.
111. Sencar-Bozic, P.; Srcic, S.; Knez, Z.; Kerc, J. *Int. J. Pharm.* **1997**, *148*, 123.
112. Shine, A.; Gelb, J. WO 98/15348, 1997
113. Chen, J.; Zhang, J. L.; Liu, D. X.; Liu, Z. M.; Han, B. X.; Yang, G. Y. *Colloids Surf., B* **2004**, *33*, 33.
114. Zhang, H. F.; Lu, J.; Han, B. X. *J. Supercrit. Fluids* **2001**, *20*, 65.
115. Zhang, J. L.; Han, B. X.; Liu, J. C.; Zhang, X. G.; He, J.; Liu, Z. M. *Chem. Commun.* **2001**, 2724.
116. Anand, M.; McLeod, M. C.; Bell, P. W.; Roberts, C. B. *J. Phys. Chem. B.* **2005**, *109*, 22852.
117. McLeod, M. C.; Anand, M.; Kitchens, C. L.; Roberts, C. B. *Nano Lett.* **2005**, *5*, 461.
118. McLeod, M. C.; Kitchens, C. L.; Roberts, C. B. *Langmuir* **2005**, *21*, 2414; Liu, J.; Anand, M.; Roberts, C. B. *Langmuir* **2006**, *22*, 3964
119. Subramaniam, B.; Saim, S.; Rajewski, R.; Stella, V. J. U. S. 5,833,891; 1998 (CAN 129:347344).
120. Niu, F.; Subramaniam, B. *Ind. Eng. Chem. Res.*, **2007**, *46*, 3153.
121. Tomasko, D. L.; Li, H.; Liu, D.; Han, X.; Wingert, M. J.; Lee, L. J.; Koelling, K. W. *Ind. Eng. Chem. Res.* **2003**, *42*, 6431.
122. Nalawade, S. P.; Picchioni, F.; Janssen, L. P. B. M. *Progr. Polym. Sci.* **2006**, *31*, 19.

36

123. McHugh, M. A.; Wang, J. D.; Mandel, F. S. In *Supercritical Fluid Technology in Materials Science: Synthesis, Properties, and Applications*; Sun, Y. P., Ed.; Marcel Dekker: New York, 2002.
124. Cooper, A. I. *J. Mater. Chem.* **2000**, *10*, 207.
125. Yeo, S.-D.; Kiran, E. *J. Supercrit. Fluids* **2005**, *34*, 287
126. Teramoto, G.; Oda, T.; Saito, H.; Sano, H.; Fujita, Y. *J. Polym. Sci., Part B: Polym. Phys.* **2004**, *42*, 2738.
127. Shieh, Y.-T.; Yang, H.-S. *J. Supercrit. Fluids* **2005**, *33*, 183.
128. Kropp, D.; Hermann Berstorff Maschinenbau Gmbh, Germany: Eur. Pat. EP 765724 A2, 1997, Application number EP 96-113789 19960829. Priority: DE 95-19536711.Eur. Pat. Appl., 1997.
129. Nagata, T.; Areerat, S.; Ohshima, M.; Tanigaki, M. *Kagaku Kogaku Ronbunshu* **2002**, *28*, 739.
130. Yeo, S.-D.; Kiran, E. *J. Supercrit. Fluids* **2005**, *34*, 287.
131. Maeda, T.; Otake, K.; Nakayama, K.; Horiuchi, S.; Sanyo Electric Co., Ltd., Japan; National Institute of Advanced Industrial Science and Technology: Jpn. Kokai Tokkyo Koho, Pat., JP 2003211483 A2, 2003, application number JP 2002-14410 200201232003.
132. Verreck, G.; Decorte, A.; Li, H.; Tomasko, D.; Arien, A.; Peeters, J.; Rombaut, P.; Mooter, G. V. d.; Brewster, M. E. *J. Supercrit. Fluids* **2006**, 38, 383.
133. Howdle, S. M.; Watson, M. S.; Whitaker, M. J.; Popov, V. K.; Davies, M. C.; Mandel, F. S.; Wang, J. D.; Shakesheff, K. M. *Chem. Commun.* **2001**, 109.
134. Shishikura, A.; Kanamori, K.; Takahashi, H.; Kinbara, H. *J. Agric. Food Chem.* **1994**, *42*, 1993.
135. Chang, C. J.; Randolph, A. D. *Biotechnol. Progr.* **1991**, *7*, 275.
136. B. Rajagopalan, Ph.D. Dissertation, University of Kansas, 2007
137. Wei, M.; Musie, G. T.; Busch, D. H.; Subramaniam, B. *J. Am. Chem. Soc.* **2002**, *124*, 2513.
138. West, K. N.; Hallett, J. P.; Jones, R. S.; Bush, D.; Liotta, C. L.; Eckert, C. A. *Ind. Eng. Chem. Res.* **2004**, *43*, 4827.
139. Lu, J.; Lazzaroni, M. J.; Hallett, J. P.; Bommarius, A. S.; Liotta, C. L.; Eckert, C. A. *Ind. Eng. Chem. Res.* **2004**, *43*, 1586.
140. Ablan, C. D.; Hallett, J. P.; West, K. N.; Jones, R. S.; Eckert, C. A.; Liotta, C. L.; Jessop, P. G. *Chem. Commun.* **2003**, 2972.
141. McHardy, J. *Supercritical Fluid Cleaning*; William Andrew Publishing, 1998; O'Neil, A.; Watkins, J. J. *MRS Bulletin* **2005**, *30*, 967.
142. Mount, D. J.; Rothman, L. B.; Robey, R. J. *Solid State Technol.* **2002**, *45*, 103.
143. Fang, J.; Jin, H.; Ruddy, T.; Pennybaker, K.; Fahey, D. and Subramaniam, B. *Ind. Eng. Chem. Res.* **2007**, *46*, 8687.
144. Melhem, G.A., *Process Safety Progress* **1997**, *16*, 203.

145. *Guide on Fire hazards in oxygen-enriched atmospheres* in *National Fire Protection Association Guides*, 1994

146. Dwyer, J., Jr.; Hansel, J. G.; Philips, T. in *Heat Treating and Surface Engineering, Proceedings of the 22nd Heat Treating Society Conference and the 2nd International Surface Engineering Congress* 2003 Indianapolis, United States..

147. Zabetakis, M.G.; Bulletin - United States Bureau of Mines **1965**, *7*, 121.

148. Jin, H.; Subramaniam, B. *Chem. Eng. Sci.* **2003**. *58*, 1897.

Thermodynamics
and Transport Properties:
Experiment and Modeling

Chapter 2

Phase Equilibrium, Structure, and Transport Properties of Carbon-Dioxide Expanded Liquids: A Molecular Simulation Study

Brian B. Laird[1], Yao A. Houndonougbo[2], and Krzysztof Kuczera[1,3]

[1]Department of Chemistry, University of Kansas, Lawrence, KS 66045
[2]Center for Environmentally Beneficial Catalysis, University of Kansas, Lawrence, KS 66047
[3]Department of Molecular Biosciences, University of Kansas, Lawrence, KS 66045

We review our recent work on phase equilibrium, transport properties and local structure in carbon-dioxide-expanded liquids (CXL's). In addition, we present new data on the effect of increasing CO_2 mole fraction on hydrogen bonding in CO_2-expanded methanol and acetic acid. Phase equilibrium and structure data were calculated using Gibbs Ensemble Monte Carlo techniques and transport properties were determined from molecular-dynamics simulation. We show, using existing force fields for pure CO_2 and a variety of pure organic solvents and standard Lorentz-Berthelot combination rules, that molecular simulation gives results for CXL vapor-liquid phase equilibria and transport properties that are in good agreement with experiment. Also, our phase equilibrium results were found to be at least as good as, and often superior to, predictions using the Peng-Robinson Equation of State (PR-EOS) and have the advantage over empirical PR-EOS predictions in that no experimental input data from the mixtures are required. Thus, molecular simulation is shown to be a potentially important tool in determining CXL properties required for design and optimization of CXL media in catalytic processes.

Introduction

The use of traditional organic solvents (e.g., acetonitrile, acetone and toluene) in industrial catalysis poses a number of challenges to responsible corporate environmental stewardship. When organic solvents enter the waste stream, great care must be taken in their collection, handling and disposal. In addition, traditional organic solvents are generally of high volatility and pose a threat to air quality and contribute to global climate change. In recent years, scientists and engineers, including many at the Center for Environmentally Beneficial Catalysis (CEBC) where our work is based, have been actively seeking environmentally friendly alternative solvents for catalytic processes, such as supercritical carbon dioxide and water, room-temperature ionic liquids and CO_2-expanded liquids (CXL's), which are the subject of the present work.

The application of supercritical CO_2 ($scCO_2$) as a benign medium in catalytic chemistry and reaction engineering satisfies several green chemistry and engineering principles such as pollution reduction, lower toxicity and use of an abundant resource with no increase in environmental burden. However, the reaction benefits are often marginal with $scCO_2$. In many cases, $scCO_2$-based reactions are limited by high process pressures (hundreds of bars) and inadequate solubilities of preferred homogeneous catalysts. Additionally, CO_2 is non-polar, often leading to low reaction rates. The combination of high pressures and low reaction rates increase energy consumption and reactor volumes, both of which adversely affect process economics and severely reduce the environmental advantages.

In contrast, the use of traditional organic solvents offers important reaction benefits. For example, solvents are typically chosen with dielectric properties that help solubilize preferred homogeneous catalysts and favor the desired reaction. However, the concerns with conventional organic solvents are toxicity and environmentally deleterious vapor emissions that may also form explosive mixtures with air.

A relatively new class of solvent media, CXL's are a mixed solvents composed of CO_2 condensed into an organic solvent. [CXL's are referred to as Gas-Expanded Liquids (GXL's) by some workers in the field.] The properties of CXL's can be continuously tuned from those of the neat solvent to those of $scCO_2$ by adjusting the CO_2 partial pressure above the liquid phase, thereby increasing the CO_2 mole fraction in the CXL. For example, a large amount of CO_2 favors gas solubility and the presence of polar organic solvents enhances metal catalyst solubility.

Over the past several years, work at the CEBC and elsewhere [1–14] has demonstrated that CXL's represent a promising class alternative catalytic media. For example, CEBC researchers have recently demonstrated that CXLs are optimal solvent media for a variety of homogeneous catalytic oxidations [15], hydroformylation [16] and solid acid catalyzed-reactions [17]. Reaction advantages include higher gas miscibility compared to organic solvents at ambient conditions; enhanced transport rates due to the properties of dense CO_2; and between one to two orders of magnitude greater TOFs (turnover frequency) than in neat organic solvent or $scCO_2$. Environmental and economic advantages include substantial (up to 80 vol.%) replacement of organic solvents with environmentally benign dense-phase CO_2 and milder process pressures (tens of bars) compared to $scCO_2$ (hundreds of bars). Thus, CXLs have emerged as important components in the optimization of catalytic systems. Moreover, the possibility of controlling CO_2 pressure in CO_2-expanded organic solvents gives an alternative to temperature controlled crystallizations by avoiding slowness of the temperature loop, cooling, and heat transfer [18].

To optimize CXLs solvents for industrial reactor use, knowledge of their vapor-liquid phase equilibria (VLE) and transport properties is essential. Experimental data on phase equilibria in CXLs are available in the literature [19–25]. However, because experiments are time-consuming, the data in the literature on CO_2 expanded solvents are sparse. For the transport properties, very little experimental data can be found in the literature. Sassiat and Morier [26] reported that the diffusion coefficients of benzene in CO_2-expanded methanol increased between 4 and 5 fold with CO_2 addition. Kho et al. [27] measured viscosities of CO_2-expanded fluorinated solvents using an electromagnetic viscometer. In the 25-35°C temperature range and at pressures from 8-72 bar, Kho et al. report up to a 4-5 fold decrease in viscosity with CO_2 addition [27]. Given these experimental difficulties, the development of protocols for the determination of these quantities from either molecular or thermodynamic modeling would be extremely useful for the optimization and design of CXL media for catalytic processes. Molecular modeling has the additional advantage of providing molecular-level structural detail - for example, local solvation structure - that can increase our fundamental understanding of this new and potentially important class of solvents.

To date, there have been relatively few molecular simulation studies of CO_2 expanded liquids using both Monte Carlo and molecular-dynamics (MD) [28]. Our efforts complement several existing molecular-modeling studies by other groups - many of these are described later in this volume. Li and Maroncelli [29] have performed MD simulations of CO_2-expanded cyclohexane, acetonitrile, and methanol to calculate energies, local compositions, viscosities, diffusion coefficients, and dielectric constants and relaxation times. Aida and Inomata [30] reported molecular dynamics simulations of the self-diffusion and dielectric properties in CO_2/methanol mixtures. In addition, Shukla et al. [31] used molecular dynamics simulations to study local solvation and transport effects in CO_2-expanded methanol and acetone.

In this paper, we review our recent work on molecular modeling of phase equilibria, and transport properties and local molecular structure in a variety of CXL's. Much of this work has been previously described in Refs. [32–35]. In our work we have demonstrated, using existing force fields for pure CO_2 and several organic solvents and standard Lorentz-Berthlot combining rules, that molecular simulation can be used to determine phase equilibrium and transport data for CXL's that is in good agreement with the existing experimental data. Thus, molecular simulation is shown to have much promise in readily predicting the properties of candidate CXL solvents in the absence of accurate experimental data. We have chosen to utilize only existing literature force fields for the pure substances (together with standard mixing rules) because force-field development and optimization would be slow and would require the availability of experimental data for the very CXL properties that we are trying to predict - limiting the usefulness of molecular modeling for CXL optimization and design. With respect to phase equilibrium, we have shown that our molecular modeling protocol generates VLE data that is as good as, and often superior to, standard PR-EOS modeling, and has the advantage that we only require data from single component CO_2 and solvent systems - that is, no experimental or literature data from the studied mixtures is necessary as input. In addition, we also report new structural results for hydrogen bonding in CO_2-expanded methanol and acetic acid.

Force Field Models

Here we discuss the force field models and simulation methods that we have used in our studies. To make such studies simple and easily extendable to other systems, we have concentrated on using standard simulation techniques with existing interaction potentials from the literature. More information can be found in our previous papers [32–35]. All molecules studied in this work are represented using a united-atom model in which the CH_n groups (with $0 \leq n \leq 4$) are represented as a single interaction site located on the carbon atom. The interaction between nonbonded sites is parameterized by pairwise additive Lennard-Jones (LJ) 12-6 potentials and Coulombic interactions between partial charges:

$$U(r_{ij}) = 4\epsilon_{ij} \left[\left(\frac{\sigma_{ij}}{r_{ij}} \right)^{12} - \left(\frac{\sigma_{ij}}{r_{ij}} \right)^{6} \right] + \frac{q_i q_j}{4\pi\epsilon_0 r_{ij}} \tag{1}$$

Where r_{ij}, ϵ_{ij}, σ_{ij}, q_i, and q_j are the separation, LJ well depth, LJ size, and partial charges, respectively, for interacting atoms i and j. Interaction parameters (ϵ_{ij} and σ_{ij}) between unlike atoms were calculated using standard Lorentz-Berthelot combination rules [36]. The use of fixed, standard combination rules in our work is designed to make for facile and consistent treatment of a wide class of molecules with off-the-shelf potentials. For all simulations, an Ewald sum method [36, 37] was used to calculate the long-range electrostatic interactions.

The CO_2 and acetonitrile molecules used in the phase equilibrium and structure studies (Monte Carlo) were modeled as rigid three-site molecules. The EPM2 potential parameters developed by Harris and Yung [38] and optimized for VLE were used for CO_2. Acetonitrile interaction parameters were taken from the work of Hirata [39]. For the phase equilibrium and structure Monte Carlo, all the other molecules intramolecular interactions were treated in the following way: the bond length between two neighboring pseudo-atoms was fixed. A harmonic potential is used to control bond angle bending

$$U_{bend} = \frac{k_0}{2}(\theta - \theta_0)^2 \qquad (2)$$

where θ is the actual bending angle, θ_0 is the equilibrium bending angle, and k_0 is the force constant. The dihedral rotations were controlled by a cosine series

$$U_{torsion} = c_0 + c_1\left[1 + \cos(\phi + f_1)\right] + c_2\left[1 - \cos(2\phi)\right] + c_3\left[1 + \cos(3\phi)\right] \qquad (3)$$

where f_1, ϕ, and c_i are a phase factor, dihedral angle, and the i-th expansion coefficient, respectively.

For the transport property simulations (MD) of CO_2/acetonitrile, the CO_2 and acetonitrile molecules were treated by flexible models in which the original force fields were modified by introducing bond stretching and angle bending potentials. Since flexible models are needed for larger molecules such as octene, this approach allows systematic treatment of molecules of any size. More details about the models can be found in Ref. [34].

The parameters for the nonbonded interactions for all molecules studied in this work are listed in Table I. For the intramolecular interaction parameters the reader should consult Refs. [32–34].

Simulation Results for Phase Equilibrium

Phase equilibrium studies for CO_2-expanded acetonitrile (CH_3CN), acetone (CH_3COCH_3), methanol (CH_3OH), ethanol (CH_3CH_2OH), acetic acid (CH_3COOH), toluene ($CH_3C_6H_5$), and 1-octene (C_8H_{16}) were performed using Gibbs Ensemble Monte Carlo (GEMC) simulation [23, 44, 45] and the MCCCS Towhee program [46]. The coexistence curve of the pure-component systems were simulated in the NVT ensemble and the calculations of the binary mixtures were performed within the NPT ensemble.

Table I. Parameters for Non-bonded Interactions (Refs. [38–43])

Substance	atom/group	ϵ/k_b (K)	σ (Å)	$q(e)$	Substance	atom/group	ϵ/k_b (K)	σ (Å)	$q(e)$
Carbon-dioxide	C	28.129	2.757	0.6512	Acetic acid	CH$_3$	98	3.75	0.12
	O	80.507	3.033	-0.3256		C	41.0	3.90	0.42
Acetonitrile	CH$_3$	90.6	3.8	0.269		O(=C)	79	3.05	-0.45
	C	105	3.0	0.129		O(-H)	93	3.02	-0.46
	N	48.8	3.4	-0.398		H	0	0	0.37
Methanol	CH$_3$	98	3.75	0.265	Toluene	CH$_3$	98	3.75	0
	O	93	3.02	-0.7		CH(aro)	50.5	3.695	0
	H	0	0	0.435		C(aro)	21.0	3.88	0
Ethanol	CH$_3$	98	3.75	0	1-Octene	CH$_3$	98	3.75	0
	CH$_2$	46	3.95	0.265		CH$_2$(-C)	46	3.95	0
	O	93	3.02	-0.7		CH	47.0	3.73	0
	H	0	0	0.435		C	40.0	3.82	0.424
Acetone	CH$_3$	98	3.75	0		CH$_2$(=C)	85.0	3.675	0
	O	79	3.05	-0.424					

Single Component Systems

In order to validate the molecular models and the simulation method for phase equilibrium, we calculated the single-component vapor-liquid coexistence curves for all substances considered. These results from Ref. [32] are summarized in Figure 1. The saturated liquid and vapor densities are in good agreement with experimental data [47–51] for carbon-dioxide, methanol, ethanol, acetone, toluene, and 1-octene. This good agreement is unsurprising because these potentials were optimized to give good vapor-liquid coexistence. The simulation results for pure acetonitrile are only in qualitative agreement with experiment at high temperatures - for example, the critical point is underestimated by the interaction model we used. As we have previously discussed [32, 33], this deviation from experimental data for the neat acetonitrile will not affect our CO_2/actenonitrile mixture results significantly at the much lower temperatures that we consider these studies.

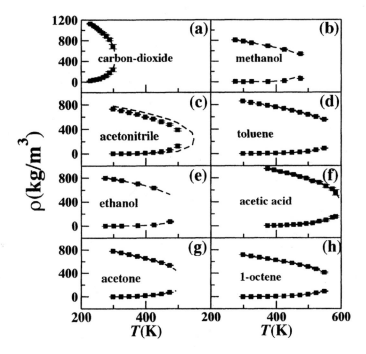

Figure 1. Vapor-liquid coexistence curves for all single component systems: (a)carbon-dioxide; (b) methanol; (c) acetonitrile; (d) toluene; (e) ethanol; (f) acetic acid; (g) acetone; and (h) 1-octene. The filled squares represent the coexistence data from our simulations and the dashed lines are the experimental results from Refs. [47–51].

Mixtures: CO_2-Expanded Solvents

As a metric for phase equilibrium, we calculate the CXL volume expansion, defined as the ratio

$$\frac{V(T,P)}{V_0(T,P_0)} \tag{4}$$

where V represents the total volume of the mixture liquid phase at temperature T and pressure P and V_0 is the total volume of the pure solvent at the same temperature and at 1 bar pressure (P_0). At selected temperatures, the volume expansion for a variety of CO_2 expanded solvents were determined for pressure up to 5 MPa using two-phase GEMC simulations in a NVT ensemble. Specifically, we have examined the binary systems CO_2 + acetonitrile at 25°C, acetone at 30°C, methanol at 30°C, ethanol at 25°C, acetic acid at 25° C, toluene at 30°C, and 1-octene at 60°C. Our results are summarized in Figure 2 and compared with both experiments [32] and predictions using the Peng-Robinson Equation of State (see the Appendix).

For the binary mixtures of CO_2 with acetonitrile, acetone, methanol, ethanol, and toluene, the results are in good agreement with experimental results. For the binary mixtures of CO_2 with acetic acid and 1-octene, the agreement with the experimental data is good at low pressure (< 4 MPa). For the highest pressures in the study (between 4 and 5 MPa), the calculated values are underestimated in CO_2-expanded acetic acid and overestimated in CO_2-expanded 1-octene. The Peng-Robinson-equation of-state volume-expansion calculations for CO_2/ methanol, CO_2/toluene, and CO_2/1octene are in close agreement with our simulation results. However, the PR-EOS results in CO_2/acetonitrile, CO_2/acetone, and CO_2/acetic acid tend to underestimate the volume expansion, relative to simulation and experiments. In these cases our simulations give superior agreement with experimental results than PR-EOS predictions.

In Ref. [35], we examined the pressure-composition diagrams for our collection of CO_2-expanded solvent systems. In Figure 3 we compare with experiment [19] our simulation and PR-EOS predictions for the vapor and liquid phase compositions of CO_2 expanded acetonitrile. For this system, both the simulation and Peng-Robinson results are in good agreement with the experimental data. Pressure-composition diagrams for the other six CO_2-expanded solvents are shown in Figure 4 to give good agreement between simulation results and the PR-EOS modeling. For all seven binary systems, the CO_2 mole fractions in the vapor phase are close to unity indicating a CO_2 rich phase.

In the liquid phase, the solubility of carbon-dioxide approximatively increased linearly with pressure in the low pressure region (below 5 MPa). Carbon-dioxide solubility up to 65 mole % is achieved in the range of pressures

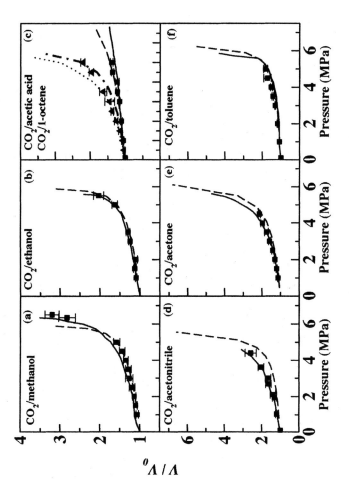

Figure 2. Volume expansion for CO_2-expanded (a) methanol (30°C); (b) ethanol (25°C); (c) acetic acid (25°C) and 1-octene (60°C); (d) acetonitrile (25°C); (e) acetone (30°C); (f) toluene (30°C). In panels (a), (b), (d), (e) and (f), the filled squares represent our simulation results, the solid lines are the experimental results and the dashed lines represent predictions from the PR EOS. In panel (c), the same scheme applies to the data for 1-octene. For the acetic acid data in panel (c) the filled triangles represent our simulation results, the dot-dashed lines are the experimental results and the dotted lines represent predictions from the PR EOS.

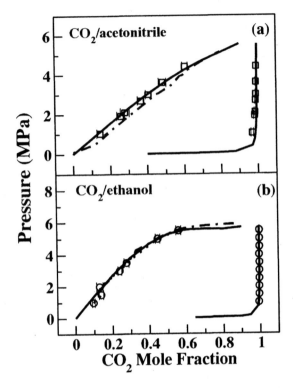

Figure 3. Pressure-composition diagrams for CO_2-expanded (a) acetonitrile at 25°C) and (b) ethanol at 25°C: Our simulation data is shown as open symbols, whereas the PR-EOS predictions are shown as solid lines. Dot-dashed lines represent the experimental data from Ref [19].

studied. Pressure-composition data for CO_2-expanded acetone is not shown, but can be found in Ref. [35]. In this system, the agreement between simulation and PR-EOS is very good for both vapor and liquid phases.

Figure 5 shows the densities of the coexistence phases in CO_2-expanded acetonitrile and ethanol. In addition to our simulation data and PR-EOS predictions, this figure shows experimental values from Ref. [19]. In both of these systems, the simulation values and PR-EOS predictions for the vapor-phase densities agree well with the corresponding experimental results. This level of agreement is also seen in the liquid-phase densities in ethanol. In CO_2-expanded acetonitrile, however, the simulated values for the liquid-phase density are in excellent agreement with experiment, but the PR-EOS predicts a value of the liquid density that is significantly lower than either experiment or simulation. For CO_2-expanded methanol, acetic acid, toluene and 1-octene, the

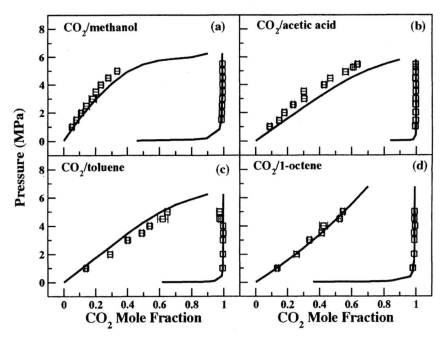

Figure 4. Pressure-composition diagrams for CO₂-expanded (a) methanol at 30°C) and (b) acetic acid at 25°C, (c) toluene at 30°C and (d) 1-octene at 60°C: Our simulation data is shown as open symbols, whereas the PR-EOS predictions are shown as solid lines.

simulation and PR-EOS predictions for the coexisting phase densities are shown in Figure 6. Agreement between these methods is very good in 1-octene and fair in methanol, but in both acetic acid and toluene the PR-EOS prediction for the liquid-phase density is significantly lower than the simulation results. Density data for CO_2-expanded acetone is not shown, but can be found in Ref. [35]. In acetone, the agreement between simulation and PR-EOS is excellent for the vapor phase, but only fair for the liquid phase, where the PR-EOS predicts a slightly smaller density than simulation for the liquid phase.

Transport Properties

In a recent paper [34], we studied the translational diffusion coefficients, the rotational correlation times, and the shear viscosities in CO_2-expanded acetonitrile liquid. In this section, we summarize the *NVT* simulation results of these investigations. These molecular-dynamics simulations were carried out

52

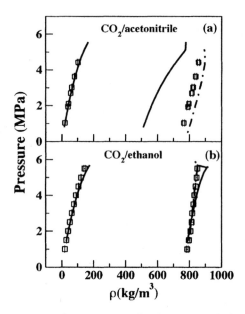

Figure 5. Pressure versus density curves for CO₂-expanded (a) acetonitrile at 25°C) and (b) ethanol at 25°C: Our simulation data is shown as open symbols, whereas the PR-EOS predictions are shown as solid lines. Dot-dashed lines represent the experimental data from Ref. [19].

using the program CHARMM [52]. Both constant volume (*NVT*) and constant pressure (*NPT*) were employed in our recent work [34] for calculations in pure liquids and binary mixtures. The *NVT* simulations were performed using the densities and compositions consistent with the experimental coexistence curve, whereas the *NPT* simulations were performed along the coexistence curve appropriate for the model potential. In this review, we report only the *NVT* results. The Nosé-Hoover algorithm was used to maintain a constant temperature at 25°C in all cases [36].

The translational diffusion coefficients were calculated from the slope of the mean-squared displacement as a function of time [28] and are plotted as functions of CO_2 mole fractions (x_{CO_2}) in Figure 7(a). The available experimental values for the pure acetonitrile [53] (x_{CO_2} =0) and pure CO_2 [54] (x_{CO_2} =1) are also plotted for comparison. For the pure liquid, the deviations between our simulation results and experimental values are in 19-23% range. These deviations are reasonable since the model parameters used in this work were not optimized for transport properties. In the mixtures, the translational diffusion coefficients for both acetonitrile and CO_2 increase with CO_2 mole fraction. The simulation results show an inverse linear variation of the translational diffusion coefficients with x_{CO_2} in CO_2-expanded acetonitrile.

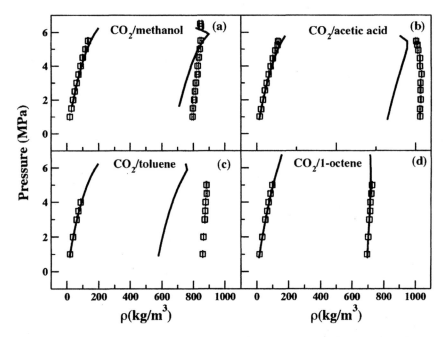

*Figure 6. Pressure versus density curves for CO₂-expanded (a) methanol at
30°C) and (b) acetic acid at 25°C, (c) toluene at 30°C and (d) 1-octene at 60°C:
Our simulation data is shown as open squares, whereas the PR-EOS
predictions are shown as solid lines.*

Figure 7(b) shows the mole fraction dependence of the calculated values of
the rotational correlation times, which were determined from the exponential
decal of the second-order rotational correlation function [34]. The deviation
between experimentally determined and simulated values fall in the 11-16%
range indicating a quite good agreement since the potential models used were
not optimized for these properties. The simulation data in Figure 7(b) exhibit
smooth variations of the correlation time with composition for both CO_2 and
acetonitrile. The data can be well fitted using a linear equation in x_{CO_2}. With the
diffusion coefficient being inversely proportional to the correlation time, Figure
7(b) shows faster rotational diffusion of both CO_2 and acetonitrile with
increasing CO_2 mole fraction for the CO_2/acetonitrile mixtures.

In Figure 8, the simulation results for the shear viscosities are presented
with the experimental values of the pure components for comparison. The
viscosities were determined using a Green-Kubo formula involving the
integration of the time autocorrelation function of the off-diagonal elements of
the stress tensor [34]. For pure acetonitrile, the simulation overestimates the
experimental result by 32%. However, considering the statistical uncertainties

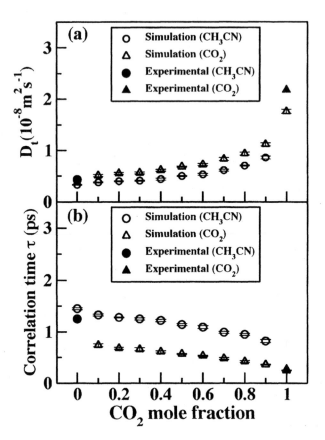

Figure 7. (a) Translational diffusion constants in CO_2/acetonitrile (25°C) mixtures. The values from MD simulation (experiment [53, 54]) are represented as open (filled) circles (acetonitrile) and triangles (CO_2) and filled diamonds. (b) Rotational correlation times for the system in (a).

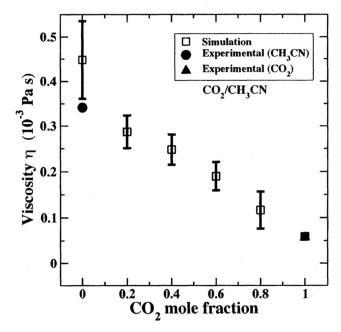

Figure 8. Shear viscosity in CO₂/acetonitrile (25°C) mixtures: Open circles show our simulation results and the filled diamonds are the experimental values for the pure solvents taken from Refs. [55, 56].

(20%), the agreement with the experimental measurement is quite good. For carbon dioxide, the difference between the calculated and measured values is only 3%. Overall, the simulated viscosities of the pure solvents are in reasonable agreement with experimental data. The predicted values of the viscosities of the mixtures presented in Figure 8 decrease with increasing of x_{CO_2}. A viscosity decrease by a factor of 8 compared to pure acetonitrile is predicted in the studied composition range. The trend can be correlated with a linear equation. Similar results were also found by Li and Maroncelli [29].

Structural Properties of CO₂-Expanded Solvents

In order to study the effect of increasing mole fraction of CO₂ on the local molecular structure of the liquid phase in CXL's, we have calculated, the radial distribution functions (RDFs) for our simulated CXL systems as a function of pressure. The RDF $g_{ij}(r)$ represents the probability, relative to an ideal gas, of finding an atom of type j on one molecule at a distance r away from an atom of type i on another molecule. The RDFs are the primary measures of local molecular structure in liquid mixtures. For this review, we focus on two aspects of our results: orientational order in CO_2-expanded acetonitrile and hydrogen bonding in methanol, ethanol and acetic acid CXL's.

Orientational Correlations in CO_2-expanded acetonitrile

In Ref. [32], we analyzed the structure of CO_2-expanded acetonitrile as a function of pressure. In this work, only the N-N RDF between two CH_3CN molecules was found to show any significant change as the pressure was changed from 0.1 to 5 MPa (1 to 50 bar) at 25°C. Figure 9 shows the $g_{NN}(r)$ RDF for 10.34, 30.96 and 50.40 bar. As the pressure (and, thus, the CO_2 mole fraction) is increased, the shape of the double nearest neighbor peak changes significantly with the first peak decreasing in amplitude and the second peak becoming more pronounced. The first maximum in the double peak was reported by Böhm, et al. [57] to be dominated by antiparallel molecular pairs. Based on this, one might conclude from Figure 9 that the fraction of antiparallel neighbor pairs decreases with increasing pressure. However, we have determined in Ref. [32] that this is not the case.

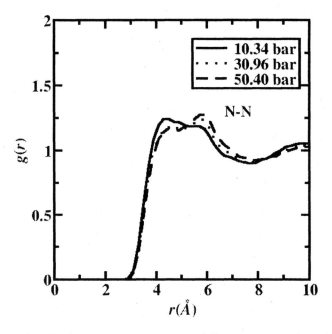

Figure 9. The RDF between N atom sites on different acetonitrile molecules for CO_2-expanded acetonitrile at 10.34, 30.96 and 50.40 bar pressure and 25°C.

To show this, we defined a set of dipole-partitioned RDFs as follows: For all MeC-NMeCN molecular pairs that go into calculation of the RDF, we define an orientation angle given by

$$\cos\theta = \hat{\mu}_1 \cdot \hat{\mu}_2,$$

where $\hat{\mu}_1$ and $\hat{\mu}_2$ are the dipole-moment unit vectors for the two molecules in the pair. The values of $\cos\theta$ for antiparallel, perpendicular and parallel orientations are 1, 0 and +1, respectively. The antiparallel dipole-partitioned RDF, $g(r)$, is defined by restricting the calculation of the RDF only to neighbor pairs for which $1 \le \cos\theta < 1/3$. Similarly, the perpendicular and parallel dipole-partitioned RDFs, $g_0(r)$ and $g_+(r)$, are defined using the ranges $1/3 \le \cos\theta < +1/3$ and $+1/3 \le \cos\theta \le +1$, respectively. The full RDF can be obtained from the sum of the three dipole-partitioned RDFs. In addition, because the orientations become uncorrelated at large separations, each of the dipole-partitioned RDFs go to the value of $+1/3$ at large r. These dipole-partitioned RDFs for the NN sites of MeCN are shown in Figure 10 as functions of pressure. From these plots, we see, in agreement with Ref. [57], that the dominant orientation to the first

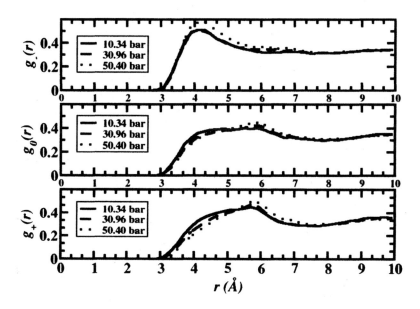

Figure 10. The dipole-partitioned NN RDFs for CO_2-expanded acetonitrile at various pressures: (top) antiparallel aligned molecules ($1 \le \cos\theta < 1/3$), (middle) perpendicular molecules ($1/3 \le \cos\theta < +1/3$) and (bottom) parallel aligned molecules ($+1/3 \le \cos\theta \le 1$). The quantity $\cos\theta$ is the dot product of the dipole unit vectors on the two acetonitrile molecules see text.

maximum of the double neighbor peak is antiparallel, whereas the second maximum is predominantly perpendicular or parallel. However, we also see that the decrease in the first maximum, and corresponding increase in the second maximum, as pressure is increased is due, not to a decrease in antiparallel orientations, but instead to a decrease of the fraction of parallel orientations in the first maximum. In fact, the fraction of antiparallel orientations in the first maximum actually increases with increasing pressure. In other words, the change in the N-N RDF is due to an increase in the degree of orientational segregation between the first and second peak.

Hydrogen Bonding in CO_2-expanded Methanol and Acetic Acid

To examine the effect of CO_2-expansion on hydrogen bonding in methanol, we have calculated via simulation the O(OH)H(OH) RDF for this system at 25°C in this system for CO_2 pressures up to 5 MPa (50 bar). Our data is plotted in Figure 11. The first peak in this RDF represents the hydrogen bond. As the pressure (and thus, the CO_2 mole fraction) is increased from 10 to 50 MPa, the height of the first peak increases from about 3.4 to about 4.5, indicating a significant increase in the degree of self-association of methanol in the presences of increasing CO_2 mole fraction. This increase is far less pronounced than that reported by Shukla et al. [31] in similar simulations, in which the first peak of the O(OH)-H(OH) RDF in methanol is calculated to reach a very large peak height of 24 at a CO_2 mole fraction of 0.884. The Shukla et al. results indicate a far larger degree of clustering in methanol at high CO_2 concentrations than we see in our results. There are a couple of possible explanations for this discrepancy. First, our highest pressure simulations (50 bar) correspond to a CO_2 mole fraction of 75%, which is lower than the 88.4% maximum CO_2 mole fraction examined in Shukla et al. We are currently examining CO_2-expanded methanol at higher pressures to see if we see a significant enhancement of clustering. Second, we use a slightly different force field for the CO_2 and CH_3OH interactions than are used in Shukla et al. Finally, whereas we perform simulations at pressures and densities (and thus, mole fractions) along the calculated coexistence curve appropriate to our model, Shukla et al. perform their simulations at densities and mole fractions consistent with the experimental coexistence curve for methanol and are thus not likely to be on (or near) the actually coexistence curve for the model that they use. Clearly, more work needs to be done to resolve the issue of methanol clustering in CO_2-expanded methanol.

Similarly, we have calculated the O(CO)H(OH) and O(OH)-H(OH) RDF's in CO_2 expanded acetic acid, shown in shown in Figures 12(a) and 12(b), respectively. Figure 12 shows strong peaks at approximately 1.8 Å for all H(OH)-O(CO) RDFs and two peaks at approximately 1.9 Å and 3.9 Å for all H(OH)- O(OH) similar to structures that we see in neat CH_3COOH. Hydrogen

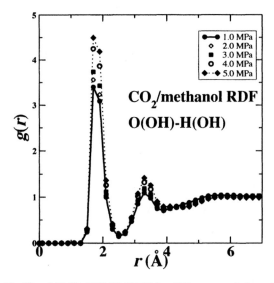

Figure 11. The O(OH)-H(OH) RDF for CO₂-expanded methanol for several pressures

bonds H(OH)-O(OH) form at approximately 3.9 Å and suggests dimer formation between CH_3COOH molecules. The effect of pressure on the concentration of dimers appears negligible. The weak first peaks seen in H(OH)-O(OH) RDFs indicate that hydrogen bonding to the hydroxl is less prevalent than that to the carboxyl oxygen in these acetic acid mixtures. While the effect of increasing pressure (increasing CO_2 mole fraction) is minimal for H(OH)-O(OH) hydrogen bonds, the H(OH)-O(CO) hydrogen bonds are enhanced with increasing pressure.

Conclusion

We have studied the phase equilibrium, structure and transport properties of a variety of carbon-dioxide expanded solvents using Monte Carlo and molecular-dynamics simulation. The interaction potentials were constructed from single-component force fields adopted from the literature together with standard Lorentz-Berthelot combining rules for the mixtures. Using both Gibbs Ensemble Monte Carlo (GEMC) molecular simulation and standard Peng-Robinson Equation of State (PR-EOS) modeling, we have determined the volume expansion, pressure-composition and pressure-density diagrams are obtained for carbon-dioxide expanded acetonitrile, acetone, methanol, ethanol, acetic acid, toluene, and 1-octene. In addition, molecular-dynamics simulation

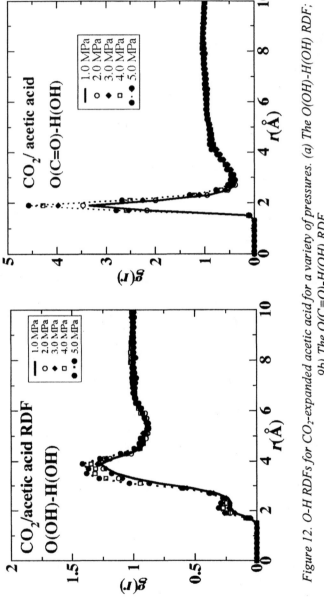

Figure 12. O-H RDFs for CO₂-expanded acetic acid for a variety of pressures. (a) The O(OH)-H(OH) RDF;
9b) The O(C=O)-H(OH) RDF.

results for translational diffusion coefficients, rotational correlation times, and shear viscosities in carbon-dioxide expanded acetonitrile were also reported. We show that, using only off-the-shelf interaction potentials for the pure systems, our GEMC results for the liquid-vapor phase equilibrium give very good agreement with experiment that is at least as good as, and often superior to, standard PR-EOS thermodynamic modeling. In addition, these simulations give detailed information as to the local solvation structure of CXL's that is unobtainable from equation-of-state modeling. Finally, for CO_2-expanded acetonitrile, we demonstrate that molecular simulations yield good agreement with existing experimental for translational and rotational diffusion constants, as well as for viscosity.

Acknowledgments

This work is supported by the National Science Foundation grant EEC-0310689 providing for the University of Kansas Center for Environmentally Beneficial Catalysis.

Appendix

In engineering, the standard method for modeling vapor-liquid equilibria in multicomponent systems is to use empirical equations of state (EOS), such as the Peng-Robinson (PR) EOS [58, 59]. The Peng-Robinson equation of state is a two-parameter extension of the van der Waals equation of state. For mixtures, the parameters and mixing rules of the PR-EOS are determined empirically from experimental data and must be optimized for each mixture studied. Thus, accurate PR-EOS modeling requires existing experimental data on the very mixtures to be studied - a series drawback to exploring potential solvent candidates. For complex systems, the mixing rules necessary for accurate modeling of the VLE are often quite complicated.

The Peng-Robinson Equation of State (PR-EOS) is given by

$$P_{\text{PR-EOS}} = \frac{RT}{v - b} - \frac{a(T)}{v + 2bv - b^2} \tag{5}$$

where T, P, R, and v are the temperature, pressure, the ideal gas constant, and molar volume. For a component i, parameters a and b are given by

$$a_i = 0.457235 \frac{(RT_{ci})^2}{P_{ci}} \left[1 + \kappa_i \left(1 - \sqrt{\frac{T}{T_{ci}}} \right) \right]^2 \tag{6}$$

$$b_i = 0.077796 \frac{RT_{ci}}{P_{ci}}, \tag{7}$$

where T_{ci} and P_{ci} are the critical temperature and pressure of compound i. The κ_i term has the form

$$\kappa_i = 0.37464 + 1.54226\omega_i - 0.26992\omega_i^2 \tag{8}$$

where ω is Pitzer's acentric factor. For this equation of state we determine the mixture parameters a and b using the conventional Van der Waals mixing and combining rules

$$a = \sum_i \sum_j x_i x_j a_{ij} \tag{9}$$

$$a_{ij} = (1 - k_{ij}) \sqrt{a_{ii} a_{jj}} \tag{10}$$

$$b = \sum_i \sum_j x_i x_j b_{ij} \tag{11}$$

$$b_{ij} = (1 - l_{ij}) \frac{b_{ii} + b_{jj}}{2} \tag{12}$$

The calculations of VLE were carried out by equating the fugacities ϕ_i of each component in the liquid phase and in the vapor phase and summing the mole fraction x_i and y_i in each phase to unity

$$\phi_i^L x_i P = \phi_i^V y_i P \tag{13}$$

$$\sum_{i=1}^{2} x_i = 1 \; and \; \sum_{i=1}^{2} y_i = 1 \tag{14}$$

The Newton-Raphson method [60] (NR) was used to solve the nonlinear equations 13 and 14.

For the binary systems CO_2/methanol, CO_2/acetic acid, and CO_2/1-octene, the simplex optimization method was used to determined the cross interaction parameters k_{ij} and l_{ij} by minimization of the objective function

$$F = \sum_{m=1}^{n} \left| P_{calc,m} - P_{exp,m} \right|^2 \tag{15}$$

where n is the number of experimental data points. In the case of $CO_2/1$-octene system, *VLE* data from our simulations were used since we could not find experimental vapor-liquid equilibrium data.

More details about the PR-EOS calculations, as well as the pure component and mixing parameters used, can be found in Ref. [32].

References

1. DeSimone, J. M. *Science* **2002**, *297*, 799.
2. Gordon, C. M.; Leitner, W. *Chimica Oggi* **2004**, *22*, 39.
3. Leitner, W. In *Green Chemistry Using Liquid and Supercritical Carbon Dioxide*; DeSimone, J. M., Tumas, W., Eds.; Oxford University Press, Inc.: New York, 2003; page 81.
4. Eckert, C.; Liotta, C. L.; Bush, D.; Brown, J.; Hallett, J. P. *J. Phys. Chem. B* **2004**, *108*, 18108.
5. Beckman, E. J. *Environ. Sci. Technol.* **2002**, *36*, 347A.
6. Tundo, P.; Anastas, P.; Black, D. S.; Breen, J.; Collins, T.; Memoli, S.; Miyamoto, J.; Polyakoff, M.; Tumas, W. *Pure Appl. Chem.* **2000**, *72*(7), 1207.
7. Morgenstern, D. A.; LeLacheur, R.; Morita, D.; Borkowsky, S.; Feng, S.; Brown, G. H.; Luan, L.; Gross, M.; Burk, M.; Tumas, W. In *ACS Symposium Series No 626 (Green Chemistry)*; 1996; page 132.
8. Jessop, P. G.; Heldebrant, D.; Li, X.; Eckert, C.; Liotta, C. L. *Nature* **2005**, *436*, 1102.
9. Jessop, P. G.; Stanley, R. R.; Brown, R. A.; Eckert, C. A.; Liotta, C. L.; Ngo, T. T.; Pollet, P. *Green Chem.* **2003**, *5*, 123.
10. Hunter, S. E.; Savage, P. E. *Chem. Eng. Sci.* **2004**, *59*, 4903.
11. Licence, P.; Poliakoff, M. In *Multiphase Homogeneous Catalysis*; Cornils, B., Ed.; Wiley-VCH Verlag GmbH and Co. KGaA: Weinheim, 2005; page 734.
12. Amandi, R.; Hyde, J.; Poliakoff, M. In *Carbon Dioxide Recovery and Utilization*; Aresta, M., Ed.; Kluwer Academic Publishers: Dordrecht, 2003; page 169.
13. Hyde, J.; Leitner, W.; Poliakoff, M. In *High Pressure Chemistry*; Eldik, R. V., Klaerner, F.-G., Eds.; Wiley-VCH Verlag GmbH and Co. KGaA,: Wilhelm, 2002; page 371.
14. Subramaniam, B.; Busch, D. H. In *ACS Symposium Series No. 809, CO_2 Conversion and Utilization*; C. Song, A. M. G., Fujimoto, K., Eds.; American chemical Society: Washington, 2002; page 364.
15. Wei, M.; Musie, G.; Busch, D.; Subramaniam, B. *Green Chem.* **2004**, *6*, 387.
16. Jin, H.; Subramaniam, B. *Chem. Eng. Sci.* **2004**, *59*, 4887.

17. Lyon, C.; Sarsani, V.; Subramaniam, B. *Ind. Eng. Chem. Res.* **2004**, *43*, 4809.
18. Eckert, C.; Bush, D.; Brown, J.; Liotta, C. *Ind. Eng. Chem. Res.* **2000**, *39*, 4615.
19. Kordikowski, A.; Schenk, A.; Nielen, R. V.; Peters, C. *J. Supercrit. Fluids* **1995**, *8*, 205.
20. Bamberger, A.; Sieder, G.; Maurer, G. *J. Supercrit. Fluids* **2000**, *17*, 97.
21. Byun, H.; Kim, K.; McHugh, M. A. *Ind. Eng. Chem. Res.* **2000**, *39*, 4580.
22. Day, C.; Chang, C.; Chen, C. *J. Chem. Eng. Data* **1996**, *41*, 839.
23. Panagiotopoulos, A. *Mol. Phys.* **1987**, *61*, 813.
24. Chen, J.; Wu, W.; Han, B.; Gao, L.; Mu, T.; Liu, Z.; Jiang, T.; Du, J. *J. Chem. Eng. Data* **2003**, *48*, 544.
25. Wu, J.; Pan, Q.; Rempel, G. *J. Chem. Eng. Data* **2004**, *49*, 976.
26. Sassiat, P. R.; Mourier, P.; Caude, M. H.; Rosset, R. H. *Anal. Chem.* **1987**, *59*, 1164-1170.
27. Kho, Y. W.; Conrad, D. C.; Knutson, B. L. *Fluid Phase Equilib.* **2003**, *206*, 179-183.
28. Frenkel, D.; Smit, B. *Understanding Molecular Simulation*; Academic Press: San Diego, 2001.
29. Li, H.; Maroncelli, M. *J. Phys. Chem. B* **2006**, *110*, 21189.
30. Aida, T.; Inomata, H. *Mol. Phys.* **2004**, *30*, 407-412.
31. Shukla, C.; Hallett, J.; Popov, A.; Hernandez, R.; Eckert, C. L. C. *J. Phys. Chem. B* **2006**, *47*, 24101-24111.
32. Houndonougbo, Y.; Jiu, H.; Rajagopalan, B.; Wong, K.; Kuczera, K.; Subramaniam, B.; Laird, B. *J. Phys. Chem. B* **2006**, *110*, 13195-13202.
33. Houndonougbo, Y.; Guo, J.; Lushington, G.; Laird, B. *Mol. Phys.* **2006**, *104*, 2955-2960.
34. Houndonougbo, Y.; Laird, B.; Kuczera, K. *J. Chem. Phys.* **2007**, *126*, 074507.
35. Houndonougbo, Y.; Kuczera, K.; Laird., B. *Mol. Sim.* **2007**, *33*, 861-869.
36. Allen, M.; Tildesley, D. *Computer Simulation of Liquids*; Oxford Science Press: Oxford, 1987.
37. Sagui, C.; Darden, T. A. *Annu. Rev. Biophys. Biomol. Struct.* **1999**, *28*, 155-179.
38. Harris, J.; Yung, K. *J. Phys. Chem.* **1995**, *99*, 12021.
39. Hirata, Y. *J. Phys. Chem. A* **2002**, *106*, 2187.
40. Stubbs, J. M.; Potoff, J. J.; Siepmann, J. I. *J. Phys. Chem. B* **2004**, *108*, 17596.
41. Chen, B.; Potoff, J. J.; Siepmann, J. I. *J. Phys. Chem. B* **2001**, *105*, 3093.
42. Kamath, G.; Cao, F.; Potoff, J. *J. Phys. Chem. B* **2004**, *108*, 14130.
43. Wick, C. D.; Martin, M. G.; Siepmann, J. I. *J. Phys. Chem. B* **2000**, *104*, 8008.
44. Panagiotopoulos, A.; Quirke, N.; Stapleton, M.; Tildesley, D. *J. Mol. Phys.* **1988**, *63*, 527.

45. Panagiotopoulos, A. Z. *Mol. Simul.* **1992**, *9* (1), 1.
46. Martin, M. G.; Chen, B.; Wick, C. D.; Potoff, J. J.; Stubbs, J. M.; Siepmann, J. I. *http://towhee.sourceforge.net* **1994**.
47. Newitt, D. M.; Pai, M. U.; Kuloor, N. R.; Huggill, J. A. W. In *Thermodynamic Functions of Gases*; Dim, F. Ed.: Butterworth: London, 1956; Vol. 123.
48. Francesconi, A.; Franck, E. U.; Lentz, H. *Bunsenges. Phys. Chem.* **1975**, *68*, 897.
49. Smith, B. D.; Srivastava, R. *Thermodynamic Data for Pure Compounds: Part B Hydrocarbons and Alcohols*; Elsevier: Amsterdam, 1986.
50. Vargaftik, N. B. *Handbook of Physical Properties of Liquids and Gases: Pure Substances and Mixtures, 3rd Augm. and Rev. Ed.*; New York : Begell House, 1996.
51. Smith, B. D.; Srivastava, R. *Thermodynamic Data for Pure Compounds: Part A Hydrocarbons and Ketones*; Elsevier: New York, 1986.
52. Brooks, B. R.; Bruccoleri, R.; Olafson, B.; States, D.; Swaminathan, S.; Karplus, M. *J. Comp. Chem.* **1983**, *4*, 187-217.
53. Hurle, R. L.; Woolf, L. A. *J. Chem. Soc. Faraday Trans. 1* **1982**, *78*, 2233–2238.
54. Etesse, P.; Zega, J. A.; Kobayashi, R. *J. Chem. Phys.* **1992**, *97*, 2022-2029.
55. Weast, R., Ed. *CRC Hand Book of Chemistry and Physics, 81st Ed.*; CRC Press: Cleveland, OH, 2001.
56. Fenghour, A.; Wakenham, W. A.; Vesovic, V. *J. Phys. Chem. Ref. Data* **1998**, *27*, 31-44.
57. Bohm, H.; Lynden-Bell, R.; Madden, P.; McDonald, I. *Mol. Phys.* **1984**, *5*, 761.
58. Orbey, H.; Sandler, S. *Modeling Vapor-Liquid Equilibria: Cubic Equations of State and Their Mixing Rules*; Cambridge University Press: Cambridge, 1998.
59. Peng, D.; Robinson, D. *Ind. Eng. Chem. Fundam.* **1976**, *15*, 59.
60. Press, W.; Teukolsky, S.; Vetterling, W.; Flannery, B. *Numerical Recipies in Fortran*; Cambridge University Press: New York, 1992.

Chapter 3

On the Molecular-Based Modeling of Dilute Ternary Systems in Compressible Media: Formal Results and Thermodynamic Pitfalls

Ariel A. Chialvo[1], Sebastian Chialvo[2], and J. Michael Simonson[3]

[1]Chemical Sciences Division, Aqueous Chemistry and Geochemistry Group, Oak Ridge National Laboratory, Oak Ridge, TN 37831–6110
[2]Department of Chemical Engineering, University of Florida, Gainesville, FL 32611–6005
[3]Center for Nanophase Materials Sciences Division and Chemical Sciences Division, Macromolecular Structures Group, Oak Ridge National Laboratory, Oak Ridge, TN 37831–6493

Truncated series expansions for the species fugacity coefficients in ternary dilute systems are derived for the systematic study of mixed solutes in highly compressible media. Then, explicit molecular-based expressions for the expansion coefficients are drawn in terms of direct and total correlation function integrals associated with the actual microstructure of the reference infinite dilute system. Finally, these self-consistent formal expressions are used (a) to derive the corresponding expressions for special systems, (b) to highlight, and discuss with examples from the literature, some frequent pitfalls in the molecular modeling of these mixtures leading to serious thermodynamic inconsistencies, and (c) to illustrate how the proposed expressions reduce exactly, in the zero-density limit, to those for the properties of mixtures obeying the 1st-order truncated virial equation of state.

Introduction

Solvation of dilute solutes in compressible media plays a pivotal role in green chemistry and engineering processes, where the success with which we can control or manipulate them depends on our ability to model those systems accurately. It is no longer a matter of having a more accurate EOS or a more successful correlation, but rather a crucial requirement to developing fundamentally based and thermodynamically consistent formalisms able to connect microscopic details of the systems with the desired macroscopic properties of interest.

One solvation phenomenon relevant to processes conducted in green media is the solubility enhancement of non-volatile solutes in near critical solvents and the potential effect on it by other species in solution. These species are usually co-solutes and/or co-solvents which might give rise to synergistic effects on the solute's solubility, known as co-solute or mixed-solute (*1, 2*) and co-solvent or entrainer effects (*3, 4*), respectively. These synergistic solvation phenomena immediately suggest a few relevant questions including: (a) the identification of the intermolecular asymmetry that originates the effect, (b) the connection between this asymmetry and the system's microscopic structural manifestation, and (c) the translation of the microstructural evidence to a measurable macroscopic counterpart that would be the target of any modeling effort. The answer to these questions will provide the required molecular picture of the underlying mechanism to tackle more appropriately the actual modeling process, without invoking or speculating about the nature and type of the involved intermolecular interactions.

From a modeling viewpoint, these questions must be put in the context of the two main features that characterize these systems, namely, the high dilution of the species in solution, and the high compressibility of the medium at the process' conditions (*5*). In principle, these two simultaneous features could become rather problematic in the modeling process, though we can also turn them to our advantage as we have done it elsewhere (*6-9*). In fact, we can facilitate the modeling by a judicious separation of contributions to the properties according to their length scales, *i.e.*, into short- and long-range, as solvation and compressibility-driven contributions, respectively, and can simultaneously take advantage of the condition of infinite dilution as a reference state for the development of free energy composition expansions, based on fugacity or activity coefficients (*10, 11*).

While truncated composition expansions have been frequently used in fluid phase equilibrium calculations, especially for binary mixtures (*12*) for which compliance with the Gibbs-Duhem equation might also imply exactness of the corresponding differentials, their multicomponent counterparts might suffer from thermodynamic inconsistencies if the truncated expressions do not fulfill the condition of exactness, *aka* the Maxwell relations in the thermodynamic jargon. In other words, as the derived truncated expansions might not describe state functions, their predictions will depend on the path used for the integration of the differential expressions, and consequently, the validity of the theoretical developments and corresponding conclusions will become questionable (see Appendix F of Ref. (*13*)).

Another relevant issue in the use of truncated composition expansions within the context of the above questions hinges around our ability not only to assess accurately the corresponding expansion coefficients, but also to link them to the solvation behavior of the species at infinite dilution. In this regard, and for more than a decade, we have placed emphasis on the development of rigorous solvation formalisms for dilute supercritical solutions, to link unambiguously the solvent microscopic (local) behavior around species in solutions (as descriptors of the solvation process) and the resulting system's thermodynamic properties (*6-9, 14-17*). We achieved this goal by splitting the thermophysical properties of infinitely dilute solutions into two contributions according to the unambiguous separation between direct and indirect correlation functions via the Ornstein-Zernike equation (*18*); a short-ranged (solvation) part, and a long-ranged (compressibility-driven) part, where the first one becomes associated with the re-arrangement of the solvent around the species in solution (local solvent density perturbation), while the second, linked to the propagation of the local density perturbation across the system.

Since the first true molecular-based analysis of solute synergism in near critical solutions appeared in the literature (*6*), numerous publications have emerged dealing with different aspects of this phenomenon (*19-22*). One common feature in all these studies is the use of a 1st-order truncated composition expansion for the species fugacity coefficients, according to the analysis proposed by Jonah and Cochran (JC)(*23*) as a multicomponent generalization of the work of Debenedetti and Kumar (DK) for binary systems (*24*). Soon after JC's publication, it became clear that their expansions were not completely consistent from a thermodynamic sense, as first discussed elsewhere (*25, 26*), yet these expressions have been used since then, without any additional discussion (*22, 27-29*).

Our goal here is to address and argue explicitly some relevant issues behind the modeling of dilute solutes in compressible solvents, placing emphasis on the thermodynamic consistency and the molecular interpretation of the formal results toward the development of successful engineering correlations. For that purpose, we first derive thermodynamically consistent truncated composition-expansions for the species fugacity coefficients of dilute ternary systems, analyze binary systems as special cases of the ternary counterparts, and then discuss the thermodynamic inconsistencies encountered in currently available 1st-order truncated expansions by pinpointing the causes and the consequences of their use. Finally, we illustrate how these newly derived expressions reduce exactly, in the zero-density limit, to the corresponding equations for the fugacity coefficients of ternary mixtures obeying the 1st-order truncated virial equation of state $Z = 1 + BP/RT$.

Truncated composition-expansions for the fugacity coefficients of dilute ternary systems

In what follows we will develop a self-consistent 2nd-order truncated expansion of the fugacity coefficients of ternary systems comprising a compressible solvent and two solute species at high (but finite) dilution (a more detailed and general analysis comprising quaternary systems and involving co-solutes and co-solvents will appear elsewhere (30, 31)). For that purpose, let us take a ternary mixture at constant pressure P and temperature T, composed of n_1 solvent molecules, n_2 solute molecules of species 2, and n_3 co-solute molecules of species 3, respectively, for which the corresponding fugacity coefficients at finite composition are 2nd-order expanded around the infinite dilution condition for the two solutes as follows,

$$\ln \hat{\phi}_1(x_2, x_3) = \ln \phi_1^o + a_1 x_2 + b_1 x_3 + c_1 x_2^2 + d_1 x_3^2 + e_1 x_2 x_3 \tag{1}$$

$$\ln \hat{\phi}_2(x_2, x_3) = \ln \hat{\phi}_2^\infty + a_2 x_2 + b_2 x_3 + c_2 x_2^2 + d_2 x_3^2 + e_2 x_2 x_3 \tag{2}$$

$$\ln \hat{\phi}_3(x_2, x_3) = \ln \hat{\phi}_3^\infty + a_3 x_2 + b_3 x_3 + c_3 x_2^2 + d_3 x_3^2 + e_3 x_2 x_3 \tag{3}$$

where we have chosen x_2 and x_3 as the independent variables, and use their link to the corresponding residual partial molar Gibbs free energy $\bar{g}_i^r$, i.e., $\bar{g}_i^r / RT = \ln \hat{\phi}_i = \left[\partial(N \ln \phi)/\partial n_i \right]_{T,P,n_j}$, (32). The goal here is to find the values of the coefficients in Eqns. (1)-(3) that satisfy simultaneously the Gibbs-Duhem equation and the exactness of the mixture's residual Gibbs free energy function $G^r(P, T, n_1, n_2, n_3)$, and later, to identify their thermodynamic meaning as well as to interpret them from a molecular viewpoint.

The Gibbs-Duhem requirement can be written as,

$$(1 - x_2 - x_3)\left(\frac{\partial \ln \hat{\phi}_1}{\partial x_2} \right)_{P,T,x_3} + x_2 \left(\frac{\partial \ln \hat{\phi}_2}{\partial x_2} \right)_{P,T,x_3} + x_3 \left(\frac{\partial \ln \hat{\phi}_3}{\partial x_2} \right)_{P,T,x_3} = 0 \tag{4}$$

and,

$$(1-x_2-x_3)\left(\frac{\partial \ln \hat{\phi_1}}{\partial x_3}\right)_{P,T,x_2} + x_2\left(\frac{\partial \ln \hat{\phi_2}}{\partial x_3}\right)_{P,T,x_2} + x_3\left(\frac{\partial \ln \hat{\phi_3}}{\partial x_3}\right)_{P,T,x_2} = 0 \qquad (5)$$

so that, after invoking Eqns. (1)-(3), we have two polynomials equated to zero, i.e.,

$$a_1 + x_2(2c_1 - a_1 + a_2) + x_3(e_1 - a_1 + a_3) + x_2^2(2c_2 - 2c_1) + x_3^2(e_3 - e_1)$$

$$+ x_2 x_3(2c_3 - 2c_1 - e_1 + e_2) = 0 \qquad (6)$$

and,

$$b_1 + x_2(e_1 - b_1 + b_2) + x_3(2d_1 - b_1 + b_3) + x_2^2(e_2 - e_1) + x_3^2(2d_3 - 2d_1)$$

$$+ x_2 x_3(2d_2 - 2d_1 - e_1 + e_3) = 0 \qquad (7)$$

whose coefficients are also null. Likewise, the relevant Maxwell relation for $G'(T,P,n_1,n_2,n_3) = RT\left(n_1 \ln \hat{\phi_1} + n_2 \ln \hat{\phi_2} + n_3 \ln \hat{\phi_3}\right)$ reduces to (*13*),

$$\left(\partial \ln \hat{\phi_2}/\partial n_3\right)_{PTn_2} = \left(\partial \ln \hat{\phi_3}/\partial n_2\right)_{PTn_3} \qquad (8)$$

so that, after recalling that $\left(\partial \ln x_i/\partial n_j\right) = (\delta_{ij}/x_i - 1)/n$, the Maxwell condition (8) reads,

$$(1-x_3)\left(\partial \ln \hat{\phi_2}/\partial x_3\right)_{PTx_2} - x_2\left(\partial \ln \hat{\phi_2}/\partial x_2\right)_{PTx_3} =$$

$$(1-x_2)\left(\partial \ln \hat{\phi_3}/\partial x_2\right)_{PTx_3} - x_3\left(\partial \ln \hat{\phi_3}/\partial x_3\right)_{PTx_2} \qquad (9)$$

Moreover, by invoking again Eqns. (1)-(3), Eqn. (9) reduces to the following polynomial expression involving null coefficients (*13*),

$$b_2 - a_3 + x_2(a_3 - a_2 - e_2 - 2c_3) + x_3(b_3 + 2d_2 - b_2 - e_3)$$

$$+ x_2^2(2c_3 - 2c_2) + x_3^2(2d_3 - 2d_2) + x_2 x_3(2e_2 - 2e_3) = 0 \qquad (10)$$

The conditions expressed in Eqns. (6), (7), and (10) constitute a homogeneous system of linear equations, $A \, c = 0$, where A is a (15x15) sparse matrix and c is a (15x1) vector. Because the matrix A has a rank of 12, leaving 3 variables free, A was row-reduced to solve the system in terms of the three free variables or equating them to zero (33). We have solved the system in terms of the coefficients of the second species, i.e., a_2, b_2, c_2, d_2 and e_2 as follows (13),

$$
\begin{aligned}
&a_1 = 0 & &a_3 = b_2 & &a_2 = -2c_2 \\
&b_1 = 0 & &b_3 = -2d_2 & &b_2 = -e_2 \\
&c_1 = -0.5a_2 & &c_3 = -0.5a_2 & & \quad\quad (11)\\
&d_1 = d_2 & &d_3 = d_2 & & \\
&e_1 = -b_2 & &e_3 = -b_2 & &
\end{aligned}
$$

Therefore, we now need to identify the thermodynamic meaning of those coefficients, and for that purpose we recall that,

$$
\mu_i(T,P,n_2,n_3) = \mu_i^{IG}(T,P,n_2,n_3) + RT \ln \hat{\phi}_i(T,P,n_2,n_3) \quad (12)
$$

$$
\mu_i^{IG}(T,P,n_2,n_3,) = \mu_i^{o}(T,P) + RT \ln x_i \quad (13)
$$

and then, we define the coefficient $k_{ij}(P,T)$ as follows,

$$
k_{ij}(P,T) \equiv -\left(\partial \ln \hat{\phi}_i / \partial x_j \right)_{P,T,x_k}^{\infty} \quad (14)
$$

Consequently,

$$
\left(\partial \ln \hat{\phi}_2 / \partial x_2 \right)_{P,T,x_3}^{\infty} = -k_{22} = a_2 \quad (15)
$$

$$
\left(\partial \ln \hat{\phi}_2 / \partial x_3 \right)_{P,T,x_3}^{\infty} = -k_{23} = b_2 \quad (16)
$$

$$
\left(\partial^2 \ln \hat{\phi}_2 / \partial x_2^2 \right)_{P,T,x_3}^{\infty} = -a_2 \quad (17)
$$

$$
\left(\partial^2 \ln \hat{\phi}_2 / \partial x_2 \partial x_3 \right)_{P,T}^{\infty} = -b_2 \quad (18)
$$

$$
\left(\partial^2 \ln \hat{\phi}_2 / \partial x_3^2 \right)_{P,T,x_2}^{\infty} = 2d_2 \quad (19)
$$

$$
\left(\partial \ln \hat{\phi}_3 / \partial x_3 \right)_{P,T,x_2}^{\infty} = -k_{33} = -2d_2 \quad (20)
$$

$$\left(\partial^2 \ln \hat{\phi}_3 / \partial x_3^2\right)^\infty_{P,T,x_2} = 2d_2 \qquad (21)$$

$$\left(\partial^2 \ln \hat{\phi}_3 / \partial x_2^2\right)^\infty_{P,T,x_3} = -a_2 \qquad (22)$$

$$\left(\partial^2 \ln \hat{\phi}_3 / \partial x_2 \partial x_3\right)^\infty_{P,T} = -b_2 \qquad (23)$$

Finally, according to the above conditions we have that Eqns. (1)-(3) become,

$$\ln \hat{\phi}_1(x_2, x_3) = \ln \phi_1^o + 0.5k_{22} x_2^2 + 0.5k_{33} x_3^2 + k_{23} x_2 x_3 \qquad (24)$$

$$\begin{aligned}\ln \hat{\phi}_2(x_2, x_3) = \ln \phi_2^\infty &- k_{22} x_2 - k_{23} x_3 \\ &+ 0.5k_{22} x_2^2 + 0.5k_{33} x_3^2 + k_{23} x_2 x_3\end{aligned} \qquad (25)$$

$$\begin{aligned}\ln \hat{\phi}_3(x_2, x_3) = \ln \phi_3^\infty &- k_{23} x_2 - k_{33} x_3 \\ &+ 0.5k_{22} x_2^2 + 0.5k_{33} x_3^2 + k_{23} x_2 x_3\end{aligned} \qquad (26)$$

or in the alternative form,

$$\ln \hat{\phi}_2(x_2, x_3) = \ln \hat{\phi}_2^\infty - k_{22} x_2 - k_{23} x_3 + \ln\left[\hat{\phi}_1(x_2, x_3)/\phi_1^o\right] \qquad (27)$$

$$\ln \hat{\phi}_3(x_2, x_3) = \ln \hat{\phi}_3^\infty - k_{23} x_2 - k_{33} x_3 + \ln\left[\hat{\phi}_1(x_2, x_3)/\phi_1^o\right] \qquad (28)$$

so that,

$$\begin{aligned}g^r(P, T, x_2, x_3) &= G^r(P, T, n_1, n_2, n_3)/(n_1 + n_2 + n_3) \\ &= RT\left(x_1 \ln \phi_1^o + x_2 \ln \hat{\phi}_2^\infty + x_3 \ln \hat{\phi}_3^\infty\right) - \\ &\quad RT\left(0.5k_{22} x_2^2 + 0.5k_{33} x_3^2 + k_{23} x_2 x_3\right)\end{aligned} \qquad (29)$$

In summary, Eqns. (24)-(29) are a set of thermodynamically consistent expressions to describe the behavior of dilute solutions, whose coefficients are rigorously connected to the structure of the system at infinite dilution, *i.e.*, the reference system. Consequently, the corresponding pressure and temperature derivative counterparts, *i.e.*, the partial molar volumetric, enthalpic, and entropic behavior will automatically be consistent (the expressions for these properties are presented and discussed elsewhere (*13*)).

In what follows, and after extracting the corresponding expressions for binary systems, we discuss the thermodynamic inconsistencies behind currently used 1st-order composition expansions for the properties of dilute solutes in binary and ternary systems, their causes as well as their consequences.

Dilute binary systems as special cases of dilute ternary systems

Having derived the expressions for the ternary system, it becomes straightforward to obtain the corresponding binary counterpart. In fact, a binary solution of a dilute solute(2) can be thought as a special case of a ternary system comprising a solute(2) and a co-solute(3), where the solute becomes identically the same to the co-solute. For this condition we have that $k_{22} = k_{33} = k_{23}$, and therefore, from Eqns. (24)-(26) we have that,

$$\ln \hat{\phi}_1(z = x_2 + x_3) = \ln \phi_1^o + 0.5 k_{22} \left(x_2^2 + x_3^2 + 2 x_2 x_3 \right)$$
$$= \ln \phi_1^o + 0.5 k_{22} z^2 \tag{30}$$

$$\ln \hat{\phi}_2(z) = \ln \hat{\phi}_2^\infty - k_{22} z + 0.5 k_{22} z^2$$
$$= \ln \hat{\phi}_2^\infty - k_{22} z + \ln \left[\hat{\phi}_1(z) / \phi_1^o \right] \tag{31}$$

where z now denotes the mole fraction of the solute, $i.e.$, $z = 1 - x_1$. Note that here we keep z, instead of the usual x_2, as the mole fraction of the solute simply to remind the reader on how the expression was derived.

Comparison between the consistent expressions and others from the literature

Now, we illustrate the thermodynamic inconsistencies behind the currently used 1st-order truncated expansions for the finite-composition fugacity coefficients of solutes in binary and ternary systems, comprising a solvent(1), a solute(2) and a co-solute(3). Let us start the discussion by considering binary dilute systems, according to the approach of Debenedetti and Kumar (24) who derived a thermodynamically consistent expression for $\ln \hat{\phi}_1(x_2)_{TP}$ by enforcing the Gibbs-Duhem relation (GDR) to a 1st-order expansion of $\ln \hat{\phi}_2(x_2)_{TP}$, $i.e.$,

$$\ln \hat{\phi}_1(x_2) = \ln \phi_1^o - k_{22} \left[\ln(1 - x_2) + x_2 \right] \tag{32}$$

$$\ln \hat{\phi}_2(x_2) = \ln \hat{\phi}_2^\infty - k_{22} x_2$$

where $k_{22} = K$ (see Eqns. (28)-(30) of Ref. (24)). Note that, because $x_1 = 1 - x_2 \rightarrow 1$, then $\ln(1 - x_2)$; $-x_2 - 0.5x_2^2$, and from a mathematical viewpoint we can recast the original DK expressions in the following approximated (truncated) form, identified hereafter as TDK, i.e.,

$$\ln \hat{\phi}_1(x_2) = \ln \phi_1^o + 0.5 k_{22} x_2^2$$

$$\ln \hat{\phi}_2(x_2) = \ln \hat{\phi}_2^\infty - k_{22} x_2$$

(33)

Expressions (33) correspond to B11-B12 in the appendix of Ref. (24), and have been widely used in the literature (21, 22, 27-29, 34-37). Unfortunately, in contrast to the original DK, the TDK expressions do not obey the GDR, i.e.,

$$(1 - x_2)\left(\partial \ln \hat{\phi}_1/\partial x_2\right) + x_2 \left(\partial \ln \hat{\phi}_2/\partial x_2\right) = k_{22} x_2 \left(1 - 2x_2\right) = 0 \qquad (34)$$

Equation (34) is satisfied only at infinite dilution, $x_2 = 0$, a condition that defeats the purpose of developing a composition expansion in the first place.

The situation of truncated expansions for ternary (or multi-component for that matter) systems becomes a bit more troublesome because the resulting expressions must satisfy not only the GDR but also, the Maxwell relations (MR) given by Eqn. (9). Note that Jonah and Cochran (23) generalized the TDK for multi-component systems based only on the compliance of the GDR, and derived the corresponding 1st-order truncated expressions for the fugacity coefficients of the dilute species, i.e., they did not propose any expression for the corresponding solvent property. This was followed by a series of papers by Chialvo (6), Chimowitz et al. (20), and Ruckenstein et al. (27-29) where these expressions were used to study ternary systems and to propose engineering correlations. All these studies involve the truncated expressions of the fugacity coefficients of all species but the solvent, i.e., invariably they appear as,

$$\ln \hat{\phi}_2(x_2, x_3) = \ln \hat{\phi}_2^\infty - k_{22} x_2 - k_{23} x_3$$

$$\ln \hat{\phi}_3(x_2, x_3) = \ln \hat{\phi}_3^\infty - k_{23} x_2 - k_{33} x_3$$

(35)

Obviously, according to our derivation, Eqns. (35) are either missing the solvent contribution $\ln\left[\hat{\phi}_1(x_2, x_3)/\phi_1^o\right]$ that is quadratic in (x_2, x_3), or they are implicitly assuming that $\ln\left[\hat{\phi}_1(x_2, x_3)/\phi_1^o\right] = 0$. In either case, the consequences of this missing term is twofold (13):

a) The MR is satisfied only when $k_{22} = k_{33} = k_{23}$, *i.e.*, when the two solutes behave identically the same, *i.e.*, the system becomes effectively binary as described by Eqns. (30)-(31).

b) Under the above MR conditions $k_{22} = k_{33} = k_{23}$, we also have that $\ln\left[\hat{\phi}_1(x_2,x_3)\big/\phi_1^o\right] = 0.5 k_{22}(1-x_1)^2 = 0$, *i.e.*, $x_2 = x_3 = 0$ since $k_{22} \neq 0$. These two sets of conditions force Eqn. (35) to satisfy the GDR.

Therefore, the preceding analysis indicates that the 1st-order truncation scheme given by Eqns. (35) is thermodynamically consistent only at infinite dilution! The non-compliance with the MR makes these equations path-dependent as discussed in detail in Ref. (*13*), and consequently, they cannot describe properly the Gibbs free energy of the system, which by definition, is a function of state (*11*). Obviously, the inconsistencies found in the 1st-order composition expansions will spill over the corresponding temperature and pressure derivatives, *i.e.*, the partial molar entropic, enthalpic, and volumetric expressions (*13*).

Molecular-based interpretation of the expansion coefficients

Finally, we make contact between the macroscopic coefficients k_{ij} (T, P), defined by Eqn. (14), and the microstructural details of the reference system, *i.e.*, the solvation behavior of the infinitely dilute system. For that purpose we invoke the general molecular-based solvation formalism for infinite dilute solutions in compressible solvents (*6, 7, 9*), *i.e.*,

$$k_{ij}(T,P) = \left(G_{11}^o + G_{ij}^\infty - G_{1i}^\infty - G_{1j}^\infty\right)\big/\upsilon_1^o \qquad (36)$$

Thus, the solvation behavior of the species in solution becomes explicitly described by the Kirkwood-Buff integral $G_{ij}^\otimes$, *i.e.*,

$$G_{ij}^\otimes = 4\pi \int_0^\infty h_{ij}^\otimes(r)r^2\,dr \qquad (37)$$

where $h_{ij}^\otimes(r) \equiv g_{ij}^\otimes(r) - 1$ represents the total correlation function, and $g_{ij}^\otimes(r)$ is the corresponding radial distribution function (microstructure) involving the interactions between species i and j, and the superscript $\otimes$ denotes either a pure component, o, or an infinite dilution condition, ∞, respectively.

As discussed extensively elsewhere (*7, 9*), by definition $g_{ij}^\otimes(r)$ comprises direct correlations associated with the perturbation of the solvent structure due to the presence of the infinitely dilute solute, and indirect correlations caused by the propagation of the structural perturbation. While the first one characterizes the solvation process (short-range direct correlations), the latter describes the solvent's correlation length, $i.e.$, it scales as the isothermal compressibility. For that reason it is always advantageous to split these two quantities in order to make a clear interpretation of their contributions to the phenomenon under study. For example, according to the argument developed previously (7, 9) we can write,

$$G_{ij}^{\otimes} = C_{ij}^{\otimes} + RT\kappa_1^o C_{1i}^{\otimes} C_{1j}^{\otimes} / v_1^{o2} \qquad i,j = 2,3$$

$$G_{1j}^{\otimes} = RT\kappa_1^o C_{1j}^{\otimes} / v_1^o \qquad\qquad j = 1,3$$

(38)

where $C_{ij}^{\otimes}$ is the direct correlation function integral counterpart (38), v_1^o denotes the molar volume of the pure solvent and κ_1^o is the corresponding isothermal compressibility so that (6),

$$k_{ij} = \left(C_{ij}^{\infty} - v_1^o\right)/v_1^o + \kappa_1^o RT \left(C_{1i}^{\infty} - v_1^o\right)\left(C_{1j}^{\infty} - v_1^o\right)/v_1^{o3} \quad i,j = 2,3 \qquad (39)$$

Once more, we can recast k_{ij} in terms of measurable macroscopic properties, through the interpretation of the two parenthetical expressions in Eqn. (39). In fact, the second and third parenthetical expressions are the structural manifestation of the isothermal-isochoric pressure perturbation caused by the replacement of a solvent molecule by a solute molecule (6), $i.e.$,

$$\left(C_{1i}^{\infty} - v_1^o\right)/v_1^o = \kappa_1^{oIG}\left[1/\kappa_1^o + \left(\partial P/\partial x_i\right)_{Tv}^{\infty}\right] \quad i = 2,3 \qquad (40)$$

where $\kappa_1^{oIG} = (RT\rho)^{-1}$ is the compressibility of the solvent at the ideal gas condition. Useful modeling alternative to Eqn. (39) are discussed in Refs. (13) and (31).

Final remarks

There is yet another way to illustrate formally the thermodynamic consistency of the truncated expansions, by invoking a rigorous zero-density limiting behavior for the radial distribution function $g_{ij}^{\otimes}(r)$, $i.e.$, $\lim_{\rho \to 0} g_{ij}^{\otimes}(r) = \exp\left(-\varphi_{ij}(r)/kT\right)$ (18). In fact, under these conditions, the Kirkwood-Buff integrals $G_{ij}(T,P,x_k)$ become proportional to the second virial coefficient $B_{ij}(T)$ for the corresponding pair interactions (39), $i.e.$,

$$\lim_{\rho \to 0} G_{ij}(P,T,x_k) = -2B_{ij}(T) \tag{41}$$

Moreover, imperfect gases, *i.e.*, those whose behavior can be described accurately at the level of their second virial coefficients, become a precisely defined model to illustrate further the consistency of the derived expressions (*13*). Accordingly, the coefficients $k_{ij}(T,P)$ reduce to the ratio between combinations δ_{kl} of second virial coefficients introduced by Van Ness (*11*) as $\delta_{kl} = 2B_{kl} - B_{kk} - B_{ll}$, and the ideal gas molar volume of the pure solvent.

To be more precise we analyze the expressions for binary and ternary systems of imperfect gases. In the case of a binary mixture of imperfect gases, and after invoking Eqns. (36) and (41), k_{22} can be written in terms of the three possible pair interactions as follows (*13*),

$$k_{22}(T,P) = 2\delta_{12}P/RT \tag{42}$$

and consequently,

$$\ln \hat{\phi}_1(x_2) = \ln \phi_1^o + (P/RT)\delta_{12}x_2^2 \tag{43}$$

$$\ln \hat{\phi}_2(x_2) = \ln \hat{\phi}_2^\infty - (2P/RT)\delta_{12}\,x_2 + (P/RT)\delta_{12}\,x_2^2$$

After some straightforward algebraic manipulations, Eqns. (43) become identical to the expressions (4-120a)-(4-120b) derived by Van Ness and Abbott (*11*) for a mixture of gases obeying the 1st-order truncated virial equation of state $Z = 1 + BP/RT$.

Likewise, for the ternary mixtures of imperfect gases, for which we have that,

$$k_{23}(P,T) = (P/RT)(\delta_{12} + \delta_{13} - \delta_{23}) \tag{44}$$

$$k_{33}(T,P) = 2\delta_{13}P/RT \tag{45}$$

resulting in the following,

$$\ln \hat{\phi}_1(x_2,x_3) = \ln \phi_1^o + (P/RT)\Big[0.5\delta_{12}\,x_2^2 + 0.5\delta_{13}\,x_3^2 +$$
$$(\delta_{12} + \delta_{13} - \delta_{23})x_2 x_3 \Big]$$

$$\ln \hat{\phi}_2(x_2, x_3) = \ln \hat{\phi}_2^\infty - (P/RT)\Big[\delta_{12}x_2 + (\delta_{12} + \delta_{13} - \delta_{23})x_3 - 0.5\,\delta_{12}\,x_2^2 - 0.5\delta_{13}\,x_3^2 - (\delta_{12} + \delta_{13} - \delta_{23})x_2x_3 \Big] \tag{46}$$

$$\ln \hat{\phi}_3(x_2, x_3) = \ln \hat{\phi}_2^\infty - (P/RT)\Big[(\delta_{12} + \delta_{13} - \delta_{23})x_2 + \delta_{13}x_3 - 0.5\,\delta_{12}\,x_2^2 - 0.5\delta_{13}\,x_3^2 - (\delta_{12} + \delta_{13} - \delta_{23})x_2x_3 \Big]$$

A close inspection of Eqns. (46) indicates that they are identical to Van Ness and Abbott's (11) expressions (4-121a)-(4-121c). Obviously, neither the 1st-order truncated Eqns. (33) nor Eqns. (35) can reduce to Eqns. (43) or Eqns. (46), respectively, in the meaningful zero-density limit.

In conclusion, we have set up a thermodynamically consistent composition expansion of the fugacity coefficients of species in dilute ternary (and binary as special case) systems, and consequently, for the corresponding molar Gibbs free energy, enthalpy, entropy, and volume of these systems. Moreover, the expansion coefficients comprise explicit and precisely-defined connections to the microstructure of the system at the infinite dilution limit, an attribute that makes the link between the microscopic behavior of the system and the resulting macroscopic properties unambiguous. The derived expressions are valid regardless of the types of intermolecular forces or the state conditions of the system under consideration. However, their application becomes more in-sightful when dealing with highly compressible media, (30, 31), and allow us to tackle the interpretation of experimental data and subsequent development of engineering correlations on a sounded molecular-based foundation.

Acknowledgements

This research was sponsored by the Division of Chemical Sciences, Geosciences, and Biosciences, Office of Basic Energy Sciences under contract number DE-AC05-00OR22725 with Oak Ridge National Laboratory, managed and operated by UT-Battelle, LLC. One of us (AAC) acknowledges fruitful discussions with Dr. Hank D. Cochran (ORNL) and Prof. Daniel A. Jonah (University of Sierra Leone) after they kindly made available their manuscript before publication (Ref. 23) in the summer of 1992.

References

1. Kwiatkowski, J.; Lisicki, Z.; Majewski, W., *Berichte Der Bunsen-Gesellschaft-Physical Chemistry Chemical Physics* **1984**, *88*, (9), 865-869.

2. Kurnik, R. T.; Reid, R. C., *Fluid Phase Equilibria* **1982**, *8*, (1), 93-105.
3. Brunner, G., *Fluid Phase Equilibria* **1983**, *10*, (2-3), 289-298.
4. Vanalsten, J. G.; Eckert, C. A., *Journal of Chemical and Engineering Data* **1993**, *38*, (4), 605-610.
5. Bruno, T. J.; Ely, J. F., *Supercritical Fluid Technology.* CRC Press: Boca Raton, 1991.
6. Chialvo, A. A., *Journal of Physical Chemistry* **1993**, *97*, 2740-2744.
7. Chialvo, A. A.; Cummings, P. T., *Aiche Journal* **1994**, *40*, 1558-1573.
8. Chialvo, A. A.; Cummings, P. T.; Kalyuzhnyi, Y. V., *Aiche Journal* **1998**, *44*, (3), 667-680.
9. Chialvo, A. A.; Cummings, P. T.; Simonson, J. M.; Mesmer, R. E., *Journal of Chemical Physics* **1999**, *110*, 1075-1086.
10. Prausnitz, J. M.; Lichtenthaler, R. N.; de Azevedo, E. G., *Molecular Thermodynamics of Fluid Phase Equilibria.* 2nd. ed.; Prentice-Hall: Englewood Cliffs, 1986.
11. Van Ness, H. C.; Abbott, M. M., *Classical Thermodynamics of Nonelectrolyte Solutions.* McGraw Hill: New York, 1982.
12. O'Connell, J. P.; Haile, J. M., *Thermodynamics: Fundamentals for Applications.* Cambridge University Press: New York, 2005.
13. Chialvo, A. A.; Chialvo, S.; Kalyuzhnyi, Y. V.; Simonson, J. M., *Journal of Chemical Physics* **2008**, (Submitted for publication).
14. Chialvo, A. A.; Debenedetti, P. G., *Industrial & Engineering Chemistry Research* **1992**, *31*, 1391-1397.
15. Debenedetti, P. G.; Chialvo, A. A., *Journal of Chemical Physics* **1992**, *97*, 504-507.
16. Chialvo, A. A.; Kusalik, P. G.; Cummings, P. T.; Simonson, J. M.; Mesmer, R. E., *Journal of Physics-Condensed Matter* **2000**, *12*, (15), 3585-3593.
17. Chialvo, A. A.; Kusalik, P. G.; Cummings, P. T.; Simonson, J. M., *Journal of Chemical Physics* **2001**, *114*, 3575-3585.
18. Hansen, J. P.; McDonald, I. R., *Theory of Simple Liquids.* 2nd. ed.; Academic Press: New York, 1986.
19. Munoz, F.; Chimowitz, E. H., *Journal of Chemical Physics* **1993**, *99*, (7), 5438-5449.
20. Munoz, F.; Chimowitz, E. H., *Journal of Chemical Physics* **1993**, *99*, (7), 5450-5461.
21. Li, T. W.; Chimowitz, E. H.; Munoz, F., *Aiche Journal* **1995**, *41*, (10), 2300-2305.
22. Munoz, F.; Li, T. W.; Chimowitz, E. H., *Aiche Journal* **1995**, *41*, (2), 389-401.
23. Jonah, D. A.; Cochran, H. D., *Fluid Phase Equilibria* **1994**, *92*, 107-137.
24. Debenedetti, P. G.; Kumar, S. K., *Aiche Journal* **1986**, *32*, 1253-1262.
25. Jonah, D. A.; Cochran, H. D., Personal Communication to Hank Cochran, Chem. Tech Div. at ORNL (Oak Ridge, TN) from UVA (Charlottesville, VA) on Aug 9, 1993.

26. Chialvo, A. A.; Cummings, P. T. In *Solute Synergism in Dilute Near-Critical Ternary Mixtures*, AIChE Spring Meeting, paper 93a, session 93: Thermodynamics of Simple Molecular Fluids II, Atlanta, GA, April 18, 1994, 1994; Atlanta, GA, 1994.
27. Ruckenstein, E.; Shulgin, I., *Fluid Phase Equilibria* **2001**, *180*, (1-2), 345-359.
28. Ruckenstein, E.; Shulgin, I., *Fluid Phase Equilibria* **2002**, *200*, (1), 53-67.
29. Shulgin, I.; Ruckenstein, E., *Industrial & Engineering Chemistry Research* **2002**, *41*, (25), 6279-6283.
30. Chialvo, A. A.; Kalyuzhnyi, Y. V.; Chialvo, S.; Simonson, J. M., *Journal of Chemical Physics* **2008**, (Manuscript in preparation, Part III).
31. Kalyuzhnyi, Y. V.; Chialvo, S.; Simonson, J. M.; Chialvo, A. A., *Journal of Chemical Physics* **2008**, (Manuscript in preparation. Part II).
32. Smith, J. M.; Van Ness, H. C., *Introduction to Chemical Engineering Thermodynamics*. 4th Edition ed.; McGraw Hill: New York, 1987.
33. Bronshtein, I. N.; Semendyayev, A. K., *Handbook of Mathematics*. 3rd ed.; Springer: 1997.
34. Liu, H. Q.; O'Connell, J. P., *Industrial & Engineering Chemistry Research* **1998**, *37*, (8), 3323-3330.
35. O'Connell, J. P.; Liu, H. Q., *Fluid Phase Equilibria* **1998**, *144*, (1-2), 1-12.
36. Ruckenstein, E.; Shulgin, I., *Journal of Physical Chemistry B* **2000**, *104*, (11), 2540-2545.
37. Ruckenstein, E.; Shulgin, I., *International Journal of Pharmaceutics* **2004**, *278*, (2), 221-229.
38. O'Connell, J. P., *Molecular Physics* **1971**, *20*, 27-33.
39. Ben-Naim, A., *Molecular Theory of Solutions*. Oxford University Press: Oxford, 2006.

Chapter 4

Viewing the Cybotactic Structure of Gas-Expanded Liquids

John L. Gohres[1,3,4], Rigoberto Hernandez[2,3,4,*], Charles L. Liotta[1,2,4], and Charles A. Eckert[1,2,4,*]

[1]School of Chemical and Biomolecular Engineering, [2]School of Chemistry and Biochemistry, [3]Center for Computational Molecular Science and Technology, and [4]Specialty Separations Center, Georgia Institute of Technology, Atlanta, GA 30332–0100

Spectroscopic experiments and molecular dynamics (MD) simulations provide a synergistic approach for studying solvation in gas-expanded liquids (GXLs). UV/Vis and fluorescence spectroscopy results indicate preferential organic solvation around the laser dye Coumarin 153 in methanol and acetone gas-expanded liquids regardless of CO_2 composition. Simple modeling techniques are presented as a viable way to extract local compositions from spectroscopic data. MD simulations confirm the presence of organic enrichment and provide an atomic-level picture of the cybotactic region.

The combined use of computer simulations and experiments is becoming commonplace for studying a variety of problems in solution chemistry. Several studies have taken advantage of this approach in a variety of solvent systems including organic liquids(*1,2*), supercritical fluids (*3,4*), and gas-expanded liquids (*5-7*) (GXLs). Both techniques have strengths and weaknesses associated with their use, but their combined use can serve as a basis for comparison and allow the development of better simulation models and tuning of experiments. Computational investigations into structural and transport properties of CO_2/organic mixtures have revealed the heterogeneous nature of these fluids and the degree to which these properties can be modified with CO_2 concentration (*8-12*). The resulting information demonstrates the versatile nature of these tunable solvents. A variety of bulk solvent properties can be tailored to fit the needs of the researcher and molecular-level solvation structures can be tuned by bulk cosolvent or CO_2 changes.

The current work uses spectroscopic experiments and molecular dynamics (MD) simulations to study solvation in GXLs by examining the local solvent environment surrounding a solute probe. This local solvent domain is known as the cybotactic region and its structure can have implications on bulk phenomena like spectroscopy, solubility, and reactivity. The existence of a cybotactic region distinct from the bulk is a well known phenomenon (*13*) and it has been studied in a wide range of solvents with spectroscopy (*14-17*) and simulation (*4,5,7*). The impact of local solvent structure has sparked the interest of many to determine the solvent composition in the cybotactic region with solvent design for reactions (*18,19*), separations (*20*), and materials processing applications (*21,22*) as an overall goal.

Absorption and emission spectroscopy probe the cybotactic region surrounding the chromophore Coumarin 153 (C153) in the ground and first excited-state. C153 is shown in the ground state and first excited-state in Figure 1. Excited-state structure is a symbolic representation of charge distribution as taken from Kumar and Maroncelli (*2*). CO_2/Methanol and CO_2/acetone mixtures were chosen as solvents because the organic species offer a variety of intermolecular interactions. The spectra are analyzed using simple spectroscopy modeling techniques to yield local compositions. MD simulations seek to compare the spectroscopic results and provide atomic-scale information regarding the solvent structure in the cybotactic region. From local solvent information, interesting chemistry can be predicted; in this case MD simulations are used to explore the effects of local solvation on spectroscopy.

Background

The cybotactic region is defined as that part of a solvent perturbed by the presence of a solute molecule. It is sometimes difficult to systematically ascertain the exact cybotactic region, so it is common practice to use a consistent

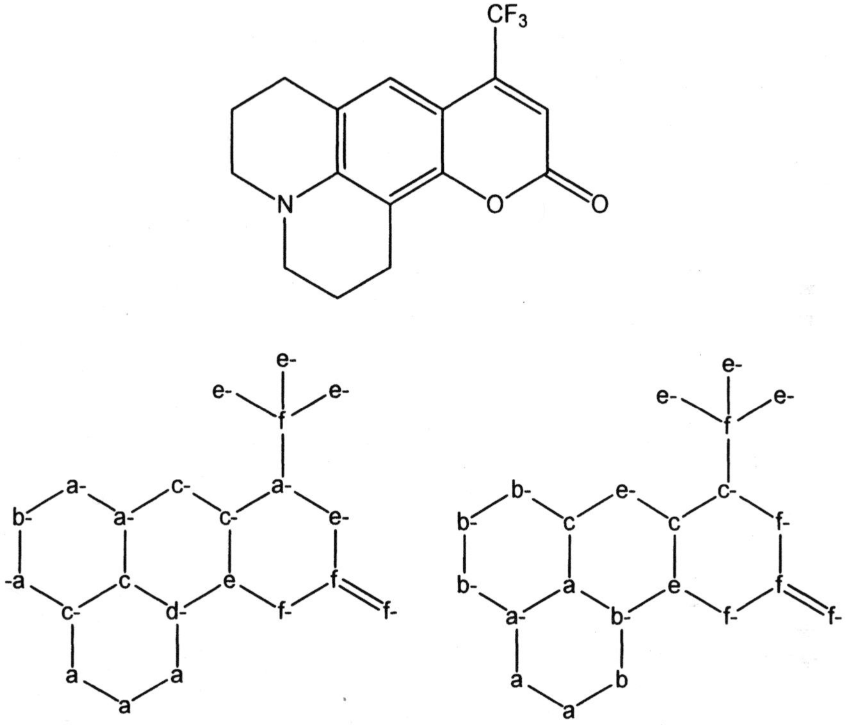

Figure 1. The molecular structure of Coumarin 153 is shown at the top. The charge distribution in the dominant molecular orbital of the ground- (S_0, bottom left) and lowest-excited- (S_1, bottom right) electronic states (C153) are also indicated schematically. Lettering scale is used to illustrate partial charges extracted from Ref. 2. **a**: $|q| < 0.1$, **b**: $0.1 < |q| < 0.2$, **c**: $0.2 < |q| < 0.3$, **d**: $0.3 < |q| < 0.4$, **e**: $0.4 < |q| < 0.5$, **f**: $|q| > 0.5$.

domain that is clearly a subset of the cybotactic region to infer properties of the latter (5). In the computational aspects of this work, a spherical region centered on the solute center of mass was defined as the cybotactic region. A sphere is the simplest model that can be used to quantify an arbitrarily-defined region. Because experiments and simulations provide important information and different perspectives into the cybotactic region, their use has been the focus of much research.

Spectroscopic probes offer an *in situ* method to probe molecular-scale interactions in the cybotactic region of GXLs. The underlying principle behind solvatochromism is the shifting of the maximum absorption or fluorescence wavelength, λ_{max}, due to short-range solvent-solute interactions. As a result, the degree of the solvatochromic shift can be related to the solvent shell surrounding the chromophore. It is assumed that the solvatochromic shift is affected primarily by solvent molecules within the cybotactic region.

In order to quantify the local compositions, a relationship between the solvatochromic shift and local polarity must be created. Onsager's Reaction Field Theory (23) (ORFT) given by Equation 1 is a relationship between the observed solvatochromic shift and solvent polarity.

$$\Delta v = A\left(\frac{\varepsilon-1}{\varepsilon+2} - \frac{n^2-1}{n^2+2}\right) + B\left(\frac{n^2-1}{2n^2+1}\right) \qquad (1)$$

The calculated solvatochromic shift (Δv) is defined as the difference between the absorption or fluorescence maximum and to that of the pure chromophore in vacuum (v_0). Solvent polarity is represented by refractive index (n) and dielectric constant (ε). Parameters A and B account for solute electronic and spatial properties. In the present study, A and B are treated as adjustable parameters to match the spectroscopy of the pure organic and CO_2 which was necessary for local compositions to approach unity at the two extremes.

In the absence of refractive index and dielectric constant data, mixing rules involving the pure organic and CO_2 values can be used to estimate those of the GXL. The Lorentz-Lorenz equation (24) was used to estimate refractive indices.

$$\frac{(n_{mix}-1)^2}{(n_{mix}+2)^2} = \sum_i \Phi_i \frac{(n_i-1)^2}{(n_i+2)^2} \qquad (2)$$

Where n is the refractive index and Φ is the volumetric fraction. The Bruggeman equation (25) was originally used for dielectric constants because it is an asymmetric formula accounting for the heterogeneities that can exist in self-associating solvents like CO_2/MeOH mixtures. Subsequently, several reports (*26,27*) have shown that this equation provides the best fit to all the currently available literature data for CO_2/MeOH systems. Simplified combining rules such as a linear expression can introduce error into the system, but have been used extensively in the literature (*15*).

$$\frac{\varepsilon_{s,mix} - \varepsilon_{s,d}}{\varepsilon_{s,m} - \varepsilon_{s,d}} \left(\frac{\varepsilon_{s,m}}{\varepsilon_{s,mix}} \right)^{1/3} = 1 - \Phi_d \tag{3}$$

Here ε is the dielectric constant and Φ is the volumetric fraction. The subscripts d and m refer to the dispersed and continuous species respectively. In this study, the organic constituent was chosen as the dispersed species because the calculated dielectric constants agreed better with literature values.

Calculated and experimental solvatochromic shifts plotted simultaneously reveal slightly different behavior. The calculated shifts are higher in energy than the experimental counterparts, meaning the chromophore senses a more polar environment than the bulk solvent. The micro-environment around the probe has increased organic presence. Preferential organic solvation would give a different absorption/emission spectrum (calculated by ORFT) than the experimental spectra. The effective micro-environment around C153 can be estimated based on the differences between experimental and calculated spectral maxima. The solvent composition that gives a calculated solvatochromic shift equal to the experimental shift is taken to be the local composition in the cybotactic region. This estimation procedure is illustrated graphically in Figure 2. It should be noted that the maxima in vacuum, v_0, is taken from the literature (*28*) which has been extrapolated from spectral data using Equation 1. The absolute solvatochromic shift will vary with this value, but the calculation procedure will yield identical results regardless of the value chosen.

Molecular Dynamics simulations are another means to estimate local compositions, but they also provide a map of organic and CO_2 distribution around the chromophore. Distribution functions are a means to explore the most probable locations of atoms or molecules relative to a chosen type of atom or molecule. A 3-dimensional axial distribution function (ADF) shown in Equation 4 was used to study solvation patterns around the Coumarin solute. 3D distribution functions provide more detailed spatial information than 1D radial distribution functions, but require more statistics and computational time. The ADF around a chosen atom i located at the origin of a Cartesian axis, is defined as

$$g_i(x,y,z) = \frac{\langle n_i(x,y,z) \rangle}{\rho_i \cdot dV} \tag{4}$$

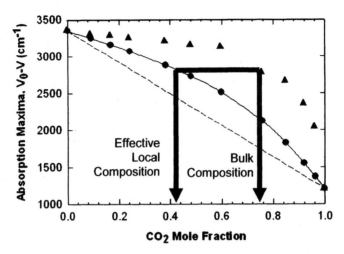

Figure 2. Experimental solvatochromic shifts for C153 absorption in CO_2-expanded MeOH (triangles) and calculated solvatochromic shifts from ORFT (circles). Linear local composition approximation (dashed line) is shown for comparison. At identical spectral shift values, the experimental composition is the bulk CO_2 composition while the composition corresponding to calculated shift represents the local or "effective" CO_2 composition in the cybotactic region.

where g is the magnitude of the ADF, ρ is the bulk number density of the atom or molecule of interest, dV is the volume of a discrete element, and n is the average number density of an atom type in a discrete element. To perform a meaningful ADF calculation, the simulation box is rotated relative to a new coordinate system based on the natural axes of the solute, *e.g.*, the principle moment of inertia or a dipole direction. The box is divided into discrete elements and statistics are recorded in a consistent manner.

Local compositions can be determined from the statistics by determining the number of each molecular species within a spherical element with an origin at the solute center of mass. The radius of the sphere is determined from the ADFs as the distance from the center of mass to the edge of the first solvent shell. The cybotactic region is a complex and non-rigidly defined region that is difficult to quantify consistently because of its dynamic nature. A sphere of fixed radius around the solute's origin is a simple subset of the cybotactic region that can be used to estimate the local composition. This radius must be large enough to encompass a significant part of the solvent located within the cybotactic region, but not so large that the properties of the sphere reach the bulk limit.

Molecular Dynamics Simulations

All molecules were modeled as rigid bodies with a Lennard-Jones plus Coulombic potential given by the equation,

$$u_{ij} = 4\varepsilon_{ij}\left\{\left(\frac{\sigma_{ij}}{r_{ij}}\right)^{12} - \left(\frac{\sigma_{ij}}{r_{ij}}\right)^{6}\right\} + \frac{q_i q_j}{r_{ij}} \tag{5}$$

where i and j are the interaction sites on two separate molecules, r_{ij} is the distance between two sites and q_i are the site charges. The Lennard-Jones parameters ε_{ij} and σ_{ij} are the site-site interactions between atoms obtained by the Lorentz-Berthelot combining rules $\varepsilon_{ij} = (\varepsilon_{ii}\,\varepsilon_{jj})^{1/2}$ and $\sigma_{ij} = 0.5(\sigma_{ii} + \sigma_{jj})$. Carbon dioxide pair interactions were modeled with the TrAPPE (Transferable Potentials for Phase Equilibria Force Field) potential (29). MeOH and acetone pair interactions were modeled with the J2 (30) and OPLS (Optimized Potential for Liquid Systems)-derived potentials (31), respectively. LJ interaction parameters for C153 were obtained from the OPLS model (32) and were assumed to be the same for ground and excited electronic states. Partial charges for the S_0 and S_1 excited states were taken from the literature (2).

All MD simulations were performed using the DL_POLY source code (33). Coulombic interactions were handled by the Ewald summation method with automatic parameter optimization handled internally by DL_POLY and rigid bodies were handled by the SHAKE algorithm. Equilibrium simulations were run at 298K in the NVT ensemble using a Nose-Hoover thermostat (34) with relaxation time of 5 ps. The system consisted of 1000 solvent molecules and one Coumarin 153 solute in a periodic box with dimension determined by experimental data (27) Simulated GXLs included 2, 5, or 20% acetone or MeOH cosolvent and both ground and first-excited state C153 molecules. These particular GXLs were selected based on spectroscopic results to obtain a range of local compositions. Initial configurations were allowed to equilibrate for 400 ps before 500 ps of NVT simulation during which statistics were recorded for ADF generation.

A representative view of the cybotactic region is provided in Figure 3 which shows the ADF of acetone around ground-state C153 in a 5% acetone GXL. Figure 3 displays the cybotactic region from two vantage points, a bird's-eye-view of the C153 plane and an in-plane perspective that captures acetone accumulation above and below the C153 plane. The results indicate a high concentration of acetone near the carbonyl and trifluoro groups of the C153 molecule. Also evident is a ring-like first solvation shell around the molecule with a symmetric acetone clustering pattern above and below the C153 plane. Interested readers are referred to the main article (5) to view organic and CO_2

solvation patterns of the other GXLs for both excited and ground state C153. Local compositions determined by MD simulation will be presented in the proceeding section alongside the experimentally determined counterpart.

Spectroscopic Experiments

HPLC grade MeOH, acetone and Coumarin 153 were obtained from Sigma Aldrich and used as received. CO_2 was purchased from Airgas and dried over molecular sieves prior to use. Absorption and emission spectra were recorded with a UV detector (Hewlett Packard 1050 Series) and an Ocean Optics USB200 fiber optic detection system. The light source for fluorescence spectroscopy was a Kratos LH151 N/2 short arc lamp with 1000-Watt power. Incident wavelength was controlled with a monochrometer. All solvatochromic experiments were performed in a high-pressure optical cell equipped with sapphire windows and cooling jacket, as described by Lu (14). CO_2 was metered into the cell with an Isco syringe pump and pressure was measured with a Druck pressure transducer with ±1 psi. Temperature was measured using an Omega J-type thermocouple (0.1°C precision) in contact with the liquid phase. Temperature was maintained at 298K by pumping an externally coupled ethylene glycol/water solution through the cooling jacket. Samples were allowed to equilibrate for several hours before recording absorption or emission spectra.

Absorption and emission maxima (λ_{max}) are presented as a function of CO_2 composition in Figures 4 and 5. A moderate blue shift occurs in the absorption spectra as CO_2 is added to the more polar organic solvents. After 80% CO_2 there is a rapid decrease in the absorption maxima as pure CO_2 is approached. This behavior is consistent with preferential solvation as the maxima resemble more closely the pure organic absorption spectra. Beyond the 80% CO_2 point the cybotactic region starts to the effects of more CO_2, but is still dominated by the organic species. This is evident in the large solvatochromic shift that occurs between 95% CO_2 and pure CO_2 in the MeOH case.

Fluorescence spectra resemble organic emission over the entire range of CO_2 concentration with a sudden decrease in between 95% CO_2 and pure CO_2. Emission solvatochromism measures the solvent's influence on solute relaxation pathways: radiative and non-radiative shown pictorially in Figure 6. In the case of C153, polar solvents can better solvate the excited state and thus support radiative decay by lowering the energy level of the excited state. Consequently the fluorescent light has a lower wavelength reflecting the decreased relaxation required to return to ground state. Strong specific interactions could dominate the fluorescence response as in the case of MeOH GXLs; however, the similar trend seen in acetone GXLs, which are limited to dipolar interactions, supports the conclusion that dipolar interactions from an organic-like cybotactic region dominate the emission spectroscopy.

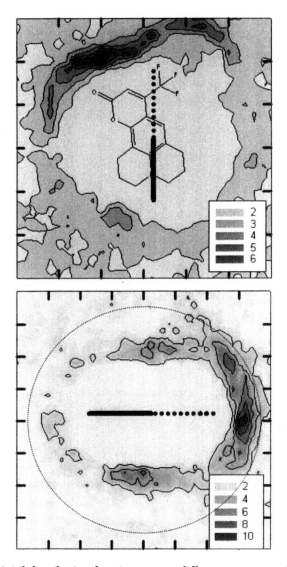

Figure 3. Axial distribution functions at two different vantage points around ground-state (S₀) C153 in a 5% acetone GXL. Scale represents values of the ADF for the carbonyl carbon of Acetone. Depicted plane is coplanar to the C153 molecule (top). Vertical line through the probe molecule represents the sample plane used to show solvent accumulation above and below the C153 plane (bottom). The solid and dotted parts of this line are used to differentiate between the two poles of the C153 molecule as the vantage point is rotated. Tick marks are spaced ~3Å.

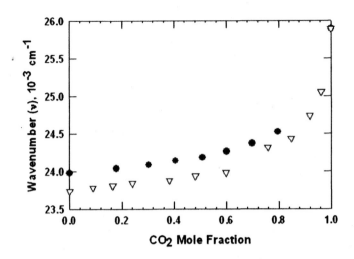

Figure 4. C153 absorption maxima at varying CO₂ concentrations.
Wavenumbers have been subtracted from gas-phase absorption values. Circles
represent CO₂-expanded MeOH absorption and triangles represent CO₂-
expanded acetone absorption. Error bars are within size of data points.

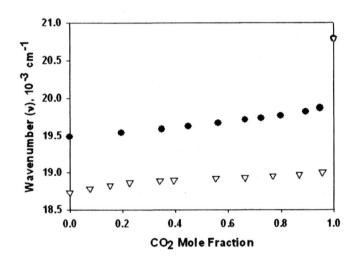

Figure 5. C153 emission maxima at varying CO₂ concentrations. Wavenumbers
have been subtracted from gas-phase absorption values. Circles represent CO₂-
expanded MeOH emission and triangles represent CO₂-expanded acetone
emission. Error bars are within size of data points.

Table I. Local compositions determined from spectroscopic experiments and MD simulations at each simulated bulk-fluid methanol concentration. Experimental values were interpolated to match the simulated counterpart.

Bulk Methanol	State	Local Methanol, Experiment	Local Methanol, Simulation
20%	S_0	59.5	57.6
5%	S_0	20.4	30.5
2%	S_0	12.0	18.0
20%	S_1	73.7	32.9
5%	S_1	69.8	13.6
2%	S_1	20.0	11.0

Table II. Local compositions determined from spectroscopic experiments and MD simulations at each simulated bulk-fluid acetone concentration. Experimental values were interpolated to match the simulated counterpart.

Bulk Acetone	State	Local Acetone, Experiment	Local Acetone, Simulation
20%	S_0	40.3	36.0
5%	S_0	10.9	5.6
2%	S_0	4.4	2.6
20%	S_1	71.3	26.0
5%	S_1	63.0	10.0
2%	S_1	20.0	8.3

Local compositions determined by simulation and experiment are presented in Tables I and II. Experimental values were interpolated to match the simulated CO_2 concentration. Enrichment around the ground-state probe agrees very well with their experimental counterparts, particularly at the 20% organic concentration. Excited state results do not match although organic enrichment is indicated. The inability of MD simulations to match the local compositions extracted from fluorescence experiments is an area of concern. Reasons offered to explain this disparity might revolve about physical phenomenon and/or shortcomings with the C153 model. First, the higher energy emission spectra seen in pure CO_2 may be the result of increased CO_2 presence in the excited state. While the organic concentration in the cybotactic region increases upon excitation, the CO_2 concentration increases as well. This dilutes the local organic composition, where the enhancements in the excited state are localized to small areas in the cybotactic region. The distribution functions show both CO_2 and organic enrichments in the cybotactic region, but CO_2 is more uniformly distributed while the organics are localized. Local CO_2 density enhancements in supercritical fluids have been demonstrated with fluorescence spectroscopy (*16*). A literature source (*35*) cites the C153 force field as another source of error. A key assumption in the models used in this work lies in the use of the same Lennard-Jones potentials for the ground and excited-state C153. This assumption is based on the fact that the atomic positions are similar in both cases and changes in the interaction are primarily manifested in the effective charge redistribution. It can therefore lead to erroneous results that are amplified by the small distances between the solvent molecules and solute-probe in the cybotactic region. The construction of more accurate models is beyond the scope of this work, and is also unnecessary to capture most of the changes in the cybotactic region observed in this work.

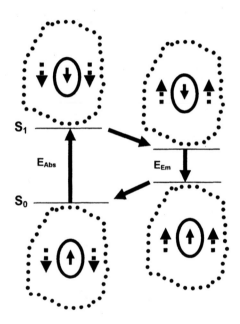

Figure 6. Schematic representation of an energy diagram. Absorption of energy (a, E_{Abs}) excites chromophore (oval) to an excited-state (S_1) with a new dipole moment (arrow inside oval). Solvent molecules in the cybotactic region (dotted region) are initially in a perturbed state and realign to solvate the chromophore's excited dipole moment (b). Chrmophore can relax to the ground-state electronic state via non-radiative or radiative pathways where light is emitted (c, E_{Em}). The cybotactic region relaxes to its equilibrium state (d).

Summary

The synergistic combination of simulations and experiments has provided a powerful method for the study of solvation in many solvent systems. This work demonstrates the ability of spectroscopic experiments to predict reasonable values for local compositions in GXLs, in particular, based on limited data and simple spectroscopic modeling techniques. The structure of the ground-state solvation of C153 has been shown to be in good agreement between experiment and computational results and exhibits a similar solvation pattern regardless of CO_2 composition. These results suggest great promise for the use of MD simulations to explore the solvent-solute and solvent-solvent interactions that are important in structure-property relationships. Also of importance is the prediction of local compositions from spectroscopic techniques for use in solvent design. While excited-state simulations underpredicted the local compositions, they can offer insight into solvation mechanisms and intermolecular interactions that are most important in excited-state dynamics. Thus the current study of the cybotactic region provides important information useful for researchers wishing to design complex systems such as separation technologies, nanomaterials processing schemes, and homogeneous catalysis for reactions.

References

1. Horng, M. L.; Gardecki, J. A.; Papazyan, A.; Maroncelli, M. *J. Phys. Chem.* **1995**, *99*, 17311.
2. Kumar, P. V.; Maroncelli, M. *J. Chem. Phys.* **1995**, *103*, 3038.
3. Patel, N.; Biswas, R.; Maroncelli, M. *J. Phys. Chem. B* **2002**, *106*, 7096.
4. Song, W.; Biswas, R.; Maroncelli, M. *J. Phys. Chem. A* **2000**, *104*, 6924.
5. Gohres, J.; Kitchens, C.; Hallett, J.; Popov, A.; Hernandez, R.; Liotta, C.; Eckert, C. *J. Phys. Chem B, In Press. (doi:10.1021/jp077552p)* **2008**, *112*, XXXX.
6. Hallett, J. P.; Kitchens, C. L.; Hernandez, R.; Liotta, C. L.; Eckert, C. A. *Acct. Chem. Res.* **2006**, *39*, 531.
7. Li, H.; Arzhantsev, S.; Maroncelli, M. *J. Phys. Chem. B* **2007**, *111*, 3208.
8. Aida, T.; Inomata, H. *Mol. Simul.* **2004**, *30*, 407.
9. Chatzios, G.; Samios, J. *Chem. Phys. Lett.* **2003**, *237*, 187.
10. Houndonougbo, Y.; Jin, H.; Rajagopalan, B.; Wong, K.; Kuczera, K.; Laird, B. B. *J. Phys. Chem. B* **2006**, *110*, 13195.
11. Li, H.; Maroncelli, M. *J. Phys. Chem. B* **2006**, *110*, 21189.
12. Shukla, C. L.; Hallett, J. P.; Popov, A. V.; Hernandez, R.; Liotta, C. L.; Eckert, C. A. *J. Phys. Chem. B* **2006**, *110*, 24101.

13. Eckert, C. A.; Ziger, D. H.; Johnston, K. P.; Kim, S. *J. Phys. Chem.* **1986**, *90*, 2738.
14. Lu, J.; Eckert, C. A.; Liotta, C. L. *J. Phys. Chem. A* **2003**, *107*, 3995.
15. Phillips, D. J.; Brennecke, J. F. *Ind. Eng. Chem. Res.* **1993**, *32*, 943.
16. Rice, J. K.; Niemeyer, E. D.; Dunbar, R. A.; Bright, F. V. *J. Am. Chem. Soc.* **1995**, *117*, 5832.
17. Yonker, C. R.; Smith, R. D. *J. Phys. Chem.* **1988**, *92*, 2374.
18. Wei, M.; Musie, G. T.; Busch, D. H.; Subramaniam, B. *J. Am. Chem. Soc.* **2002**, *124*, 2513.
19. Eckert, C. A.; Liotta, C. L.; Bush, D.; Brown, J. S.; Hallett, J. P. *J. Phys. Chem. B* **2004**, *108*, 18108.
20. Eckert, C.; Liotta, C.; Ragauskas, A.; Hallett, J.; Kitchens, C.; Hill, E.; Draucker, L. *Green Chem.* **2007**, *9*, 545.
21. McLeod, M. C.; Anand, M.; Kitchens, C. L.; Roberts, C. B. *Nano Lett.* **2005**, *5*, 461.
22. Kitchens, C. L.; Roberts, C. B. *Ind. Eng. Chem. Res.* **2006**, *45*, 1550.
23. Onsager, L. *J. Am. Chem. Soc.* **1936**, *58*, 1486.
24. Hadrich, J. *Appl. Phys.* **1975**, *7*, 209.
25. Bruggeman, D. A. G. *Ann. Phys.* **1935**, *24*, 636.
26. Roskar, V.; Dombro, R. A.; Prentice, G. A.; Westgate, C. R.; McHugh, M. A. *Fluid Phase Equilib.* **1991**, *77*, 241.
27. Weikel, R. R.; Hallett, J. P.; Liotta, C. L.; Eckert, C. A. *Topics in Catalysis* **2006**, *37*, 75.
28. Horng, M. L.; Gardecki, J. A.; Papazyan, A.; Maroncelli, M. *J. Phys. Chem.* **1995**, *99*, 17311.
29. Potoff, J.; Siepmann, J. I. *AIChE J.* **2001**, *47*, 1676.
30. Jorgensen, W. *J. Phys. Chem.* **1986**, *90*, 1276.
31. Jorgensen, W.; Briggs, J.; Contreras, M. *J. Phys. Chem.* **1990**, *94*, 1683.
32. Pranata, J.; Wierschke, S. G.; Jorgensen, W. *J. Am. Chem. Soc.* **1991**, *113*, 2810.
33. Smith, W.; Forester, T. R. *J. Molecular Graphics* **1996**, *14*, 136.
34. Allen, M. P.; Tildesley, D. J. *Computer Simulations of Liquids*; Oxford Science, London, 1987.
35. Cichos, F.; Brown, R.; Bopp, P. A. *J. Chem. Phys.* **2001**, *114*, 6834.

Chapter 5

Solvatochromism and Solvation Dynamics in CO_2-Expanded Liquids

Chet Swalina[1], Sergei Arzhantzev[1], Hongping Li[2], and Mark Maroncelli[1,*]

[1]Department of Chemistry, The Pennsylvania State University, University Park, PA 16803
[2]Department of Chemistry, Zhengzhou University, No. 100 Science Road, Zhengzhou, Henan 450001, China

Molecular dynamics (MD) simulations and electronic spectroscopy are used to investigate solvatochromism and solvation dynamics in CH_3CN+CO_2 mixtures. Experimental results for the probe solute *trans*-4-(dimethylamino)-4'-cyanostilbene (DCS) reveal a nonlinear dependence of the solvatochromic shifts upon composition, suggesting substantial preferential solvation of DCS by CH_3CN. MD simulations show that analyzing these shifts using the common assumption of spectral additivity leads to grossly exaggerated estimates of the extent of this preference. Solvation dynamics has been measured using Kerr-gated emission spectroscopy and calculated using MD simulations. Simulation-experiment comparisons offer an interpretation of the observed dynamics in terms of nonspecific preferential solvation.

Introduction

Gas-expanded liquids (GXLs) are mixtures of a liquid organic solvent and a near-critical gaseous component such as CO_2 (1-4). A surprising attribute of GXLs, shown in the top panel of Figure 1, is that gas expansion leads to an increase in mass density compared to the original liquid (5). But, as shown in the bottom panel, simulations reveal that this increased density is accompanied by a decrease in the packing fraction, the fraction of the total volume occupied by the van der Waals volumes of the constituent molecules. At high x_{CO_2}, the packing fractions decrease to values ~60% of those original liquids. Thus, there is substantially more "free-volume" in the CO_2-expanded liquids than in the original liquids. The unique capabilities of GXLs such as enhanced solubilities of gases and increased fluidity arise from this extra space. Combined with the ability to readily pressure-tune the amount of expanding gas in the liquid, these features make GXLs potentially useful green solvent alternatives (1,2,4).

The focus of this paper is on exploring molecular aspects of solvation in these rarified solvents. One objective is to describe preferential solvation in GXLs, the enrichment of one solvent component in the cybotactic region relative to the bulk. The extent of preferential solvation in mixtures can be determined from knowledge of how gas-to-solution spectral shifts depend on local compositions. Typically, spectral shifts in homogeneous mixtures are assumed to be linearly related to the shifts observed in the pure component solvents, weighted by their mole fractions in the mixture,

$$v = x(1)v(1) + x(2)v(2) \tag{1}$$

Inversion of this equation provides an approximate expression for the local mole fraction of a particular solvent component (6),

$$x_u^v(1) = [v - v(2)]/[v(1) - v(2)] \tag{2}$$

Throughout this paper, the subscript u denotes the cybotactic region of the solute and (1) and (2) refer to the liquid and gaseous components of a GXL mixture, respectively. The linear interpolation in eq 1 implicitly assumes that contributions to the spectral shift from each component are additive with respect to the numbers of solvent molecules in the cybotactic region,

$$v = N_u(1)\delta(1) + N_u(2)\delta(2) \tag{3}$$

Here, $\delta(i)$ is the per-molecule or per-atom contribution to the spectral shift and $N_u(i)$ is the number of molecules or atoms in the first solvation shell associated

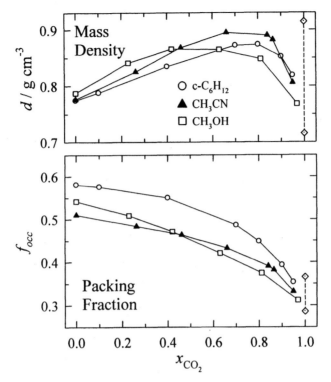

Figure 1. Volumetric properties of some CO_2-expanded liquids. The top panel contains mass densities and the bottom panel displays the packing fractions. For both panels, the two diamonds connected by a dashed line signify the values of these quantities in pure CO_2 at 298 K and at pressures of 6.5 (lower) and 20 MPa (upper point). (Reproduced with permission from reference (5). Copyright 2006 American Chemical Society.)

with solvent component i. (In this paper, analysis is conducted via per-atom quantities.) Figure 2 shows spectral shift data representative of what is observed in the present work and other recent experiments in GXLs (*3,7-9*). As illustrated here, analysis according to eqs 1 and 2 suggests very large differences between local and bulk compositions. Typically, a 5-10-fold enrichment of the liquid component is implied for the region around the solute.

To what extent are these simple assumptions useful for inferring preferential solvation in GXLs? Molecular dynamics (MD) simulations in GXLs provide a means to answer this question and a number of GXL simulations have already been performed (*5,10-21*). In previous work (*20*), we reported experiment-simulation comparisons of absorption and emission shifts of two solutes in the set of CO_2-expanded liquids presented in Figure 1. Comparison of the extent of preferential solvation estimated using spectral shifts to that directly observed in MD simulations demonstrated that the enrichment of the cybotactic region determined spectroscopically was 2-5 times larger than the correct values.

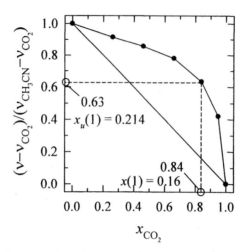

Figure 2. Gas-to-solution emission frequency shift data modeled for CH_3CN+CO_2 mixtures. The dashed lines demonstrate how substantial preferential solvation by the CH_3CN component could be inferred when its bulk mole fraction is 0.16.

In the present work, we provide an analogous comparison of preferential solvation using the solvatochromic probe *trans*-4-(dimethylamino)-4'-cyanostilbene (DCS) (see inset to Figure 3) in several CH_3CN+CO_2 mixtures. Here too we find that use of the simplistic approaches embodied in eqs 1-3 to measure local compositions grossly exaggerates the extent of preferential solvation in this prototypical GXL.

The newer objective of this work is to explore the *dynamics* of preferential solvation in GXLs. The DCS/CH_3CN+CO_2 system is an excellent prototype for studying preferential solvation dynamics due to the similar size, shape, and inertial properties of the solvent components, and also for its lack of specific solute-solvent interactions. There have been several simulation studies of preferential solvation in conventional liquid mixtures (*22-28*), however, few have examined *nonspecific* preferential solvation dynamics. Day and Patey simulated dynamics of preferential ion solvation in binary Stockmayer mixtures (*26*). Ladanyi and Perng simulated preferential solvation in $CH_3CN+benzene$ mixtures using a model benzene-like dipolar solute (*24*). In both systems, the solvation times were found to be maximized in mixtures that were most dilute in the more polar component. The preferential solvation process in these systems was observed to occur in two distinct phases, a fast electrostriction step followed by a slower redistribution phase. The present work reports the first experimental measurements of solvation dynamics in a prototypical GXL, CH_3CN+CO_2, based on Kerr-gated emission spectroscopy. Simulations are used to provide molecular interpretations of the observed dynamics.

This paper uses steady state spectroscopy of the solute DCS to emphasize general features of solvation in GXLs which have previously been described using other solutes (*20*). The solvation dynamics work reported is new and will be described more completely in a forthcoming paper (*29*). Here, we emphasize the use of molecular dynamics (MD) simulations to interpret experimental solvatochromic shifts and dynamic Stokes shift data in GXLs, and minimize the discussion of the experimental aspects of the work. For a detailed description of these and related experiments, interested readers are referred to refs (*20,29*).

Models and Methods

The observables of interest here are gas-to-solution spectral shifts and the solvation response functions. Absorption and emission spectral shifts are assumed to consist of independent contributions from electrostatic and dispersion interactions (*30*). The electrostatic contribution is modeled using

$$\Delta U_{el} = \sum_{\alpha} \sum_{j} \frac{\Delta q_{\alpha} q_j}{r_{\alpha j}} \qquad (4)$$

where the indices α and j denote solute and solvent atoms, respectively, the Δq_{α} are $S_1 - S_0$ differences in solute atomic charges, and the q_j are solvent charges. The contribution to the spectral shift arising from dispersion interactions is modeled by assuming that the $S_1 - S_0$ changes in solute-solvent dispersion interactions originate from solute atoms belonging to the π system of DCS and that these interactions are proportional by a factor, f, to their values in the ground state:

$$\Delta U_{dsp} = -f \sum_{\alpha'} \sum_{j} \frac{4\varepsilon_{\alpha' j} \sigma_{\alpha' j}^6}{r_{\alpha' j}^6} \qquad (5)$$

Here, $\varepsilon_{\alpha' j}$ and $\sigma_{\alpha' j}$ are Lennard-Jones interaction parameters and the prime signifies a sum over only solute atoms belonging to the π system. f is the single adjustable parameter used to fit the simulated absorption and emission shifts to experiment.

The normalized solvation response function, $S_v(t)$, is measured experimentally from the temporal evolution of the fluorescence Stokes shift. Employing linear response theory, we have approximated $S_v(t)$ using the time correlation function of ΔU_{el}, via

$$S_v(t) \cong C_{\Delta U_{el}}^{(n)}(t) = \frac{\left\langle \delta \Delta U_{el}(0) \delta \Delta U_{el}(t) \right\rangle_n}{\left\langle \left(\delta \Delta U_{el} \right)^2 \right\rangle_n} \tag{6}$$

In eq 6 $\delta \Delta U_{el} = \Delta U_{el} - \left\langle \Delta U_{el} \right\rangle$ is a fluctuation in ΔU_{el} and $\left\langle \cdots \right\rangle_n$ signifies an ensemble average observed when the solvent is in equilibrium with the solute in state n. Adherence to linear response can be monitored using the condition $C_{\Delta U_{el}}^{(0)}(t) = C_{\Delta U_{el}}^{(1)}(t)$. When this condition is not met, we have calculated solvation response functions, $S_{\Delta U_{el}}(t)$, directly using nonequilibrium simulations.

Molecules were represented by rigid collections of interaction sites with interactions being of the Lennard-Jones (12-6) plus Coulomb form. Solvent molecules were represented using models previously reported in the literature for the corresponding neat systems. The EPM2 model of Harris and Jung (*31*) was used for CO_2. Similarly, CH_3CN consisted of three interaction sites, one representing the CH_3 group, as parameterized by Edwards et al. (*32*). The solute, DCS, was modeled using the all-atom representation developed previously for simulations of supercritical fluoroform (*33*). The structure of DCS was assumed to be the same in S_0 and S_1 and was obtained from geometry optimization at the RHF/6-31G(d) level. The stilbene framework was constrained to be planar instead of the twisted conformation produced by this level of theory (*34*). Atomic charges for S_0 were obtained by fitting the electrostatic potential (ESP fit) to the MP2/6-311G(d,p)//RHF/6-31G(d) electronic density. $S_1 - S_0$ atomic charge differences were calculated at the AM1/CI level. In contrast to ref (*33*), the $S_1 - S_0$ charge differences used in this study were scaled by a factor of 1.72. This scaling is required due to the small $S_0 \rightarrow S_1$ change in dipole moment obtained with the AM1/CI calculations, as discussed in refs. (*29,33*). S_1 charges were obtained by adding these charge differences to the S_0 charges.

The systems simulated consisted of a single DCS molecule solvated by a total of 1000 solvent molecules. Simulations were performed using a modified version of the DL_POLY program (*35*) in the NVT ensemble at a temperature of 298 K using a Nosé–Hoover thermostat (*36*). System volumes were chosen to provide the densities of the GXLs measured experimentally (*5*). Cubic periodic boundary conditions were applied and long-range electrical interactions were treated using the standard Ewald method. For spectral shift calculations, electrical interactions described in eq 1 were treated using the Kubic harmonic Ewald approximation (37). Rigid body equations of motion were integrated using the leapfrog method with a 2 fs time step. For all mixtures, DCS was inserted into a pre-equilibrated CH_3CN+CO_2 mixture and the systems then further equilibrated for at least 500 ps prior to data collection. Trajectory data were collected over 5 ns in 500 ps blocks.

Solvatochromism and Local Composition

Solvatochromic shifts as a function of CO_2 mole fraction observed in experiment and calculated from simulation are presented in Figure 3. The optimized value of the single adjustable parameter, f, in eq 5 is 0.24. A 24% increase in dispersion interactions upon electronic excitation is reasonable. There is excellent agreement between the experimental and simulated shifts of DCS in the CH_3CN+CO_2 mixtures studied. Similarly good agreement was obtained in previous work using the solvatochromic probe Coumarin 153 (C153) in mixtures of CH_3CN+CO_2 (20). Such consistency validates the use of our simulation and solvatochromic models for examining relationships between spectral shifts and local environment.

We therefore use this model to examine to what extent the assumption of a linear relationship between spectral shift and composition provides accurate

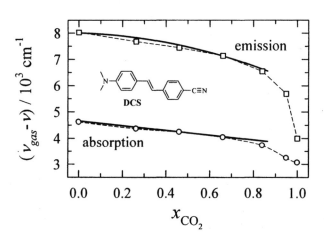

Figure 3. Observed (thick curves) and simulated (symbols) gas-to-solution shifts of absorption and emission frequencies. Uncertainties in the experimental absorption and emission shifts are on the order of ±1000 cm^{-1}. The structure of the DCS solute is provided in the inset.

measures of preferential solvation in CO_2-expanded CH_3CN. To begin, we analyze the solvation structure of DCS in CH_3CN+CO_2 mixtures by examining local compositions averaged over the first solvation shell. Spatial solvation structure can be analyzed using solvation shell distribution functions (38), $g_{ss}(r)$. This function is defined as the relative probability of finding an atom of a given solvent component at a distance r from the nearest solute atom, normalized to a random distribution of solvent molecules. $g_{ss}(r)$ functions corresponding to both DCS electronic states are depicted in Figure 4 for $x_{CO_2} = 0.46$, 0.84, and 0.95. The CO_2 distributions have a prominent first peak and the oscillations common to dense liquid packing. They are similar for all three mixtures and change little between the S_0 and S_1 solute states. In contrast, there is a substantial increase in the aggregation of CH_3CN molecules going from S_0 to S_1 as the CH_3CN component becomes more dilute. It is interesting that for S_1, as x_{CO_2} increases, the minima in the CH_3CN distributions fill in, producing a distribution more characteristic of a gas rather than a dense liquid.

We now focus on the immediate neighborhood of the solute by integrating these $g_{ss}(r)$ distributions to obtain coordination numbers, defined as the number of solvent atoms within 5 Å of any solute atom. Coordination numbers for the individual solvent components as well as the total number of atoms belonging to

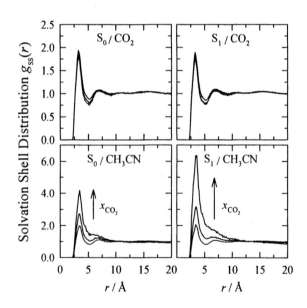

Figure 4. Solvation shell distribution functions, $g_{ss}(r)$, for $x_{CO_2} = 0.46$, 0.84, and 0.95.

any solvent molecule as a function of x_{CO_2} are presented in Figure 5. An important aspect of Figure 5 is that the total coordination number, N_u, deviates substantially from linearity. This trend arises from the buildup of mass density with x_{CO_2} depicted in Figure 1 and, as will be discussed later, it has important consequences for relating spectral shifts to local compositions. Another interesting feature in Figure 5 is that the CH_3CN coordination numbers deviate much more from ideality than the CO_2 coordination numbers. For both components, there are small, but significant changes in coordination numbers between the S_0 and S_1 solute states.

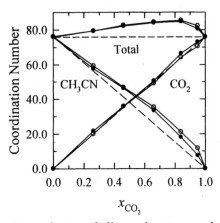

Figure 5. First solvation shell coordination numbers of DCS.
Filled (open) circles signify values observed in S_0 (S_1).

Armed with this knowledge concerning local solvation structure, we can now compare these 'real' measures of preferential solvation to those that rely on the validity of eq 1. The extent of preferential solvation is most clearly measured using "enrichment factors" defined as

$$\frac{x_u(1)}{x(1)} = \frac{N_u(1)/N_u}{x(1)} \tag{7}$$

In eq 7 $x_u(1)$ is the local mole fraction of CH_3CN in the cybotactic region, $x(1)$ is its bulk value, $N_u(1)$ is the number of atoms associated with CH_3CN and N_u the total number of atoms from any solvent component in the first solvation shell. Spectral enrichment factors, $x_u^v(1)/x(1)$, based on eq 2 are compared to those based on coordination numbers, $x_u(1)/x(1)$, in the top portion of Figure 6.

The spectral enrichment factors deduced from simulated solvatochromic data are up to 3 times larger than the correct values.

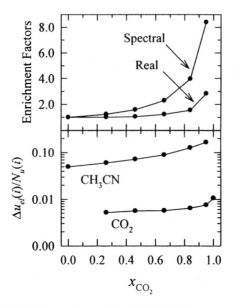

Figure 6. (Top) Enrichment factors for the CH_3CN component. (Bottom) Per-atom contributions from the two solvent components to $S_1 - S_0$ differences in solute-solvent electrical interaction energies from which spectral shifts are calculated. Units are in 1000 cm^{-1}.

Discrepancies between $x_u^v(1)$ and $x_u(1)$ originate from two effects. The bottom panel of Figure 6 depicts the primary reason for the breakdown of eqs 1 and 2 for determining local compositions in CH_3CN+CO_2 mixtures. Here we have plotted per-atom contributions to the emission shifts, $\Delta u_{el}(i)/N_u(i)$, from each component i. Not surprisingly, CH_3CN molecules (atoms) contribute much more (>10-fold) to the electrical shifts than CO_2 molecules. What is not obvious from this figure is the fact that these electrical contributions account for >90% of the composition dependence of the shifts and virtually all of the Stokes shifts except in neat CO_2. More important for explaining the difference between $x_u^v(1)$ and $x_u(1)$ is the fact that the incremental contributions to the shift depend strongly upon composition. If the assumption of spectral additivity (eq 3) was valid, the per-atom electrostatic contributions in Figure 6 would be independent of composition. The decrease in Δu_{el} with increasing polarity of the mixtures is due to the nonlinear relationship between favorable solute-solvent and unfavorable solvent-solvent interactions. The breakdown of spectral additivity is not unique to GXLs, rather it occurs in *any* mixture where the components differ significantly in polarity and where the probe solute is sensitive to electrostatic interactions.

The second effect is specific to GXLs and stems from the fact that GXLs pass through a region of maximum mass density with respect to composition (Figure 1). This density maximum causes the positive deviations from nonlinearity in the total coordination number depicted in Figure 5. Such variations in the total coordination number cause $x_u^v(1) > x_u(1)$ even when spectral additivity holds. This fact can be appreciated by examining the relationship between $x_u^v(1)$ and $x_u(1)$. Assuming that the coordination numbers at the neat solvent limits are equal, $\langle N_u \rangle_1 = \langle N_u \rangle_2 = N_u^o$, using eqs 2 and 3, one finds that

$$x_u^v(1) = x_u(1)\left(1 + \frac{\Delta N_u}{N_u^o}\right) + \frac{\Delta N_u}{N_u^o}\left(\frac{\delta_2}{\delta_1 - \delta_2}\right) \qquad (8)$$

where $\Delta N_u = \langle N_u \rangle - N_u^o$. Because ΔN_u and $\delta_2/(\delta_1 - \delta_2)$ are positive for CH_3CN+CO_2, $x_u^v(1) > x_u(1)$.

In summary, the present simulations are consistent with previous ones (*20*) for C153/CH_3CN+CO_2, indicating that use of eq 2 to measure local compositions in GXLs produces exaggerated results. Inaccuracy of eq 2 arises from two sources. First, additivity (eq 3) is a poor approximation when electrostatic interactions are important. Second, even for cases where additivity holds, the fact that the coordination number varies with composition leads to deviations from the behavior expected in simple mixtures. Dielectric continuum solvation models can treat the composition-dependent nonlinearity in electrostatic interactions (*39-41*), but for CH_3CN+CO_2 such approaches become unreliable at high CO_2 concentrations (*20*). Nevertheless, as directly observed in our simulations, preferential solvation of DCS by CH_3CN is substantial in these GXLs. We now shift attention to the dynamical aspects of this preferential solvation.

Enrichment Dynamics

Experimental spectral response functions and simulated solvation correlation functions for $x_{CO_2} = 0$, 0.46, and 0.86 are compared in Figure 7. The experimental and simulated functions both develop significant long-time components as the mole fraction of CO_2 is increased. The integral times (inset) of the experimental and simulated response functions increase systematically as the CH₃CN component becomes more dilute. The simulation data show that $\langle \tau \rangle$ increases 3-fold upon going from $x_{CO_2} = 0.86$ to 0.95. For this series of compositions, the simulated and experimental solvation times agree to within the error of the experimental values and there is at least semi-quantitative agreement between the simulated and experimental solvation response functions. This level of agreement between the simulated and experimental dynamics as well as for the spectral shifts discussed previously encourage a more detailed analysis of the observed dynamics. In particular we wish to provide a molecular-level interpretation of the long-time component which becomes pronounced in mixtures dilute in CH₃CN.

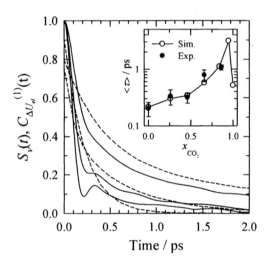

Figure 7. Comparison of simulated solvation correlation functions (S₁ state, solid curves) and experimental spectral response functions (dashed curves) at three compositions: $x_{CO_2} = 0, 0.46,$ and 0.86. The inset shows the integral times of these functions.

Figure 8 presents simulated solvation correlation functions over a range of CH₃CN+CO₂ compositions. It is evident from Figure 8 that both the short and long time regimes of the decays are affected by composition. Such trends are typically analyzed using characteristic times such as the solvation frequency (*42*) and average decay time obtained by fitting the correlation functions to a Gaussian plus exponentials function (*43*). We will not dissect these trends here, but rather, briefly offer some general observations from a more complete analysis (*29*). The solvation frequency in these mixtures tends to decrease as x_{CO_2} increases as a result of the response becoming less collective as the CO₂ concentration increases (44). Correspondingly, the fractional Gaussian character present in the response also decreases. The long-time regime of the decays, typically characterized as being diffusive in nature, becomes more prominent as the CH$_3$CN component becomes more dilute.

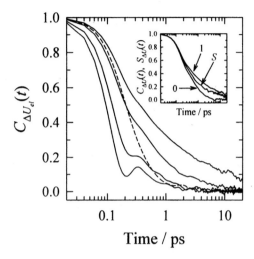

Figure 8. Simulated solvation correlation functions (S$_1$ state) at five compositions: x_{CO_2} = 0, 0.46, 0.84, 0.95, and 1. *The dashed curve is the correlation function for neat CO$_2$. The inset compares solvation correlation functions (S$_1$ and S$_0$ states) to the nonequilibrium solvation response function at* x_{CO_2} = 0.95.

We have observed that the linear response condition $C_{\Delta U_{el}}^{(0)}(t) \cong C_{\Delta U_{el}}^{(1)}(t)$ holds for all but the most dilute compositions where $x_{CO_2} \geq 0.86$. The inset in Figure 8 compares the solvation correlation functions $C_{\Delta U_{el}}^{(0)}(t)$ and $C_{\Delta U_{el}}^{(1)}(t)$ to the corresponding nonequilibrium solvation response function, $S_{\Delta U_{el}}(t)$ for $x_{CO_2} = 0.95$. Within the long-time regime of the decay, $S_{\Delta U_{el}}(t)$ more closely resembles $C_{\Delta U_{el}}^{(1)}(t)$; $C_{\Delta U_{el}}^{(0)}(t)$ does not develop the characteristic long-time component present in the other two functions.

Preferential solvation of DCS by CH$_3$CN is largest for the mixtures most dilute in CH$_3$CN. These compositions exhibit a prominent diffusive component in their response functions. In order to reveal the mechanism of this enrichment process, we performed nonequilibrium MD simulations using the most dilute mixture, $x_{CO_2} = 0.95$. The results are summarized in Figure 9, where we compare $S_{\Delta U_{el}}(t)$ to time-dependent changes in the local solvent composition. The data shown here are averaged over 1000 independent trajectories.

As shown in the upper panels of Figure 9, roughly 4 atoms associated with CH_3CN molecules enter the first solvation shell over the course of 50 ps. Concurrent with this enrichment in CH_3CN, 2 atoms associated with CO_2 exit the first solvation shell, resulting in a net gain of about 2 total solvent atoms. Thus both electrostriction and redistribution of solvent molecules occurs as a result of the excitation of DCS. The time scales for these two processes are approximately the same, about 20 ± 4 ps based on a single-exponential fit of the least noisy $\Delta N(CH_3CN)$ data. The comparison in the bottom panel of Fig. 9

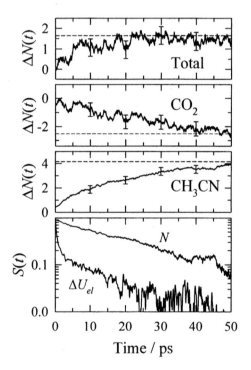

Figure 9. Time-dependent changes in the numbers of atoms of a particular solvent type in the first solvation shell compared to the solvation response at $x_{CO_2} = 0.95$. Dashed reference lines indicate values calculated using differences between the equilibrium values in S_1 and S_0. The bottom panel compares the solvation response function (ΔU_{el}) and the normalized particle response function $S_N = [N(t) - N(\infty)]/[N(0) - N(\infty)]$ of the number of CH_3CN atoms.

shows that, to within uncertainties, this decay time is the same as that of the tail of the solvation response function. Thus, as expected, the long time tail that develops in the solvation response at low CH_3CN concentrations is the result of the slow diffusive transport of CH_3CN into the first solvation shell of DCS.

Finally, it is interesting to compare what is observed here to what was previously reported in simulations of the dynamics of preferential solvation. As mentioned in the Introduction, both Day and Patey (26) and Ladanyi and Perng (24) reported clearly distinct time regimes for electrostriction and solvent redistribution. The fact that we do not observe a rapid and separable electrostriction step can be traced to two differences from these earlier simulations. First, our solute perturbation is much smaller than those simulated by Day and Patey. Rather than the pronounced changes observed upon ionization of a small solute, the $S_0 \rightarrow S_1$ change in DCS involves a much more modest change in solute – solvent interactions distributed over many more solvent molecules. As a result the electrostriction here is a rather subtle 2% change in the net density in the first solvation shell. Ladanyi and Perng (24) simulated a dipole change more similar to the one considered here. The difference in this case is that they monitored electrostriction in terms of changes in a specific solute-solvent atom-atom distribution function because of its importance to the spectral shift they simulated. There are no comparably important single-site interactions in the present system, and the overall electrostriction, that we monitor here, is probably best viewed as just a part of the solvent redistribution process.

Summary and Conclusions

MD simulations have been used to interpret composition-dependent trends in the solvatochromic shifts and time-resolved fluorescence of a prototypical GXL system, CH_3CN+CO_2. Observed fluorescence shifts together with the common assumption of spectral additivity imply pronounced preferential solvation of the DCS solute, especially for compositions dilute in the liquid component. The actual enrichment observed in simulations was found to be 3 times less than that determined by spectral analysis, but is in qualitative accord with spectral results. The inherent nonlinearity of spectral shifts with respect to composition arises primarily from the complex interplay between solute-solvent and solvent-solvent electrostatic interactions. To a lesser extent, the fact that total coordination numbers are not constant with composition also contributes to nonlinearity. These results for the DCS/CH_3CN+CO_2 system are consistent with previous results based on the solvatochromic probe C153 (20) and are likely to be general for polar + nonpolar mixtures and polarity sensitive probes.

The present work offers the first direct experiment-simulation comparisons of solvation dynamics in GXLs. Analysis of the solvation response functions reveals that both the short and long time regimes are affected by composition. As the concentration of CO_2 is increased, a diffusive tail emerges and the decay times for the mixtures surpass those observed for the neat components. Nonequilibrium simulations were used to analyze the solvation dynamics of the mixture most dilute in CH_3CN. Based on time-dependent analysis of the local solvent compositions, we have observed enrichment of the CH_3CN component at the expense of the CO_2 component as well as a net gain in solvent molecules within the first solvation shell. Thus, as reported in other systems (26,24) both electrostriction and redistribution are important, however, unlike the other cases studied we find no distinction between the timescales of these two processes.

References

1. Eckert, C. A.; Bush, D.; Brown, J. S.; Liotta, C. L. *Ind. Eng. Chem. Res.* **2000**, *39*, 4615.
2. Eckert, C. A.; Liotta, C. L.; Bush, D.; Brown, J. S.; Hallett, J. P. *J. Phys. Chem. B* **2004**, *108*, 18108.
3. Hallett, J. P.; Kitchens, C. L.; Hernandez, R.; Liotta, C. L.; Eckert, C. A. *Acc. Chem. Res.* **2006**, *39*, 531.
4. Jessop, P. G.; Subramaniam, B. *Chem. Rev.* **2007**, *107*, 2666.
5. Li, H.; Maroncelli, M. *J. Phys. Chem. B* **2006**, *110*, 21189.
6. Marcus, Y. *Solvent Mixtures, Properties and Selective Solvation*; Marcel-Dekker: New York, **2002**.
7. Kim, S.; Johnston, K. *AIChE J.* **1987**, *33*, 1603.
8. Kelley, S. P.; Lemert, R. M. *AIChE J.* **1996**, *42*, 2047.
9. Wyatt, V. T.; Bush, D.; Lu, J.; Hallett, J. P.; Liotta, C. L.; Eckert, C. A. *J. Supercrit. Fluids* **2005**, *36*, 16.
10. Pfund, D. M.; Fulton, J. L.; Smith, R. D. *ACS Symposium Series* **1993**, *514*, 158.
11. Moon, S. D. *Bull. Korean Chem. Soc.* **2002**, *23*, 811.
12. Chatzis, G.; Samios, J. *Chem. Phys. Lett.* **2003**, *374*, 187.
13. Aida, T.; Inomata, H. *Mol. Sim.* **2004**, *30*, 407.
14. Stubbs, J. M.; Siepmann, J. I. *J. Chem. Phys.* **2004**, *121*, 1525.
15. Houndonougbo, Y.; Guo, J.; Lushington, G. H.; Laird, B. *Mol. Phys.* **2006**, *104*, 2955.
16. Houndonougbo, Y.; Jin, H.; Rajagopalan, B.; Wong, K.; Kuczera, K.; Subramaniam, B.; Laird, B. *J. Phys. Chem. B* **2006**, *110*, 13195.
17. Houndonougbo, Y.; Laird, B.; Kuczera, K. *J. Chem. Phys.* **2007**, *126*, 074507.
18. Houndonougbo, Y.; Kuczera, K.; Subramaniam, B.; Laird, B. *Mol. Sim.* **2007**, *33*, 861.

19. Shukla, C. L.; Hallett, J. P.; Popov, A. V.; Hernandez, R.; Liotta, C. L.; Eckert, C. A. *J. Phys. Chem. B* **2006**, *110*, 24101.

20. Li, H.; Arzhantzev, S.; Maroncelli, M. *J. Phys. Chem. B* **2007**, *111*, 3208.

21. Skarmoutsos, I.; Samios, J. *J. Mol. Liq.* **2006**, *125*, 181.

22. Yoshimori, A. *J. Chem. Phys.* **1998**, *109*, 3222.

23. Martins, L. R.; Tamashiro, A.; Laria, D.; Skaf, M. S. *J. Chem. Phys.* **2003**, *118*, 5955.

24. Ladanyi, B. M.; Perng, B. C. *J. Phys. Chem. A* **2002**, *106*, 6922.

25. Cichos, F.; Brown, R.; Rempel, U.; Borcyskowski, C. v. *J. Phys. Chem. A* **1999**, *103*, 2506.

26. Day, T. J.; Patey, G. N. *J. Chem. Phys.* **1997**, *106*.

27. Day, T. J. F.; Patey, G. N. *J. Chem. Phys.* **1999**, *110*, 10937.

28. Skaf, M. S.; Ladanyi, B. M. *J. Phys. Chem.* **1996**, *100*, 18258.

29. Swalina, C.; Arzhantzev, S.; Li, H.; Maroncelli, M. *J. Phys. Chem. B* **2008**, *(to be submitted)*.

30. Stone, A. *The Theory of Intermolecular Forces*; Oxford University Press: New York, **1996**.

31. Harris, J. G.; Yung, K. H. *J. Phys. Chem.* **1995**, *99*, 12021.

32. Edwards, D. M. F.; Madden, P. A.; McDonald, I. R. *Mol. Phys.* **1984**, *51*, 1141.

33. Kometani, N.; Hoshihara, Y.; Maroncelli, M. *Rev. High Pressure Sci. Technol.* **2006**, *16*, 113.

34. Arzhantzev, S.; Zachariasse, K. A.; Maroncelli, M. *J. Phys. Chem. A* **2006**, *110*, 3454.

35. DL_POLY_2. In *Version 2.14*; Smith, W., Forester, T. R., Eds.; CCLRC Daresbury Laboratory: Daresbury, UK, 2001.

36. Frenkel, D.; Smit, B. *Understanding Molecular Simulation, From Algorithms to Applications*, 2nd ed.; Academic Press: New York, **2002**.

37. Adams, D. J.; Dubey, G. S. *J. Comput. Phys.* **1987**, *72*, 156.

38. Patel, N.; Biswas, R.; Maroncelli, M. *J. Phys. Chem. B* **2002**, *106*, 7096.

39. Suppan, P. *J. Photchem. Photobiol. A* **1990**, *50*, 293.

40. Schulte, R. D.; Kauffman, J. F. *J. Phys. Chem.* **1994**, *98*, 8793.

41. Khajehpour, M.; Welch, C. M.; Kleiner, K. A.; Kauffman, J. F. *J. Phys. Chem. A* **2001**, *105*, 5372.

42. Maroncelli, M.; Kumar, V. P.; Papazyan, A. *J. Phys. Chem.* **1993**, *97*, 13.

43. Kumar, P. V.; Maroncelli, M. *J. Chem. Phys.* **1995**, *103*, 3038.

44. Ladanyi, B. M.; Maroncelli, M. *J. Chem. Phys.* **1998**, *109*, 3204.

Chapter 6

Phase Behavior and Equilibria of Ionic Liquids and Refrigerants: 1-Ethyl-3-methyl-imidazolium Bis(trifluoromethylsulfonyl)imide ([EMIm][Tf₂N]) and R-134a

Wei Ren[1], Aaron M. Scurto[1,*], Mark B. Shiflett[2],
and Akimichi Yokozeki[3]

[1]Department of Chemical and Petroleum Engineering and NSF–ERC Center for Environmentally Beneficial Catalysis, University of Kansas, Lawrence, KS 66045
[2]DuPont Central Research and Development, Experimental Station, Wilmington, DE 19880–0304
[3]DuPont Fluoroproducts Laboratory, Chestnut Run Plaza, 711, Wilmington, DE 19880–0711
*Corresponding author: phone: +1 785–864–4947; fax: +1 785–864–4967; email: ascurto@ku.edu

Several potential applications of ionic liquids with refrigerant gases have recently been developed. However, an accurate understanding of phase behavior and equilibria are needed for any type of process development. Here, the global phase behavior and equilibria of 1-ethyl-3-methyl-imidazolium bis(trifluoromethylsulfonyl)imide ([EMIm][Tf₂N]), and the refrigerant, 1,1,1,2-tetrafluoroethane (R-134a) are measured from approximately 20°C to 101°C and up to 160 bar. High-pressure vapor-liquid equilibrium at 25°C, 50°C, and 75°C are reported. Regions of multiphase equilibria exist at various temperatures and pressures and include vapor-liquid (VLE), liquid-liquid (LLE), vapor-liquid-liquid (VLLE) equilibrium, mixture critical points, and lower and upper critical endpoints. The global phase behavior indicates that the system is Type V according to the Scott-van Konynenburg classification scheme. An equation of state model has been used to predict the high-pressure phase equilibrium and the global behavior based upon parameters regressed from low-pressure data with good agreement.

Introduction

The solubility and separation of refrigerant gases are important problems for several industries. Many refrigerants and their intermediates have very similar physical and chemical properties which can often render their separation costly by traditional distillation. The separation of the common azeotropic mixtures encountered among refrigerant mixtures usually requires additional components in extractive distillation. Ionic liquids have been shown to provide efficient solutions to these problems (1).

The solubility of refrigerant gases in low-volatility liquids forms the basis of absorption refrigeration. Here, the gas dissolves in a liquid at low temperatures in a unit called the *absorber* and is transported to another unit (the *generator*), which is heated and liberates (boils) the high pressure gas. The liquid is then returned to the absorber to repeat the process. The high pressure gas from the generator then proceeds to the condenser and evaporator which are common to both the more common vapor-compression system and the absorption refrigeration system. The wide-spread application of absorption refrigeration has been impeded by the often bulky equipment that is needed to purify the high-pressure gas from the generator as the presence of any absorption liquid significantly decreases the efficiency. Non-volatile ionic liquids in these systems with refrigerants may help solve these problems (2).

High-pressure phase behavior and equilibria data are the most important aspects to utilize any of the aforementioned applications. The phase behavior determines the conditions (temperature and pressure) of the transitions in equilibria, whether vapor-liquid, liquid-liquid, etc. Phase equilibria data quantify the solubility of each component in each phase in these regions. In addtion, a model to correlate and to predict data would be highly useful. Here, a model system has been chosen comprising the ionic liquid, 1-ethyl-3-methyl-imidazolium bis(trifluoromethylsulfonyl)imide ([EMIm][Tf$_2$N]), and the refrigerant, 1,1,1,2-tetrafluoroethane (R-134a) (see Figure 1), to illustrate the most common features of IL/refrigerant systems. An equation of state model has been used to correlate and predict the behavior and equilibria using various amounts of experimental data.

Background

Ionic Liquids

Ionic liquids (ILs) are organic salts that are liquid at or near room temperature ($T_m < 100°C$). There are a myriad of cation/anion combinations that can yield an organic salt.

Figure 2 illustrates a few of the common classes of ionic liquids: imidazolium, pyridinium, quaternary ammonium, and phosphonium. Ionic liquids can be molecularly engineered for specific physico-chemical properties through various "R-"groups and cation/anion selection, *e.g.* viscosity, solubility properties, density, acidity/basicity, etc. It has been estimated that $\sim 10^{14}$ unique cation/anion combinations are possible (3). Preliminary toxicology studies of ILs (4-7) indicate that ILs can have low to high toxicity depending on the cation and anion, but most ILs have no measurable vapor pressure, thus eliminating air pollution. Negligible vapor pressure also results in highly-elevated flash points, usually by the flammability of decomposition products at temperatures greater than 300°C. The immeasurable vapor pressure may facilitate more efficient absorption refrigeration processes. These properties have led researchers to claim ionic liquids as potential "green" solvents. Novel ILs and processes are continually being developed for extractions, reactions (8, 9), and material processing (10-14).

Ionic Liquids/Refrigerants

Combining ionic liquids and refrigerants has a number of advantages. Most refrigerant gases have a high solubility in ionic liquids (15). One of the challenges of ionic liquids is their viscosity which is often higher than conventional solvents (16) and leads to slower mass transport rates. In concurrent studies in our laboratory, the viscosity of the IL, 1-hexyl-3-methyl-imidazolium bis(trifyl)imide ([HMIm][Tf$_2$N]) has been measured and found to decrease approximately 80+% with up to 6 bar of R-134a at 25°C. This decrease in the mixture viscosity has been used to predict the diffusivity in the ionic liquid with approximately a 5-fold improvement over the same pressure range. This dramatic increase in the mass transport properties will lead to processes with less capital intensity as smaller equipment (e.g. heat exchangers, contactors, etc.) can be used.

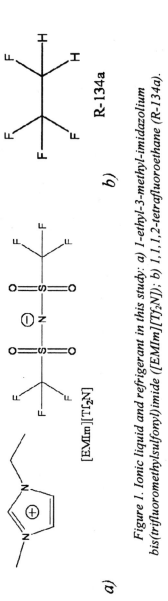

a) [EMIm][Tf₂N]

b) R-134a

Figure 1. Ionic liquid and refrigerant in this study: a) 1-ethyl-3-methyl-imidazolium bis(trifluoromethylsulfonyl)imide ([EMIm][Tf₂N]); b) 1,1,1,2-tetrafluoroethane (R-134a).

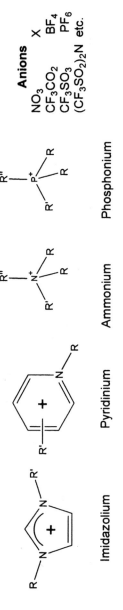

Imidazolium   Pyridinium   Ammonium   Phosphonium

Anions
X
NO₃   BF₄
CF₃CO₂   PF₆
CF₃SO₃   etc.
(CF₃SO₂)₂N

Figure 1. Illustration of common cations and anions for ionic liquids.

There are a few groups in the literature to report phase equilibria data for ionic liquids and refrigerant gases. Shiflett and Yokozeki (1, 15, 17-32) reported vapor-liquid equilibria (VLE) of a series of refrigerants, including R-23, R-125, R-41, R-134a, R-143a, R-152a, R-161, CO_2 and NH_3 in several ILs , such as 1-butyl-3-methyl-imidazolium hexaflurophosphate ([BMIm][PF$_6$]), [HMIm][Cl], 1-ethyl-3-methyl-imidazolium bis(trifluoromethylsulfonyl)imide ([EMIm][Tf$_2$N]), [HMIm][Tf$_2$N], [BMIm][BF$_4$], etc. For R-32 and R-134a, they also synthesized a series of new ILs to study the effects of anions and alkyl chain lengths on the refrigerant solubility (17, 18). Shariati and Peters compared the high-pressure phase behavior between CO_2 and trifluoromethane (CHF$_3$) in [EMIm][PF$_6$] and [BMIm][PF$_6$] (33). Kim *et. al.* (34) have studied the refractive index, heat capacity, and vapor pressure of mixtures of [BMIm][BF$_4$] and [BMIm][Br] with 2,2,2-trifloroethanol (TFE) and water (34).

Global Phase Behavior

A rich variety of phase behavior is possible with liquids and gases. For the process design of any system with a compressed gas, the global phase behavior must be characterized. The Gibbs Phase Rule indicates that even for binary systems, up to four phases may coexist in equilibrium. Thus, two phase and three phase regions are likely to exist at various temperatures, pressures and compositions. Van Konynenburg and Scott (35) were the first to provide a uniform classification scheme for binary phase behavior of liquids and gases. They classify all systems into six unique types, labeled I through VI as illustrated in Figure 3. More recently, Bolz *et al.* (36) have proposed a new nomenclature system for the International Union of Pure and Applied Chemistry (IUPAC) in which a more descriptive nomenclature is presented. The new designation conveys the topology and connectivity of critical curves. Each phase behavior type has characteristic phase transitions. For example, Type V behavior of the Scott-van Konynenburg scheme is characterized by vapor-liquid equilibrium at low temperatures, followed by a lower critical end-point (LCEP) where another liquid phase forms (thus, vapor-liquid-liquid equilibrium), followed by an upper critical end-point (UCEP) at higher temperatures in which the liquid phases merge. The mixture critical point extends from the LCEP to the pure component critical point of the less volatile component.

Type IV is similar to type V, except a low temperature region of LLE, VLLE and a second UCEP exist. Yokozeki and Shiflett (30) predicted using an equation of state, that the system of trifluoromethane and [BMIm][PF$_6$] is a Type V system. Shiflett and Yokozeki (24) measured and modeled the low-pressure vapor-liquid equilibrium of R-134a and [EMIm][Tf$_2$N] with a modified cubic equation of state. From the parameters regressed from the data, the model predicts that the R-134a and [EMIm][Tf$_2$N] is a Type V system. The present

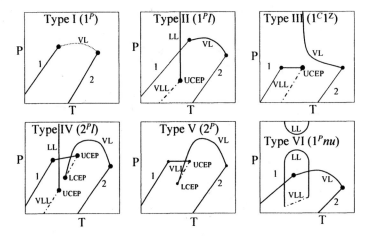

Figure 3. Common types of global phase diagrams according to the classifications of Scott and van Konynenburg (37); and Bolz et al. (38) (in parentheses).

work will measure the high-pressure global phase behavior and equilibria (VLLE, LCEP, UCEP, mixture critical points, etc.) to confirm these predictions.

Experimental

Safety

The experiments described here were performed under elevated pressures and should be handled with care. All equipment should have the proper pressure ratings and standard operating procedures as determined by trained professionals.

Phase Behavior and Equilibrium

In this study, two different apparatus were used to investigate the phase behavior and equilibria of the ionic liquid and refrigerant system. The phase behavior and the temperature and pressure of the transitions were observed in a windowed high-pressure autoclave similar to a design by Leitner and coworkers (37) and modified in our group; for further details see Schleicher (38). The ~10mL viewcell rated to 150°C and 400 bar of pressure and is stirred with a small stirbar. The temperature was maintained by a heating plate (Ika Werke,

GmbH, PN: RET Basic C) with electronic temperature control (Ika Werke, GmbH, PN: IKATRON ETS-d4-fuzzy) using a Pt-1000Ω RTD (DIN IEC 751 Class A) placed through the wall of the autoclave; the temperature precision is ±0.1°C. Pressures were measured with a high precision digital pressure gauge, Omega DPG7000-3000psi with an error band of ± 0.05% of full scale. Three different types of phase behavior transitions were observed: vapor-liquid (VLE) to vapor-liquid-liquid (VLLE); VLLE to liquid-liquid (LLE); and VLE, VLLE or LLE to critical transitions including upper-critical (UCEP) and lower-critical endpoints (LCEP). Approximately 1.5 mL of ionic liquid was loaded into the cell and heated to the desired temperature. The refrigerant was slowly introduced by a high-pressure syringe pump (Teledyne-Isco, Inc. model 260D) and vented to remove any argon or air. R-134a was slowly added to the vessel to the desired pressure, stirred and allowed to reach equilibrium (approximately 30 minutes). The pressure was slowly raised and the mixture stirred; this was repeated until the first sign of the phase transition. The vessel was partially vented to a pressure just below the transition, and allowed to re-equilibrate. The process was repeated until the transition was reproduced within ±1 bar. In addition, the temperature of the system was changed at constant pressure. These temperatures were reproducible to approximately ±0.5°C.

The solubility of the refrigerant in the ionic liquid at various temperatures and pressures was measured in a static equilibrium apparatus. Details of the apparatus are described in Ren and Scurto (39), and an overview will be given here. The system consists of a high-pressure equilibrium cell, a high-precision syringe pump filled with R-134a, a water bath and accessories to measure the temperature and pressure. As the R-134a has a high solubility even at below-ambient conditions, a vacuum was employed to approximately 0.2 bar absolute to begin the experiment. The apparatus is capable of determining the solubility (mole fraction, molarity, etc.), density of the liquid solution, molar volume, and volume expansion. These calculations are based on the mass balance, by determining R-134a delivered from the pump, the moles/mass of R-134a in the headspace above the liquid and in the tubing/lines to the equilibrium cell. This method can yield solubility data of high resolution (often better than ±0.001), pressure with accuracy of ±0.2 bar, temperature of ±0.01°C, density up to ±0.4 %, and volume expansion to ±0.05%.

Synthesis of Ionic Liquid

[EMIm][Tf$_2$N] (see Figure 1.) was synthesized by anion exchange from 1-ethyl-3-methyl-imidazolium bromide ([EMIm][Br]). [EMIm][Br] was synthesized by a quaternization reaction of 1-methyl-imidazole and bromoethane in acetonitrile at 40°C. *Caution: this reaction is highly exothermic and adequate amounts of solvent or cooling should be used.* To purify 1-ethyl-3-

methyl-imidazolium bromide, acetonitrile is added to dissolve the solid and then passed through a plug of celite and a short column of acidic alumina. The solvent was removed on a rotavap under reduced pressure and 40°C. [EMIm][Tf$_2$N] was prepared from the anion exchange of [EMIm][Br] with lithium bis(trifluoromethylsulfonyl)imide (Li[Tf$_2$N]) in deionized water as described in the literature (40). The denser hydrophobic IL phase is decanted and washed with twice the amount of water for 8-10 times. The IL is then dried under vacuum for at least 48 hours and stored in Schlenk tubes under dry argon. Water content was determined by a Mettler Toledo DL32 Karl Fisher Coulometer. ^{1}H NMR chemical shifts (relative to TMS internal standard) and coupling constants J/Hz: δ=8.59 (s, 1H), 7.42(d, 2H, J=10.3), 4.24 (q, 2H, J=7.36), 3.93(s, 3H), 1.54(t, 3H, J=7.4). Analysis calculated for C$_8$H$_{11}$N$_3$F$_6$S$_2$O$_4$: C, 24.55; H, 2.83; N, 10.74; S, 16.39. Found: C, 24.62; H, 2.84; N, 10.71; S, 15.88, Water content is 63 ppm; purity is estimated to be 99+% based on NMR and elemental analysis.

Materials

1,1,1,2-Tetrafluoroethane (R-134a , 99.99%) was purchased from Linweld, Inc and used as received. 1-Methylimidazole, (CAS 616-47-7) 99+%, lithium bis(trifluoromethylsulfonyl)imide (CAS 90076-65-6) 99.95%, acetonitrile, ≥99.9%; were purchased from Sigma-Aldrich. 1-Bromoethane (CAS 74-96-4) 99+%, was obtained from Acros. The 1-methylimidazole and bromoethane were vacuum distilled and used immediately for the synthesis.

Modeling

A modified Redlich-Kwong type of cubic equations of state (EoS) is employed in this study.

$$P = \frac{RT}{\underline{V} - b} - \frac{a(T)}{\underline{V}(\underline{V} + b)}$$

Where

$$a_i(T) = \frac{0.42748 R^2 T_{c,i}^2}{P_{c,i}} \alpha_i(T) \qquad b_i = \frac{0.08664 RT_{c,i}}{P_{c,i}}$$

where, $\underline{V}$ is the molar volume, $T_{c,i}$ is the critical temperature of the ith component; $P_{c,i}$ is the critical pressure of the ith component; x_i is the mole fraction of the ith component; and R is the universal gas constant.

The temperature dependent term, $\alpha(T)$, is modified using the following empirical form:

$$\alpha_i(T) = \sum_{k=0}^{3} \beta_i (\frac{1}{T_r} - T_r)^k \qquad T_r = T/T_{c,i}$$

The EoS parameters used are summarized in Table I. The mixture parameters (a and b) are calculated from the binary interaction parameters.

$$a = \sum_{i=1}^{N} \sum_{j=1}^{N} \sqrt{a_i a_j} f_{ij}(T)(1 - k_{ij}) x_i x_j$$

where

$$f_{ij}(T) = 1 + \frac{\tau_{ij}}{T}, \quad \tau_{ij} = \tau_{ji}, \tau_{ii} = 0 \qquad k_{ij} = \frac{l_{ji} l_{ij}(x_i + x_j)}{l_{ji} x_i + l_{ij} x_j}, \quad k_{ii} = 0$$

and

$$b = \frac{1}{2} \sum_{i=1}^{N} \sum_{j=1}^{N} (b_i + b_j)(1 - k_{ij})(1 - m_{ij}) x_i x_j, \quad m_{ij} = m_{ji}, \quad m_{ii} = 0$$

and N is the total number of components. In the above model, there are a maximum of four binary interaction parameters: l_{ij}, l_{ji}, m_{ij}, τ_{ij} for each binary pair. The optimal binary interaction parameters were determined from the low-pressure VLE data of Shiflett and Yokozeki (24).

Table I. Equation of state parameters and binary interaction parameters[a].

Compound	T_c [K]	P_c [bar]	β_0	β_1	β_2	β_3
[EMIm][Tf$_2$N]	808.82	20.28	1.0	1.35500	0	0
R-134a	374.21	40.59	1.0025	0.50532	-0.04983	0

	l_{12}	l_{21}	$m_{12} = m_{21}$	$\tau_{12} = \tau_{21}$ [/K]
R-134a/ [EMIm][Tf$_2$N]	-0.03084	-0.00416	0.00726	17.50

[a] From data of Shiflett and Yokozeki (24)

The fugacity coefficient, ϕ_i, of the ith component for a particular phase is calculated as:

$$\ln \phi_i = \ln \frac{RT}{P(V - b)} + b_i' \left(\frac{1}{V - b} - \frac{a}{RTb(V + b)} \right) + \frac{a}{bRT} \left[\frac{a_i'}{a} - \frac{b_i'}{b} + 1 \right] \ln \frac{V}{V + b}$$

where

$$a_i' = \left(\frac{\partial na}{\partial n_i} \right)_{n_{j \neq i}} = 2 \sum_{j=1}^{N} \sqrt{a_i a_j} f_{ij}(T) x_j \left(1 - k_{ij} - \frac{l_{ij} l_{ji}(l_{ij} - l_{ji}) x_i x_j}{(l_{ji} x_i + l_{ij} x_j)^2} \right) - a$$

$$b_i' = \left(\frac{\partial nb}{\partial n_i}\right)_{n_{j\neq i}} = \sum_{j=1}^{N} (b_i + b_j)(1 - m_{ij})x_j\left(1 - k_{ij} - \frac{l_{ij}l_{ji}(l_{ij} - l_{ji})x_i x_j}{(l_{ji}x_i + l_{ij}x_j)^2}\right) - b$$

The phase equilibrium boundaries for binary systems are obtained by equilibrium conditions.

For VLE: $x_i\phi_i^L = y_i\phi_i^V$ and for VLLE: $x_i^{L_1}\phi_i^{L_1} = x_i^{L_2}\phi_i^{L_2} = y_i\phi_i^V$, where x_i and y_i are the liquid and vapor mole fraction compositions, respectively.

Results and Discussion

The global phase behavior and phase equilibria of [EMIm][Tf$_2$N] and R-134a were studied between approximately 20°C to 101°C and pressures to ~160 bar. Previously, Shiflett and Yokozeki (24) studied the vapor-liquid equilibrium of [EMIm][Tf$_2$N] and R-134a at 10°C, 25°C, 50°C, and 75°C, but only until approximately 3.5 bar due to experimental limitations using a Hiden Gravimetric Microbalance (20, 24). Here, the high-pressure VLE data at 25°C, 50°C, and 75°C have been extended to just below the vapor pressure of the pure R-134a or the pressure of vapor-liquid-liquid equilibrium (VLLE). The experimental data are found in Table II and plotted in Figure 4. As shown in the figure, the experimental data at the lower pressures correlate very well with that of Shiflett and Yokozeki (24) who used a different experimental method and apparatus, *viz.* a gravimetric microbalance. The higher pressure data smoothly continue to the pure R-134a vapor pressure or VLLE region. The modeling prediction is shown with the data. The equation of state binary interaction parameters were regressed from the lower pressure data and used to predict the higher pressure data with very good accuracy.

The plot also shows experimental data and model predictions for the VLLE region at temperatures of 60, 65, 70, and 75.9°C. As the second liquid phase is nearly pure R-134a, the plot omits the VLE envelope that technically extends from the pure vapor pressure to the second liquid composition. Again, the model predicts the occurrence and composition of this 3-phase equilibrium quite well. The liquid phase compositions of the VLLE region become closer at the lower temperatures measured. This seems to indicate a lower critical endpoint (LCEP) exists. From the Scott-van Konynenburg classification schemes, only Types IV to VI have LCEPs.

The global phase behavior for [EMIm][Tf$_2$N] and R-134a was measured and is presented in Figure 5 and Table II. This contribution is one of the first complete global phase diagrams for a compressed gas and an ionic liquid. As shown, various regions of single phase and multi-phase equilibria exist. One of the interesting aspects of this behavior is the presence of a single miscible phase or critical region. This is opposed to the majority of the high-pressure equilibria

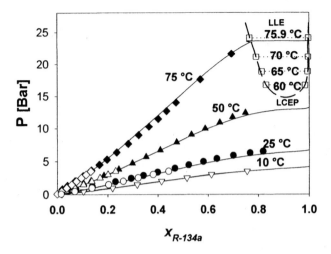

Figure 4. Vapor-liquid equilibrium data for [EMIm][Tf₂N] and R-134a. Closed symbols: experimental data in this study; Open symbols are the data of Shiflett and Yokozeki (24) for VLE and VLLE respectively; lines are equation of state predictions using only the low-pressure data.

data in the literature for ionic liquids and various gases (41-43). Shariati and Peters (33) have reported a vapor-liquid mixture critical pressure and composition for [EMIm][PF₆] and high-pressure CHF_3. The three phase vapor-liquid-liquid (VLLE) equilibrium exists between the lower and upper critical endpoints (LCEP and UCEP) and is equal to the pure component vapor pressure of R-134a within the experimental accuracy of this study. This often indicates that the second liquid phase is almost pure R-134a (x_{R-134a}<0.01) as confirmed by Shiflett and Yokozeki (21, 24) for this and other ionic liquids. As the R-134a-rich phase is nearly pure, its UCEP is nearly incident (within experimental accuracy) with the pure component critical point. Below the LCEP temperature, the system is in vapor-liquid equilibrium until the pressure is raised to the vapor pressure of the pure R-134a, at which point the system becomes miscible. These points of miscibility have been confirmed experimentally and are shown on the plot Figure 5. In contrast with CO_2, only VLE and then LLE are possible for temperatures below the pure CO_2 critical point. For [EMIm][Tf₂N] and R-134a above the LCEP but below the UCEP, four different equilibria are possible. VLE exists at the lower pressures and VLLE exists near or at the pure R-134a vapor pressure. Above the VLLE, liquid-liquid (LLE) equilibria exists until the mixture critical point and the system becomes critical/miscible. Above the UCEP temperature, VLE exists and terminates at the mixture critical point.

The global phase behavior was predicted using the model parameters regressed with only the low pressure VLE data. The model predicts the VLLE data very well and is nearly equal to the pure R-134a vapor pressure line. From convergence of the critical point and VLLE predictions, the lower critical endpoint (LCEP) is approximated as 54°C and 14.5 bar. This is very close to the experimental point of 55°C and 15 bar, despite the fact that the prediction is from lower pressure data. The UCEP is approximated at 101°C and 40.5 bar which is within a 0.5°C and 0.3 bar of the experimental point. The model provides an adequate correlation of the mixture critical pressure at 60°C. For the other critical points, the model under-predicts the critical pressure. Considering that the interaction parameters for the prediction are based solely on low-pressure data, the level of agreement is noteworthy.

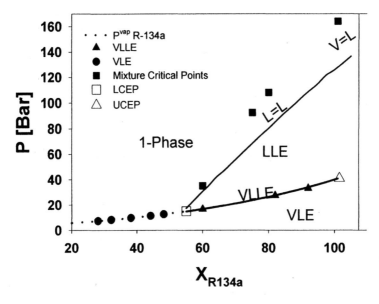

Figure 5. Experimental global phase behavior for [EMIm][Tf₂N] and R-134a with model predictions (thick solid lines).

The pressure-composition behavior of the mixture at 75°C is shown in Figure 6 with experimental VLE and mixture critical point data and the model prediction. As shown, the model correlates the experimental VLE and VLLE data very well. The model prediction for the liquid-liquid equilibrium (LLE) data and mixture critical point is also shown. The predicted composition at the critical point is close to the experimental value, but under predicts the critical pressure. The prediction of the mixture critical points using cubic equations of state often is quantitatively difficult, especially from interaction parameters regressed with lower pressure vapor-liquid equilibrium data. Here, the pressure

Table I. VLE data of [EMIm][Tf$_2$N] and R-134a

25 °C		50°C		75°C	
P [bar]	x_{R134a}	P [bar]	x_{R134a}	P [bar]	x_{R134a}
0.28	0.0161±0.0020	0.34	0.0105±0.0012	0.51	0.0087±0.0008
0.48	0.0411±0.0019	1.04	0.0616±0.0011	0.82	0.0203±0.0008
0.79	0.0754±0.0018	1.66	0.1055±0.0010	1.33	0.0409±0.0007
1.25	0.1392±0.0016	2.08	0.1345±0.0010	2.04	0.0715±0.0007
1.95	0.2304±0.0013	2.84	0.1844±0.0009	2.79	0.1035±0.0006
2.44	0.2943±0.0011	3.79	0.2437±0.0008	3.64	0.1355±0.0006
2.86	0.3428±0.0009	4.69	0.2986±0.0007	4.48	0.1670±0.0006
3.35	0.3992±0.0008	5.66	0.3531±0.0006	5.26	0.1963±0.0005
3.95	0.4681±0.0006	6.67	0.4090±0.0006	6.34	0.2354±0.0005
4.44	0.5228±0.0005	7.86	0.4745±0.0005	7.63	0.2818±0.0004
4.93	0.5812±0.0004	9.01	0.5338±0.0004	8.96	0.3235±0.0004
5.37	0.6344±0.0004	9.95	0.5884±0.0004	10.26	0.3644±0.0005
5.85	0.7004±0.0003	10.89	0.6465±0.0003	11.46	0.4014±0.0005
6.24	0.7667±0.0002	11.74	0.6974±0.0003	12.68	0.4345±0.0005
6.48	0.8183±0.0001	12.40	0.7486±0.0002	13.99	0.4722±0.0005
				17.57	0.5687±0.0003
				21.60	0.6928±0.0002
				92.29[a]	0.9628±0.0001

[a] Mixture critical point (L=L)

Table II. Global phase behavior data for [EMIm][Tf$_2$N] and R-134a.

Transition	T[°C]	P [bar]	Transition	T[°C]	P [bar]
Mixture Critical Points	60.00	35.00	LCEP	55.00	15.00
	75.00	92.29	UCEP	101.5	40.80
	80.00	108.00			
	101.0	164.00			
VLE^a	28.15	7.30	VLLE	60.00	16.60
	32.18	8.19		81.95	27.41
	38.20	9.67		92.05	33.00
	44.17	11.33			
	48.19	12.56			

^{a} confirmed conditions of liquid-vapor miscibility at the vapor pressure of R-134a.

range used to regress the interaction parameters is nearly 30 times lower than the mixture critical pressure. Even regressing interaction parameters from VLE data from low pressures to nearly the mixture critical point, researchers have described difficulty in predicting both the critical composition and pressure at a given temperature (44, 45). Often, one must fit the interaction parameters directly to the critical point data to get quantitative agreement for pressure and composition.

From the experimental data and the modeling predictions, the system of [EMIm][Tf$_2$N] and R-134a is believed to be classified as a Type V system (see Figure 3). The Type IV system is similar, but exhibits another lower temperature region of liquid-liquid immiscibility and VLLE ending at another UCEP. For reference, the system of phenyloctane and R-134a is a Type IV system, and the difference between the lower-branch of the UCEP and the LCEP is 29°C (46). The VLE region for [EMIm][Tf$_2$N] and R-134a was investigated experimentally to 0°C with no occurrence of liquid immiscibility. The phase behavior at cryogenic temperatures is beyond the capabilities of the experimental equipment. The model does not predict any liquid immiscibility below the LCEP.

Conclusion

The global phase behavior and equilibria of [EMIm][Tf$_2$N] with the refrigerant gas, R-134a, was measured and modeled with a modified cubic equation of state. R-134a is highly soluble in this ionic liquid and is miscible in certain regions of temperature and pressure. Regions of multiphase equilibria exist, *viz.* VLE, LLE, and VLLE. This phase behavior indicates that this is a

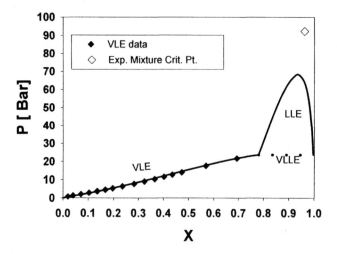

Figure 6. Pressure-composition diagram for the VLE, VLLE, and mixture critical point at 75 ℃. Lines: model prediction from low pressure data.

Type V system according to the classification of Scott and van Konynenburg. The cubic equation of state using parameters regressed at the lowest pressures was able to predict the higher pressure VLE, VLLE and the LCEP.

Acknowledgements

This work was supported by the DOT:KU Transportation Research Institute (TRI) (DOT# DT0S59-06-G-00047); NSF-ERC Center for Environmentally Beneficial Catalysis (CEBC EEC-0310689); and DuPont Central Research and Development. The author (AMS) appreciates the support of the DuPont Young Professor Award.

References

1. Shiflett, M. B.; Yokozeki, A., *Chimica Oggi/Chem. Today* **2006**, *24*, 28-30.
2. Shiflett, M. B.; Yokozeki, A. Absorption cycle utilizing ionic liquids as working fluid Patent WO 2006/08426 A1, 2006.
3. Holbrey, J. D.; Seddon, K. R., *Clean Prod. Proc.* **1999**, *1*, 223-236.
4. Bernot, R.; Brueseke, M.; Evans-White, M.; Lamberti, G., *Environ. Toxicol. Chem.* **2005**, *24*, 87-92.
5. Garcia, M.; Gathergood, N.; Scammells, P., *Green Chem.* **2005**, *7*, 9-14.

6. Ranke, J.; Stolte, S.; Stormann, R.; Arning, J.; Jastorff, B., *Chem. Rev.* **2007**, *107*, 2183-2206.
7. Stepnowski, P.; Skladanowski, A.; Ludwiczak, A.; Laczynska, E., *Hum. Exp. Tox.* **2004**, *23*, 513-517.
8. Wasserscheid, P.; Keim, W., *Angew. Chem. Int. Ed.* **2000**, *39*, 3773-3789.
9. Welton, T., *Chem Rev* **1999**, *99*, 2071-2083.
10. Barisci, J. N.; Wallace, G. G.; MacFarlane, D. R.; Baughman, R. H., *Electrochem. Commun.* **2004**, *6*, 22-27.
11. Murugesan, S.; Linhardt, R. J., *Curr. Org. Synth.* **2005**, *2*, 437-451.
12. Zhao, D. C.; Xu, H. T.; Xu, P.; Liu, F. Q.; Gao, G., *Prog. Chem.* **2005**, *17*, 700-705.
13. Gathergood, N.; Garcia, M.; Scammells, P., *Green Chem.* **2004**, *6*, 166-175.
14. Cao, J. M.; Fang, B. Q.; Wang, J.; Zheng, M. B.; Deng, S. G.; Ma, X. J., *Prog. Chem.* **2005**, *17*, 1028-1033.
15. Shiflett, M. B.; Yokozeki, A., *AIChE J.* **2006**, *52*, 1205-1219.
16. Crosthwaite, J. M.; Muldoon, M. J.; Dixon, J. K.; Anderson, J. L.; Brennecke, J. F., *J. Chem. Thermodyn.* **2005**, *37*, 559–568.
17. Shiflett, M. B.; Harmer, M. A.; Junk, C. P.; Yokozeki, A., *J. Chem. Eng. Data* **2006**, *51*, 483-495.
18. Shiflett, M. B.; Harmer, M. A.; Junk, C. R.; Yokozeki, A., *Fluid Phase Equil.* **2006**, *242*, 220-232.
19. Shiflett, M. B.; Kasprzak, D. J.; Junk, C. P.; Yokozeki, A., *J. Chem. Thermodyn.* **2008**, *40*, 25-31.
20. Shiflett, M. B.; Yokozeki, A., *Ind. Eng. Chem. Res.* **2005**, *44*, 4453-4464.
21. Shiflett, M. B.; Yokozeki, A., *J. Chem. Eng. Data* **2006**, *51*, 1931-1939.
22. Shiflett, M. B.; Yokozeki, A., *Ind. Eng. Chem. Res.* **2006**, *45*, 6375-6382.
23. Shiflett, M. B.; Yokozeki, A., *J. Phys. Chem. B* **2006**, *110*, 14436-14443.
24. Shiflett, M. B.; Yokozeki, A., *J. Chem. Eng. Data* **2007**, *52*, 2007-2015.
25. Shiflett, M. B.; Yokozeki, A., *J. Chem. Eng. Data* **2007**, *52*, 2413-2418.
26. Shiflett, M. B.; Yokozeki, A., *Fluid Phase Equil.* **2007**, *259*, 210-217.
27. Shiflett, M. B.; Yokozeki, A., *J. Phys. Chem. B* **2007**, *111*, 2070-2074.
28. Shiflett, M. B.; Yokozeki, A., *Ind. Eng. Chem. Res.* **2008**, *47*, 926-934.
29. Shiflett, M. B.; Yokozeki, A., *J. Chem. Eng. Data* **2008**, *53*, 492-497.
30. Yokozeki, A.; Shiflett, M. B., *AIChE J.* **2006**, *52*, 3952-3957.
31. Yokozeki, A.; Shiflett, M. B., *Appl. Energ.* **2007**, *84*, 1258-1273.
32. Yokozeki, A.; Shiflett, M. B., *Ind. Eng. Chem. Res.* **2007**, *46*, 1605-1610.
33. Shariati, A.; Peters, C. J., *J. Supercrit. Fluids* **2003**, *25*, 109-117.
34. Kim, K.-S.; Shin, B.-K.; Lee, H.; Ziegler, F., *Fluid Phase Equil.* **2004**, *218*, 215-220.
35. Van Konynenburg, P. H.; Scott, R. L., *Philos. Trans. R. Soc. London, Ser. A* **1980**, *298*, 495-540.
36. Bolz, A.; Deiters, U. K.; Peters, C. J.; de Loos, T. W., *Pure Appl. Chem.* **1998**, *70*, 2233-2257.

37. Koch, D.; Leitner, W., *J. Am. Chem. Soc.* **1998**, *120*, 13398-13404.
38. Schleicher, J. MS Thesis, University of Kansas, Lawrence, 2007.
39. Ren, W.; Scurto, A. M., *Rev. Sci. Instrum.* **2007**, *78*, 125104-7.
40. Nockemann, P.; Binnemans, K.; Driesen, K., *Chem. Phys. Lett.* **2005**, *415*, 131-136.
41. Aki, S. N. V. K.; Mellein, B. R.; Saurer, E. M.; Brennecke, J. F., *J. Phys. Chem. B* **2004**, *108*, 20355-20365.
42. Kumelan, J.; Perez-SaladoKamps, A.; Tuma, D.; Maurer, G., *J. Chem. Eng. Data* **2006**, *51*, 1364-1367.
43. Kumelan, J.; Perez-SaladoKamps, A.; Tuma, D.; Maurer, G., *Ind. Eng. Chem. Res.* **2007**, *46*, 8236-8240.
44. Castier, M.; Sandler, S. I., *Chem. Eng. Sci.* **1997**, *52*, 3579-3588.
45. Scurto, A. M.; Lubbers, C. M.; Xu, G.; Brennecke, J. F., *Fluid Phase Equil.* **2001**, *190*, 135-147.
46. Proot, W.; de Loos, T. W., *Fluid Phase Equil.* **2004**, *222-223*, 255-259.

Reactions

Chapter 7

In Situ Alkylcarbonic Acid Catalysts Formed in CO_2-Expanded Alcohols

Jason P. Hallett[1], Charles A. Eckert[2], and Charles L. Liotta[3]

[1]Department of Chemistry, Imperial College, London, United Kingdom
[2]School of Chemical and Biomolecular Engineering, Georgia Institute of Technology, Atlanta, GA 30332–0100
[3]School of Chemistry and Biochemistry, Georgia Institute of Technology, Atlanta, GA 30332–0400

We have developed self-neutralizing acid catalysts formed from carbon dioxide and alcohols for use in chemical synthesis. These alkylcarbonic acid catalysts are generated *in situ* and can be easily neutralized via depressurization. This technique combines a medium with good organic solubility with acid catalysts that do not require neutralization, thus completely eliminating the solid wastes associated with many acid processes. We examined the origins and characteristics of alkylcarbonic acids by comparing reaction rates of acid-sensitive probes in different CO_2-expanded alcohols, elucidating the effect of CO_2 pressure on solvent properties and acid formation, and demonstrating the use of alkylcarbonic acids for a variety of reaction types.

Introduction

Acids are the most used catalysts in industry, and they produce more than 1 x 10^8 metric ton/year of products.(*1*) Some of the most common acids include Brønstead acids such as HCl, H_2SO_4 and H_3PO_4 and Lewis acids such as $AlCl_3$ and BF_3. Unfortunately, acid-catalyzed processes are plagued by large amounts of waste associated with post-reaction acid neutralization. As an extreme example, for the Friedel-Crafts acylation of methyl benzoate with acetic anhydride, nearly 20 kg of $AlCl_3$ are used per kg of product.(*2*) Acid neutralization produces a contaminated waste salt which must be disposed of carefully (and often expensively).

As an alternative, we have developed alkylcarbonic acids, which are acid catalysts generated *in situ* through the interactions between CO_2 and alcohols in gas-expanded liquids (GXLs). Alkylcarbonic acids are a potential replacement for mineral acids which possess all of the solvation and separation advantages inherent in GXLs(*3-11*) while the acid neutralization step is eliminated in favor of simple depressurization.

It is well known that CO_2 interacts with water to form carbonic acid, as shown in Scheme 1; an analagous interaction between CO_2 and alcohols produces alkylcarbonic acids. The discovery of the catalytic properties of these acids came about as part of our work studying GXLs, when methanol-CO_2

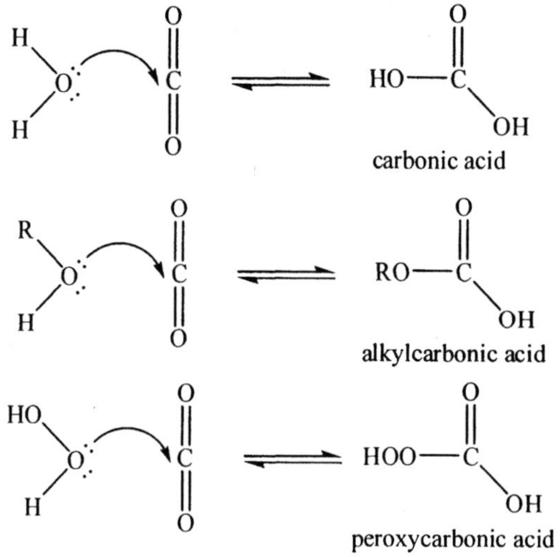

Scheme 1. Formation of carbonic acid, alkylcarbonic acid, and peroxycarbonic acid in gas-expanded liquids.

mixtures were observed to protonate Reichardt's dye. We explained this phenomenon by demonstrating the presence of alkylcarbonic acids in such systems.(*12*) These acids are reversibly formed with CO_2 pressure, and the easy removal of CO_2 forms the basis of our claim of 'self-neutralization'. We have repeatedly used diazodiphenylmethane (DDM) as a probe to measure relative reaction rates of these acids as well as carbonic acid in aqueous-CO_2 systems.(*12,13*).

In this report, we review our work with a brief summary of both the characterization (*12,13*) and utilization (*14-16*) of alkycarbonic acid as a catalyst in GXL systems. The use of GXLs as a replacement for traditional organic solvents offers many other advantages, which are detailed in previous work (*3-11*) and elsewhere in this volume. Instead, this paper will focus on the acidity of CO_2-expanded alcohols, the effect of solvent properties (CO_2 expansion, dielectric constant) on this acidity, and the application of alkylcarbonic acids to a variety of potential industrial syntheses, including the hydrolysis of β-pinene, the formation of ketals and several diazotization reactions.

Characterization of Alkylcarbonic Acids

Measurement of Relative Acidities using Diazodiphenylmethane (DDM)

The characterization of the acid properties of gas-expanded alcohols has involved several stages. We first reported preliminary finding (*12*) verifying the presence of alkylcarbonic acids in CO_2-expanded alcohols by using DDM as a reactive probe to trap the acid species. One of the advantages of the DDM probe is that it incorporates the conjugate base of any reactive acids present in the system into the reaction products. Therefore, we could confirm through NMR analysis that alkylcarbonic acids were our proton source and not just the solvent alcohol. Futher, we used the relative rates of reaction of the alkylcarbonic acids with DDM to develop a 'scale' for correlating the relative strengths of different alkylcarbonic acids. Our initial study (*12*) reported the relative reaction rates of DDM in both methanol and ethanol expanded by 20 mole % CO_2. The rate in methanol was found to be roughly 2.8 times that in ethanol, indicating that there are more available protons in the GXL-methanol system. It should be noted that we cannot distinguish between relative acid strength (pK_a) and equilibrium acid concentration using this method, because the expansion of the alcohols with CO_2 will increase the equilibrium concentration of acid species but simultaneously decrease the polarity of the solvent mixture and therefore the acid dissociation (concentration of free protons).

DDM results using acetone as a common diluent

To separate these effects, we compared the rates obtained for different alcohols (including water) using acetone as a common diluent to maintain a constant CO_2 concentration. Kinetic measurements were conducted at 40 °C using a solution containing 60 mol% acetone, 20 mol% CO_2, and 20 mol% R-OH. Our results revealed that the rates in alcohols follow a logical progression, with a monotonic decrease in rate with increasing alkyl chain length and shorter-chain alcohols yielding faster rates with the fastest rate in methanol and the relative order 1°>2°>3°. These results are consistent with steric and electronic factors affecting the equilibrium and dissociation of the alkylcarbonic acids. The rate using carbonic acid (from water and CO_2) was found to be slower than either methanol or ethanol. In our second study,(*13*) we used a wider set of alcohols, including methanol, ethylene glycol, propylene glycol, benzyl alcohol, 4-nitro benzyl alcohol, 4-chloro benzyl alcohol, and 4-methoxy benzyl alcohol. Once more, acetone was used as the dominant component in the liquid phase (60 mol%) in order to control CO_2 solubility as closely as possible. Surprisingly, reactions utilizing ethylene glycol were faster than all mono-hydroxy alcohols, including even methanol. The reaction involving propylene glycol was similarly accelerated, with rates similar to those obtained using methanol, and faster than ethanol. It is expected that the relative basicity and nucleophilicity of the oxygen in the hydroxyl functionality will control the formation and acid strength of alkylcarbonic acid, as more basic oxygens will be more likely to react with CO_2. We further believe that the second alcohol fuctonality in glycols stabilizes the alkylcarbonic acid form of the CO_2-OH interaction through hydrogen bonding. The results for the reaction of DDM with various CO_2-expanded alcohols are summarized in Figure 1.

Effect of CO_2 concentration on DDM reactions

In order to determine the effect of CO_2-expansion on the DDM reactions, we performed a series of experiments in which the pure alcohol was expanded with CO_2 (no diluent). Figure 2 displays the results for CO_2-expanded methanol. The most important result from these experiments was the observance of a maximum in rate at 60 bar CO_2. The increase in rate with CO_2 expansion at pressures below 60 bar was attributed to an increase in the equilibrium concentration of alkylcarbonic acid (at the expense of free alcohol) with increased CO_2 concentration. The decrease in rate after 60 bar results from this increase in concentration being counteracted by a large decrease in acid dissociation resulting from a drop in medium polarity. Any decrease in the ability of the solvent to solvate free protons will reduce the effective acidity of the medium. This decrease in polarity can be explored by examining the

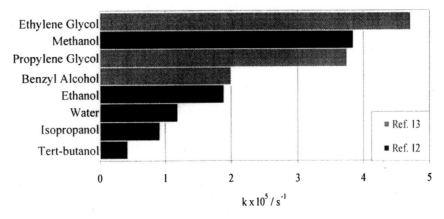

Figure 1. Rate of DDM decomposition in various CO₂-expanded solvents.
All reactions were performed in 60 mol% acetone/20 mol% ROH/20 mol% CO₂
at 40 °C. Data from references 12 and 13.

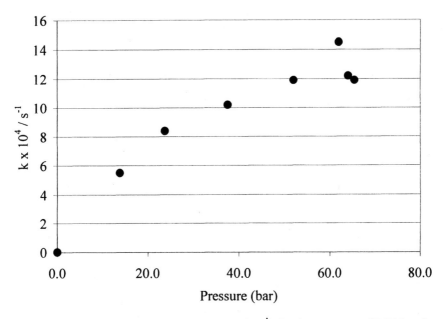

Figure 2. Pseudo first order rate constant (k, s⁻¹) for the reaction of DDM with
methylcarbonic acid as a function of CO₂ pressure at 40°C.

dielectric constant as a function of CO_2 concentration, which is discussed below. It should be noted that the maximum rate is 3 orders of magnitude faster than the rate obtained in the pure alcohol. Thus, alkylcarbonic acids demonstrate a clear catalytic effect on DDM decomposition.

Hammett-type analysis of DDM reactions

In practical terms, the use of the alcohol as a GXL solvent would necessitate an understanding of the susceptibility of the reaction to changing the electronic nature of the alcohol making up the alkylcarbonic acid. A Hammett-type analysis is a very useful method for acquiring such mechanistic information. In the case of our alkylcarbonic acids, we preformed a series of experiments using substituted benzyl alcohols, choosing sustituents for which the Hammett substituent values (σ) are known. Our Hammett correlation is log (k_x/k_0) = $\sigma*\rho$, where k_x is the rate constant for the disappearance of DDM with different substituted benzyl alcohols, k_0 is the rate constant for the reaction involving unsubstituted benzyl alcohol, σ is a constant for the ring substituent based on electron-withdrawing/donating ability, and ρ is the reaction sensitivity to electronic effects. Since unsubstituted benzyl alcohol has a moderate rate of reaction when used as a solvent, a wide range of substituents could be used. 4-nitro-, 4-chloro-, and 4-methoxy- benzyl alcohols were chosen in addition to benzyl alcohol (see Table I).(*17*) Once again, we equalized CO_2 solubility in the system by using 85 mol% acetone, 5 mol% alcohol, and 10 mol% CO_2 for all experiments. Additionally, most of our substituted alcohols are solid at room temperature (and some have overlapping bands of UV absorbance with DDM), so low concentrations of alcohol enabled us to examine the electronic effect of alcohol nature on the alkylcarbonic acid activity in the absence of bulk solvent polarity effects or CO_2 solubility concerns.

The resulting Hammett correlation yielded a ρ value for substituted benzyl alcohols of 1.71, which is very similar to the value of 1.93 reported for DDM reacting with benzoic acid in acetone at 30°C.(*18*) These relatively high values of ρ indicate that the reaction is very sensitive to electronic effects. The first step is a nucleophilic attack by a benzyl alcohol on the carbon in CO_2, which should be increased by electron-donating groups and decreased by electron-withdrawing groups. On the other hand, the second step of proton dissociation should be increased by electron-withdrawing groups and decreased by electron-donating groups. Meanwhile, the second step, proton dissociation, should be accelerated by electron-withdrawing groups (which increase the acidity of the carboxylic proton) and decelerated by electron-donating substituents. Since the fastest rate occurred with the highly electron-withdrawing nitro substitution, the second step in our reaction scheme (proton dissociation) must be the dominant effect controlling r, and therefore the most sensitive step to electronic effects.

**Table I. Rate constants used in Hammett correlation of
para-substituted benzyl alcohols**

Substituent	σ	$k \times 10^6 / s^{-1}$
NO$_2$	0.78	559
Cl	0.23	62.0
H	0.00	24.5
OCH$_3$	-0.27	9.10

Additionally, benzoic acid only undergoes the second step (proton dissociation) since there is no acid formation necessary. The ρ value for the benzoic acid reaction in acetone is similar to our alkylcarbonic acid value, so the acid formation step for alkylcarbonic acids is not likely to be limiting. The Hammett plot also revealed a two order of magnitude difference between the most electron-withdrawing and the most electron-donating groups. Electron-withdrawing substituents enhance the proton transfer step in the reaction with DDM and electron-donating groups decrease the rate of proton transfer. These results are consistent with the proton transfer step being rate-determining.

Analysis of dielectric constant of GXLs

One of our primary goals with understanding alkylcarbonic acids is to decouple the relative contributions of acid strength and acid concentration. In order to accomplish this, some simple measure of proton dissociation ability would be beneficial. We chose to do this by measuring the dielectric constant of CO_2-expanded methanol as a function of CO_2 concentration (or gas expansion). These measurements give some indication of how much the polarity of a GXL (and thus the dissociation of acid species present) is decreased when CO_2 is added to the liquid phase. The results of these experiments are displayed in Figure 3. The dielectric constant decreases in a nearly additive manner when compared to volume expansion of the system, resulting in an approximately linear decay of dielectric constant. This supports the concept that proton dissociation will be inhibited as CO_2 concentration in the GXL is increased.

Applications of Alkylcarbonic Acids

Acetal formation reactions

We have also reported several example reactions for which alkylcarbonic acids serve as catalyst. In Xie et al. (*16*) we used CO_2-expanded methanol and

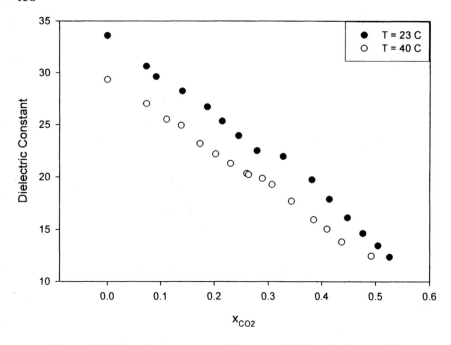

Figure 3. Dielectric constant as a function of CO_2 mole fraction for CO_2-expanded methanol at 40°C. Data from reference 13.

CO_2-expanded ethylene glycol to create ketals of cyclohexanone. Acetals are a commonly used protecting group for the carbonyl groups of aldehydes and ketones, which are susceptible to highly basic nucleophiles. Traditionally, acetals are formed from a carbonyl-containting substrate by use of a large excess of the protecting group (normally an alcohol) and a strong mineral acid catalyst. We replaced the strong mineral acid with CO_2-expansion of the protecting alcohol, forming acetals of cyclohexanone. We reported initial rates of reaction as shown in Figure 4. By replacing CO_2 with ethane, we were able to confirm that the large catalytic effect of gas-expansion is due to alkylcarbonic acid formation and not simple gas expansion of the solvent because the rate of acetal formation was similar to that in the pure alcohol. The most important aspect of Figure 4 is the existence of a maximum in the rate constant, which occurs at each temperature we studied. We observed a similar effect during the DDM reactions, with the maximum attributable to the tradeoff between increasing alkylcarbonic acid formation and decreasing acid dissociation with increased CO_2 expansion. In the acetal reactions, the maximum shifts from about 20 bar at 25 °C to 35 bar at 50 °C as more pressure is required to maintain a similar CO_2 concentration in the liquid phase. However, in the Xie *et al.* study, equation-of-state modelling revealed that the mole fraction of CO_2 at the various maxima are not identical.

Therefore, there are additional factors controlling the location of the optimum pressure for reaction. The most surprising result was the high reaction rate observed in CO_2-expanded ethylene glycol, which dissolved less than 2% CO_2 at all conditions (expansion data not shown).

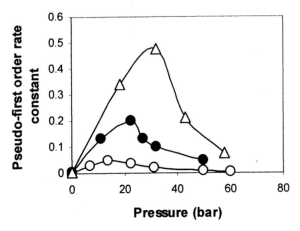

Figure 4. The pseudo-first order rate constants of cyclohexanone acetal formation in CO_2-expanded methanol at various CO_2 pressures and 25 °C (open circles), 40 °C (filled circles) and 50 °C (triangles). (Reproduced from reference 16. Copyright 2004 American Chemical Society.)

Diazotization reactions

A second application of alkylcarbonic acids as catalysts involved the synthesis of several intermediates using a diazotization route.(*14*) We used alkylcarbonic acids to catalyze the formation of a diazonium ion, which was then further converted to several products. The formation of the diazonium intermediate is considered the limiting step in many of these cross-coupling and Sandmeyer-type reactions. One of our targets, methyl yellow, is an industrial dye synthesized via a diazonium cross-coupling scheme requiring the use of a strong mineral acid such as HCl or H_2SO_4 to promote the reaction. Our synthesis involved a single pot containing aniline, sodium nitrite, *N,N*-dimethyl aniline and CO_2 with methanol as solvent, as shown in Scheme 2. Our best results for this reaction were obtained at 5°C with excess nitrite salt and a high CO_2 loading. In this optimal case, 97% conversion of aniline (with 72% yield of methyl yellow) was obtained after 72 hours. Increasing the CO_2 improved the yield up to our maximum pressure of 47 bar. This contradicts the DDM and acetal reaction systems, where an optimum CO_2 pressure was found at a variety of different

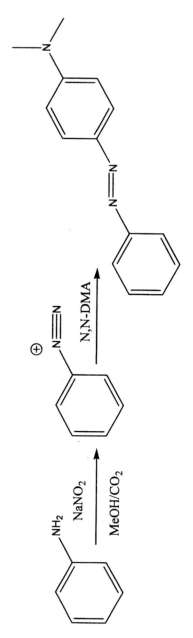

Scheme 2. Synthesis and mechanism for formation of methyl yellow from aniline.

temperatures. Apparently, the proton dissociation step is not as important for our diazotization reactions as the acid formation step is, so the effects of increased CO_2 concentration were found to be universally beneficial. This is possibly due to the instability of the diazonium cation, which often limits reaction yield, though not reaction rate. Similar cross coupling reactions, to form products such as N,N-diethyl-4-[(4-nitrophenyl)azo]aniline (DENAB), yielded similar results.

Several different contols were utilized to determine if the catalysis was due to methylcarbonic acid or a generic solvent effect. Using normal reaction conditions in the absence of CO_2 we obtained no measurable yield. When CO_2-expanded tetrahydrofuran was used in place of CO_2-expanded methanol, we obtained a minimal yield of 0.3%. We proposed that this trace catalysis is likely due to the formation of carbamic acid from aniline and CO_2 (*19*), which would be present in small quantities. When the solventless reaction was examined (basically, CO_2-expanded aniline was the "solvent") we obtained a methyl yellow yield of 14%. Our conclusion is that carbamic acid formed from aniline and CO_2 might be responsible for a minimal percentage of the reaction yield, but methylcarbonic acid clearly provides the majority of the catalysis in our system, and alkylcarbonic acids are superior to carbamic acids in this situation.

We also reported the effect of varying the reagents used and their quantities in the reaction mixture. Isoamyl nitrite was found to give half the yields of sodium nitrite at similar concentrations. We did observe a significant increase in reaction rate and yield when the amount of excess sodium nitrite was increased from 1.04 times (20%) to 2.2 times excess (63%). A range of temperatures was also employed for the coupling reaction, because diazotizations are usually carried out at low temperature due to the instability of the diazonium ion. We did observe this behavior; for example in the coupling to form N,N-diethyl-4-[(4-nitrophenyl)azo]aniline (DENAB) using methylcarbonic acid at temperatures of 5, 25, and 50 °C, the highest yields were at 5°C and yield decreased with increasing temperature.

In addition to the cross-coupling reactions, we synthesized iodobenzene in a single pot from potassium iodide and aniline using CO_2-expanded methanol. These halide addition reactions are usually catalyzed by copper salts or alkali metal salts of the nucleophilic anion. The reaction proceeds through a diazonium intermediate as the coupling reactions did, but in this case the diazonium ion reacts with the nucleophilic iodide to form iodobenzene. The most effective conditions were similar to those of the coupling reaction except in the case of temperature. Our best yield (72%) was obtained at the highest temperature (50°C) instead of the lowest. Clearly, the stability of the diazonium cation is not limiting, as it was in the case of the coupling reactions. Instead, simple Arrhenius behavior on the rate constant is dominant; therefore the yield increases uniformly with temperature.

Another difference from the coupling reactions was observed in the effect of nitrite loading. For the iodobenzene synthesis, increasing the nitrite loading produced a negligible change in overall yield. We attribute this to the much lower solubility of potassium iodide in methanol as compared to the organic *N,N*-dimethyl aniline. This hypothesis is supported by the negligible effect observed by increasing the loading of potassium iodide. This may also explain the temperature effect on the rate because iodide solubility will increase with temperature. Our results from increased CO_2 loading also agree with this hypothesis since more CO_2 (and thus more acid) did not have an effect on the yield, indicating that acid concentration is not limiting.

Solvolysis reactions

In Chamblee *et al.* (*15*) we compared the hydrolysis of β-pinene in both high temperature water and CO_2-expanded liquids. The solvents selected for CO_2 expansion were mixtures of methanol and water, providing both carbonic and alkylcarbonic acid. Literature analyses of CO_2/water systems report that the pH can reach values as low as 2.8;(*20*) however, in CO_2/water/methanol systems, a minimum in pH was reported at 4.22.(*21*) As in pure alcohol/CO_2 systems, this difference was attributed by the authors to the competition between increased acid concentration and decreased dielectric constant with CO_2 expansion. For our hydrolysis reaction, the product selectivities were much higher when using CO_2-expanded liquids when compared to hot water; in addition, we found that by using CO_2/methanol/water (1:1 methanol:water) at 75 °C, our highest conversion 71% could be obtained. Thus, a 1:1 mixture of water and methanol would seem to be the optimum solvent ratio for this hydrolysis. In the absence of methanol, essentially no reaction took place, probably due to the low solubility of the substrate. Figure 5 demonstrates that both the reaction rate and product distribution were optimal for the 1:1 volume ratio of methanol/water. The conversion drops off quickly at higher methanol ratios, perhaps again due to the lowering of dielectric constant. The conversion also drops off with increasing water content; this is perhaps due to either reduced substrate solubility or is an indication of the relative strengths of methycarbonic acid (pKa = 5.61) and carbonic acid (pKa = 6.37).

Conclusions

We have demonstrated the use of alkylcarbonic acids as self-neutralizing acid catalysts for use in chemical synthesis. These catalysts are generated *in situ* and are reversibly formed with CO_2 pressure. We reviewed the characterization of alkylcarbonic acids, which revealed that CO_2-expanded ethylene glycol and methanol form the most strongly acidic media, while alcohols decreased in strength with increasing chain length and according to the series $1° > 2° > 3°$. We

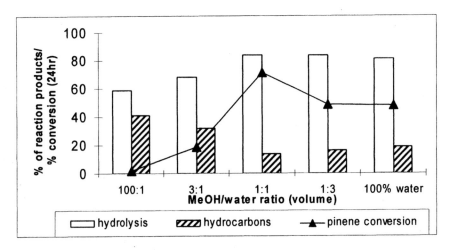

Figure 5. β-pinene conversion and product distribution variations with different methanol/water ratios (75 °C, 24 hrs, 0.127M β-pinene). (Reproduced from reference 10. Copyright 2008 American Chemical Society.

also demonstrated the use of alkylcarbonic acid for a number of synthetic reactions: the protection of ketones and aldehydes, the diazotization of aromatics, and the solvolysis of alkenes. In most cases, an optimum CO_2 pressure effect on reaction rate was found. This can generally be attributed to a tradeoff between increased acid concentration and decreased acid dissociation with CO_2 pressure.

It is not certain how many acid-catalyzed processes could be performed using alkylcarbonic acids. Our successes to date have focused on reactions involving the formation of diazonium or carbonium ions or of protecting groups such as acetals. It is evident that alkylcarbonic acids are strong enough to replace mild acids such as carboxylic acids (the pKa's are similar) and will suffice for some mineral acid catalyzed processes. However, alkylcarbonic acids are unlikely to be a suitable replacement for applications requiring strong acids, such as Friedel-Crafts alkylations, unless the starting materials are suitably activated. The full potential of these benign acids has yet to be elucidated.

References

1. Corma, A. Solid acid catalysts. *Curr. Opin. Solid State Mater. Sci.* **1997**, *2*, 63.
2. Dartt, C. B.; Davis, M. E. *Ind. Eng. Chem. Res.* **1994**, *33*, 2887.

3. Hallett, J. P.; Kitchens, C. L.; Hernandez, R.; Liotta, C. L.; Eckert, C. A. *Acc. Chem. Res.* **2006**, *39*, 531.
4. Jessop, P. G.; Subramaniam, B. *Chem. Rev.* **2007**, *107*, 2666.
5. Eckert, C. A.; Liotta, C. L.; Bush, D.; Brown, J. S.; Hallett, J. P. *J. Phys. Chem. B* **2004**, *108*, 18108.
6. Eckert, C. A.; Bush, D.; Brown, J. S.; Liotta, C. L. *Ind. Eng. Chem. Res.* **2000**, *39*, 4615.
7. Wyatt, V. T.; Bush, D.; Lu, J.; Hallett, J. P.; Liotta, C. L.; Eckert, C. A. *J. Supercrit. Fluids* **2005**, *36*, 16.
8. Li, H.; Maroncelli, M. *J. Phys. Chem. B* **2006**, *110*,
9. Shukla, C. L.; Hallett, J. P.; Liotta, C. L.; Eckert, C. A.; Popov, A. V.; Hernandez, R. *J. Phys. Chem. B* **2006**, *110*, 24101.
10. Hallett, J. P.; Pollet, P.; Liotta, C. L.; Eckert, C. A. *Acc. Chem. Res.*, **2008**, in press..
11. Musie, G.; Wei, M.; Subramaniam, B.; Busch, D. H. *Coord. Chem. Rev.* **2001**, *219-221*, 789.
12. West, K. N.; Wheeler, C.; McCarney, J. P.; Griffith, K. N.; Bush, D.; Liotta, C. L.; Eckert, C. A. *J. Phys. Chem. A* **2001**, *105*, 3947.
13. Weikel, R. R.; Hallett, J. P.; Levitin, G. R.; Liotta, C. L.; Eckert, C. A. *Top. Catal.* **2006**, *37*, 75.
14. Weikel, R. R.; Hallett, J. P.; Liotta, C. L.; Eckert, C. A. *Ind. Eng. Chem. Res.* **2007**, *46*, 5252.
15. Chamblee, T. S.; Weikel, R. R.; Nolen, S. A.; Liotta, C. L.; Eckert, C. A. *Green Chem.* **2004**, *6*, 382.
16. Xie, X.; Liotta, C. L.; Eckert, C. A. *Ind. Eng. Chem. Res.* **2004**, *43*, 2605.
17. Carroll, F. A. *Perspectives on Structure and Mechanism in Organic Chemistry*; Brooks/Cole Publishing Company: Pacific Grove, CA, 1998 p.384.
18. Buckley, A.; Chapman, N. B.; Dack, M. R. J.; Shorter, J.; Wall, H. M. *J. Chem. Soc. B* **1968**, *2*, 631.
19. Salvatore, R.N.; Flanders, V. L.; Ha, D.; Jung, K. W. *Org. Lett.* **2000**, *2*, 2797.
20. Towes, K. L.; Shroll, R.; Wai, C. M.; Smart, N. G. *Anal. Chem.* **1995**, *67*, 4040.
21. Wen, D.; Olesik, S. V. *Anal. Chem.* **2000**, *72*, 475.

Chapter 8

Catalytic Oxidation Reactions in Carbon Dioxide-Expanded Liquids Using the Green Oxidants Oxygen and Hydrogen Peroxide

Daryle H. Busch[1,3,*] and Bala Subramaniam[2,3]

Departments of [1]Chemistry and [2]Chemical and Petroleum Engineering, and [3]Center for Environmentally Beneficial Catalysis (CEBC), University of Kansas, Lawrence, KS 66047

This research focuses on sustainable catalytic oxidation chemistry for the commercial production of wide ranges of products. O_2 in the air and minimum cost H_2O_2 are the oxidants of choice because of their green footprints. Herein, the gas expanded liquid principle is applied to various kinds of reaction components: (a) media expanded liquids (MXLs) including the well known carbon dioxide expanded liquids (CXLs); (b) substrate expanded liquids (SXLs), as in our propylene to propylene oxide process, and (c) reagent expanded liquids (RXLs), as exemplified by certain ozone oxidations.

Oxidative Catalysis in Supercritical Carbon Dioxide ($scCO_2$)

A CO_2 expanded liquid (CXL) is a mixed solvent which contains a large fraction of the condensed carbon dioxide as part of the volume of the liquid, commonly 20 to 80% by volume. These versatile media are described as carbon dioxide expanded solvents because of the simple process by which they are produced; i.e., the mixing of a sample of appropriate liquid with CO_2 under pressures that cause the CO_2 to dissolve into the organic phase. In other words, the liquid phase can be "expanded" by pumping dense CO_2 into it, and contracted by decreasing the pressure to release the dissolved CO_2. In a very real sense, the development of CXLs was made possible by earlier studies on $scCO_2$. In their pioneering work on catalytic oxidation in dense carbon dioxide, Tumas and coworkers (1) first reported oxidation of cyclohexene to adipic acid using a ruthenium catalyst and NaOCl as the oxidant in a phase transfer catalytic process carried out in $scCO_2$ and water as the reaction media. In later publications, Tumas et al. (2) explored homogeneous catalytic oxidations of cyclohexene by O_2 in $scCO_2$ using halogenated porphyrins (e.g., 345 bar, 80°C), yielding multiple products, including cyclohexenol, cyclohexenone, and 4-hydroxy-2-cyclohexene-1-one, and the oxidation of several allylic alkenes, with t-BuOOH using a variety of Lewis acid catalysts in dense phase CO_2 (defined as either $scCO_2$ or subcritical, liquid CO_2 (3). The latter study also reported highly promising stereoselective epoxidation reactions in dense CO_2. In parallel research, Hass and Kolis studied the epoxidation of allylic alcohols using a vanadium Schiff base catalyst and t-BuOOH as the oxidant, in $scCO_2$ (213 bar, 45-50°C), reporting yields and selectivities similar to those obtained in organic solvents. These investigators (4) independently contributed to the understanding of epoxidations with t-BuOOH of unactivated olefins in $scCO_2$, using $Mo(CO)_6$ as the catalyst source. Indeed 1998 was a year of major growth in the understanding of catalytic oxidation reactions in $scCO_2$. Kreher and his collaborators (5) also used molybdenum and titanium catalysts to oxidize cyclooctene with t-BuOOH, and Wu and coworkers (6) made use of perfluorinated tetraphenyl porphyrin-iron and acetaldehyde as an initiator to oxidize cyclohexane with air. In virtually all these investigations, the use of CO_2 to totally replace the organic solvent medium provided environmental benefits but only marginal, if any, reaction benefit. The introduction of the concept of carbon dioxide expanded liquids, CXLs, as media for catalytic reactions, especially oxidations, follows from the lessons learned from these earlier investigations and was aimed at optimizing both the reaction and environmental benefits.

CXLs and Beyond

Early studies in our laboratories introduced CXLs for catalytic chemical processes in the course of a study of certain oxidation systems and these are discussed in a section to follow. CXLs are continua of media generated by pressure solubilization of CO_2 into any suitable solvent, but typically an organic solvent. The compositions of CXLs vary from mostly organic solvent, through all intermediate volume ratios of CO_2:organic solvent, to mostly dense phase CO_2 (7, 8). Since the properties of the two components of the CXL contribute to the properties of that mixed solvent, the tuning of solvent properties is a convenient tool for process optimization. The solubilities of substrates, catalysts and other solutes are often readily managed using this solvent tunability. For many solutes, solubilities in dense phases of CO_2 itself are limited by its nonpolar nature, and this limitation is overcome in appropriately tuned CXLs. A major advantage of CXLs is in the high concentration of the oxidatively inert carbon dioxide in solutions that might otherwise be flammable. The dominance in the vapor phase of inert CO_2, the abundant reservoir of liquid CO_2 backing up that equilibrium vapor phase, and the high heat capacity (e.g., compared to N_2) of CO_2 strongly compress the flammability envelope for CXL solutions making such media attractive for large scale industrial applications, where size hugely multiplies risk. Because of their advantages CXLs have been applied to many chemical reactions including oxidations, hydroformylation, solid acid catalysis, and hydrogenation, among others.

The unique properties of the so-called gas-expanded liquids (GXLs), of which CXLs are a sub-set, are associated with the proximity between the critical temperature (T_c) of the dissolving gas and the temperature (T) at which the CXL is generated. In general, when $T = 0.9\text{-}1.2\ T_c$, the dissolving gas is highly compressible in the vicinity of the critical pressure (P_c). The solubility parameter (δ) correspondingly increases upon compression, and the density becomes liquid-like beyond the critical pressure, resulting in increased dissolution into the liquid phase causing the liquid phase to visibly swell or expand. In this manner, CO_2 ($P_c = 73.8$ bar; $T_c = 31.1°C$) expands several conventional solvents several-fold volumetrically at room temperature ($\sim 20°C$) and at tens of bars. In the same pressure domain, transport properties such as diffusivity and viscosity are also sensitively tuned with pressure. The quantitative variations (i.e., tunability) of the physical properties of CXLs are discussed in detail in a recent review (8). As the temperature is increased beyond $1.2\ T_c$ ($\sim 90°C$), the compressibility and the solubility parameter of CO_2 diminish. Consequently, the ability to expand a liquid phase at mild pressures is severely hampered, and pressures well above 100 bars would be needed to effect any appreciable volumetric expansion of the liquid phase.

Based on the understanding of the tunability of those properties facilitating the use of GXLs, the expanding gas could either be an inert solvent medium (such as CO_2), a substrate (such as propylene or ethylene) or even an oxidant (such as O_3).   The use of CO_2 to expand liquid reaction phases and to significantly improve solubilities of reactant gases, such as O_2, H_2 and CO, has resulted in commensurate process intensification in cases where gas starvation in the reaction phase is rate-limiting.   Recently, we extended the technique to enhance the solubility of light olefins such as propylene (P_c = 46.1 bar; T_c = 91°C) and ethylene (P_c = 50.8 bar; T_c = 9.5°C)) in aqueous reaction phases containing a miscible alcohol (9).   In the 25-40°C range, the light olefins may be substantially dissolved into the liquid phase at mild pressures.   In this manner, dramatically enhanced yields of propylene oxide are attainable at mild pressures during homogeneous catalytic propylene epoxidation with hydrogen peroxide, as discussed in a later section. This amounts to a *substrate expanded liquid* (SXL) for the intensified reaction. Very recently, we have been able to demonstrate a *reagent expanded liquid* (RXL) using ozone (P_c = 55.73 bar; T_c = 260.95 K) as the reagent/oxidant and CO_2 as the liquid phase (10). Thus, it may be possible to exploit the critical properties of any gas that is a major component in a chemical reaction to increase its availability in the reaction phase by pressure intensification, thereby enhancing the reaction rate.

Principles of Catalyst Design

A simple perspective on the essential properties that make a catalyst valuable anticipates many motivating factors that have influenced the chemistry described here. First of all, <u>sustainability</u> is essential in chemical processes throughout this modern culture, whether the system be confined to a test tube in a high school laboratory or be secured in huge reactors behind fences in a commodity scale manufacturing plant.   Sustainability reflects health and safety matters, the local maintenance of the environment, the preservation of planet earth and its habitability, and economic viability. These sustainability factors ought to be pertinent to all situations and locations. Green chemistry teaches us that since risk is a function of hazard and exposure, risk is only truly minimized by elimination of the hazard; i.e., exposure is not infinitely controllable. This is the chemists' basis for choosing to work with those elements that are least likely to present hazards; i.e., use iron instead of chromium, oxygen instead of chlorine, etc. Sustainability has a strongly compelling financial component; green chemistry provides no benefit when it is not implemented. Practically, the increasingly rare precious metals will not be first choices when catalysts will be lost to the environment, or simply lost, for that matter.

In an example from one of the author's laboratories (11, 12) a homogeneous catalyst was to be designed for a consumer application that needed only a mild oxidizing agent for a commodity market. The fact that the proposed market is that of commodities suggests that *sustainability* is most likely for a catalyst based on one of the less expensive metals from the first transition element series in the periodic table. The vast literature of bio-inorganic chemistry makes it clear that nature manages much of its oxidation chemistry around the three metallic elements copper, iron and manganese. Copper presents more of a hazard than iron and manganese so these would be the two metallic elements of choice for this proposed new catalyst--call it the Tim and Daryle catalyst (T&D cat) (13, 14).

The first technical requirement for a useful catalyst is <u>efficacy</u>, the ability to accelerate the targeted reaction enough to make the catalyst system competitive with alternative catalysts, if any exist. For oxidative catalysis, this implies an oxidizing potential that is high enough to produce complete oxidation in an acceptable contact time. In a real sense, <u>selectivity</u> defines the opposite boundary that makes an oxidation catalyst useful. Such a selective oxidation catalyst accomplishes the targeted oxidation (is efficacious) without over-oxidizing or triggering parallel oxidation pathways to undesirable alternative products (is selective). It is obvious that too potent an oxidizing capability may lead to undesirable results, even including destruction of the desired product. Mechanistic sources of selectivity can be even more selective than those based on oxidizing power since chemical systems undergoing oxidation often involve two or more reaction pathways occurring in parallel and varying in mechanism and selectivity. For example, oxygen atom insertion reactions, which are fundamentally 2-electron processes, may be accompanied by the most common, and often highly non-selective, 1-electron processes, radical reactions. In nature, the activated oxidation site may be extremely powerful, and selectivity may be achieved by binding only the chosen substrates. (Figure 1). For practical applications, that may not be the most effective way to achieve selectivity. The proposed use suggested that a metal ion in a moderate oxidation state might provide selectivity due to limited oxidizing power. Since unsaturated ligands, anionic ligands, and redox active ligands favor strongly oxidizing high oxidation states, new ligands were designed by using only uncharged, tertiary nitrogen donors in the primary ligand for T&D cat (15-17).

The in-service lifetime of a catalyst may determine its value for a particular use. For example, the manufacture of a low profit margin commodity chemical may require very long catalyst life (even if it is cheap), while a medicine that catalyzes an *in vitro* process may soon be delivered to the environment, making its short in-service or environmental lifetime a necessity. It follows that, depending on the specific case, catalysts must be designed for either long or short term <u>durability</u>. The catalyst must function for the necessary in-service lifetime, and it may be very important that it not remain functional after its use has terminated. Now, returning to the T&D cat, the proposed use of the catalyst would require its presence in a highly basic solution (pH ~10) for something like an hour at a temperature that might be ~50°C. After that, the solution containing

150

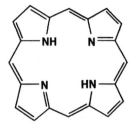  

Free porphyrin ligand
note ionizable protons

Usual porphyrin complex
note charges; n = 2 or 3

Activated iron
porphyrin
"Compound I"
note n+ = 4+ &
ligand radical

Figure 1. Activated enzyme site via highly oxidized metal ion
plus redox active ligand

the dissolved catalyst would be poured into a community sewer system. Ideally the T&D catalyst should be stable for substantially over an hour, but it should decompose in a matter of multiple hours in the environment. Recall that the central metal ions are to be either iron or manganese, so the basic cost is not a problem, nor is the return of the metal to the soil as rust in the case of iron or as mixed manganese oxides. In fact, in strong base the challenge is to assure stability of the catalyst in the basic aqueous environment for well over an hour since manganese and iron complexes are notorious for the zest with which they are precipitated into rust or mixed manganese oxides. The first choice (18, 19) would be a cage-like ligand that encompasses the metal ions (called cryptands), but that would leave no points of attack on the metal ion. To approach cryptate (complex of metal ion with cryptand) stability, bridged macrocyclic ligands (Structure 1) were used. Experiments proved that the complex was millions of times more stable in acid solution than anything short of a cryptate (20, 21). Experiment also showed that, in base, the manganese is air oxidized to manganese(III), and that complex is indefinitely stable in solution. The durability of the catalyst is well demonstrated.

Certainly the technical management of chemical processes illustrates the venerable slogan "knowledge is power", and for a given system, the definability of all those parameters and variables that determine the behavior of a chemical process is a matter of major concern. In the limit, detailed understanding of the thermodynamic and mechanistic details of the system is of the utmost value. However, one rarely understands a complex chemical process in ultimate fundamental detail. In the case of the T&D catalyst the maximum oxidation state attainable by the manganese derivative is 4+, and the oxidizing power proved to be moderate as expected (12, 16). Further, the weakly oxidizing power of this

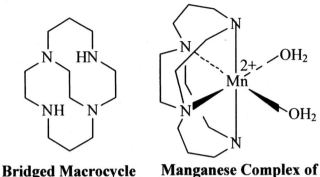

Bridged Macrocycle   **Manganese Complex of Bridged Macrocycle**

Structure 1

Mn^{4+} species was found to be accompanied by strong catalytic activation of oxygen transfer to olefins, forming epoxides. In fact the activated catalyst itself has been found to oxidize substrates, gently, by three mechanisms, the dominant mechanism being the clean oxygen atom transfer process. Thus the T&D catalyst exemplifies the goal of finding selectivity through mechanistic definability.

In summary, the value of a catalyst system may be judged by the five factors: efficacy, selectivity, durability, definability and sustainability, and as the discussion proceeds, these considerations will reappear both with and without notice.

Co(salen) Studies: First Use of CXLs in Catalytic Oxidations

Background in catalytic studies in scCO₂

The University of Kansas group began research on catalytic oxidation reactions in dense phase CO_2 with supercritical CO_2, building on the previous work mentioned above. Because O_2 is the much used and least expensive oxidant for commercial applications and because O_2 is infinitely miscible with $scCO_2$, the catalysts chosen for early studies were cobalt oxygen carriers (22). These species were already well known as moderate oxidation catalysts, but had not been examined in media providing such high concentrations of O_2. $Co^{II}(Salen)$, the parent of such a family of Schiff base catalysts, was soon found to be insoluble in $scCO_2$, but a readily available, heavily substituted derivative (Co(salen*) known as Jacobson's catalyst showed considerable solubility and

Structure 2

Structure 3

was used extensively (see Structures 2 and 3). The oxidation of phenols to quinones was chosen as the reaction to be studied, and 2,6-di-t-butylphenol (DTBP) was the substrate most studied in reactions conducted at 70°C and 207 bar in pure scCO$_2$ as the solvent. Reactions were carried out in cylindrical view-cell reactors of 10 and 17 mL volume, having sapphire windows at each end, fitted for fiber optics for UV/vis spectroscopic monitoring. Five ports, located midway between the ends of the cylinder, provide connections for gas inlet and exit, a sampling system, a pressure transducer, a thermocouple and a liquid injection/safety rupture disk. All reactions were conducted under a predetermined pressure of pure O$_2$. At the end of the batch reactions, samples were recovered, internal standards added, and analysis conducted by GC/FID and/or GC/MS. The reactions were complete in ~21 hours and the ratio of the desired product, 2,6-di-t-butylquinone to the dimeric by-product, 3,5,3',5'-tetra-t-butyl-4,4'-diphenoquinone (TTDBQ) determined the selectivity. Interestingly, varying the temperature from 50 to 90 °C showed a strong temperature dependence on the part of conversion, but a selectivity that is temperature independent (Figure 2). As Scheme 1 shows, this supports the expectations that conversion involves hydrogen abstraction producing a phenoxy radical, a process that should indeed be associated with a sizable enthalpy of activation. In contrast, both the product and by-product forming reactions involve radical-radical coupling, reactions that should involve minimal activation enthalpies. The dependence of the reaction on O$_2$ concentration is equally in agreement with the simple representation of the mechanism in Scheme 1. As O$_2$ partial pressure is increased, the conversion initially increases in proportion to the concentration, but it eventually saturates (Figure 3). This indicates that the true catalyst is the O$_2$ adduct, (salen*)(AL)Co^{III}-OO·, where the O$_2$ is represented as the superoxide radical anion that is believed to be its structure in oxygen adducts with cobalt (23, 24). When every cobalt is bound to an O$_2$ molecule there is relatively little effect on conversion if more O$_2$ is added. However, high O$_2$ concentrations enjoy an additional advantage. Selectivity in favor of the desired DTBQ is best (~90%) at very high O$_2$/DTBP ratios. AL means axial ligand and in this work

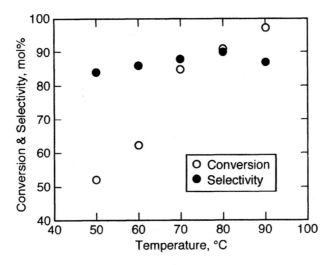

Figure 2. Co(salen*) Studies: Temperature dependence of conversion and selectivity in scCO₂. (Reproduced with permission from reference 22. Copyright 2001 American Chemical Society.)

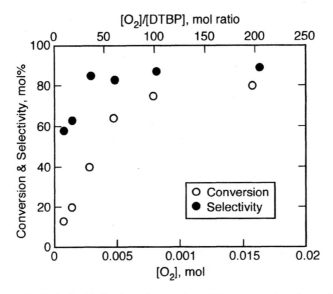

Figure 3. Co(salen*) Studies: Saturation of O₂ conversion dependence in scCO₂. [DTBP]=9.2 mM; [Co(Salen*)] = 0.46 mM; [Me-Im = 0.59 mM; pTotal = 207 bar; T = 70 °C; Reaction time = 21 h. (Reproduced with permission from reference 22. Copyright 2001 American Chemical Society.)

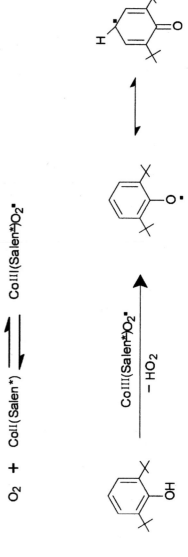

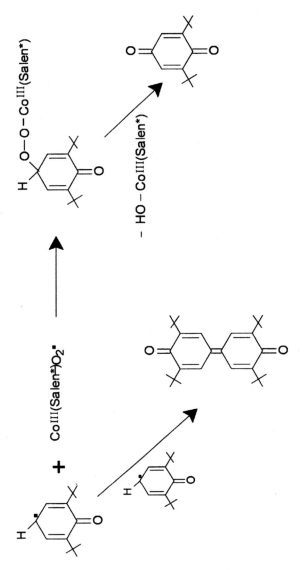

Scheme 1. Mechanism of oxidation of a phenol by Co(salen)(AL)/O₂ system

AL was usually the base N-methylimidazole, MIm. Strong O_2 binding only occurs when an axial ligand is also bound to the cobalt(II), so it is not surprising that conversion to products increased with AL concentration, but only up to the point where an equivalent of the axial ligand had been added. This dependence will be discussed later in more detail. Overall, oxidation with Co(salen*) in $scCO_2$ gave promise of more selective reactions, but the reactivity appeared not to exceed that attainable in organic solvents.

Moving into CXLs as Media for Catalytic Oxidations

Mixed solvents composed of organic liquids and CO_2 have been known and studied for many years (25), and were best known in separations for a variety of purposes (26, 27). The phase behavior associated with the use of CXLs was studied experimentally and modeled by De la Fuente Badilla (28) and more recently by Laird and coworkers (29). Shortly thereafter, CXLs were introduced into reaction chemistry in studies of catalytic oxidation reactions in these laboratories (30). In that first work, the practical enhancement of O_2 solubility over that available in other sustainable media was reported (Figure 4), and the dependence of catalyst solubility on the extent of CO_2 expansion over the original volume of organic solvent was defined in terms of the maximum homogeneous expansion limit (MHEL) for a given initial concentration of a solid solute. Using the same reaction system that is described above for $scCO_2$, except for solvent, it was easy to display the kinetic advantage of the CXL, using simple batch reactions, as shown in Figure 5 . Scheme 1 above describes the reactions in both media ($scCO_2$, CXL) equally well. CXLs exhibit both reaction and environmental advantages over either $scCO_2$ or the organic solvent for oxidation reactions (31). The behavior pattern for the reaction system is similar in CXLs to that described for $scCO_2$, except for the major advantages that are associated with the mixed solvents. For example, conversion shows a strong temperature dependence whereas selectivity is insensitive to temperature. The reaction advantages of CXLs are manifold: (1) as much as two orders of magnitude greater O_2 solubility; (2) increased solubility of the catalysts; e.g., unsubstituted Co(salen) has sufficient solubility to perform well; (3) greater turnover frequencies; (4) equal or better selectivities than those in either solvent component alone; (5) easy catalyst separation post-reaction; i.e., simply increase the amount of CO_2 until the catalyst precipitates. These reaction advantages are accompanied by the environmental and economic advantages for CXLs over both the organic liquid and $scCO_2$: (1) replacement of a large fraction of organic solvent (~20-80%) with environmentally benign CO_2; (2) operations proceed at lower total pressures (tens instead of hundreds of bars); (3) safer operations result from the dominant presence of inert CO_2 in both the gas and liquid phases,

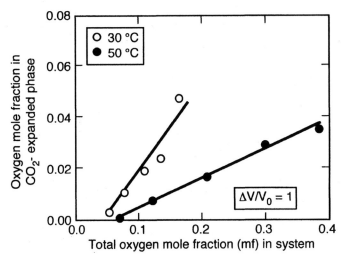

*Figure 4. Co(salen) Studies: Enhanced solubility of O_2 in CXLs. O_2 mf in CH$_3$CN (25°C & pO$_2$ = 1 bar) 5.5*10^{-4}; O_2 mf in CO$_2$ at 25°C = 0.05*, ~100X increase in O_2 solubility. *Solubility data series: Volume 7: Oxygen and ozone, Oxford ; New York : Pergamon, 1981. (Reproduced with permission from reference 30. Copyright 2002 American Chemical Society.)*

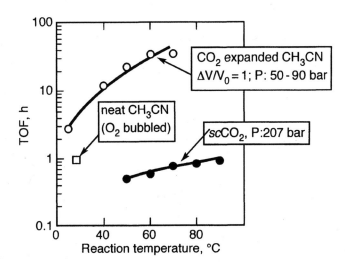

Figure 5. Co(salen) Studies: Relative efficacies of scCO$_2$ and CXL systems. Reaction conditions: catalyst: 2.5 mg; methylimidazole, 2 μL; Volume, 10 mL; O$_2$/Substrate, 4. Selectivity: CXL, 80%; CH$_3$CN, 60%; scCO$_2$, 85-90%. (Reproduced with permission from reference 30. Copyright 2002 American Chemical Society.)

the higher heat capacity of CO_2 in general, but especially in the near-critical region; (4) lower pressures and higher reaction rates favor lower operating costs.

Recent research investigated how the basicity of the axial ligand (AL) affects the activity of the catalyst (32, 33), using five axial ligands, methylimidazole, pyridine, water, acetonitrile, and triethylamine, and with 2,6-di-tert-butyl phenol as the substrate. Batch catalytic reactions were carried out with all five ligands at two different Co(salen) to AL ratios, one and ten. At the 1:1 ratio, the most powerful ligands were completely bound to the cobalt, producing the maximum rate of catalysis for that system. This is because the axial ligand is necessary in order to have effective O_2 binding to cobalt, and it is necessary for there to be a 6th vacant site available on the cobalt to bind the oxygen molecule (Figure 6). These (1:1) results are summarized in Figure 7, showing that triethylamine and N-methylimidazole are both strongly binding and similarly effective in supporting catalysis by Co(salen). In contrast, pyridine, acetonitrile and water are less effective. At the higher AL concentration, both the triethylamine and the N-methylimidazole systems are less effective reflecting the fact that the strong ligands compete for the binding site needed for O_2 binding. Also when present in excess, acetonitrile and pyridine approach the triethylamine in reactivity, a result that emphasizes the fact that, while an axial ligand is essential for good O_2 binding to the cobalt atom in Co(salen), the system is not very sensitive to the particular base that serves as the AL. In addition to substituted quinones, (34) Co(salen) has been studied as a catalyst for the O_2 oxidation of cyclooctane (35), alcohols (36) including veratryl alcohol (37, 38) and methylindole (39).

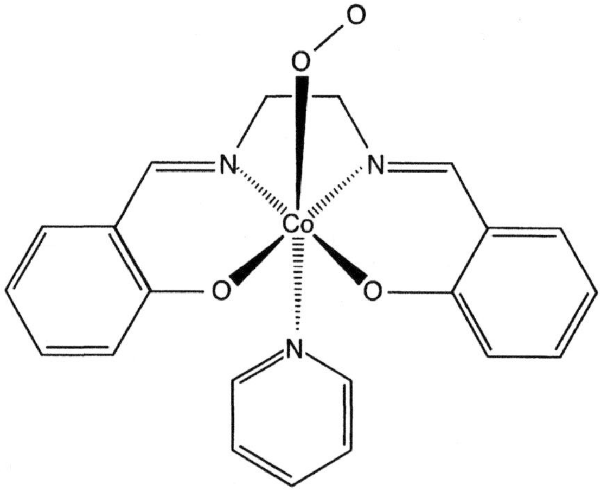

Figure 6. Co(salen) Studies: Structure of the Co(salen) complex with oxygen and pyridine as axial ligands.

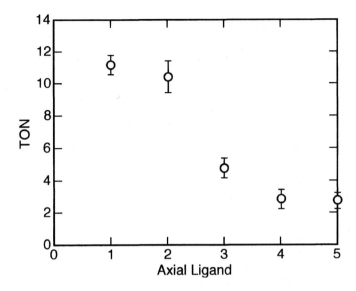

Figure 7. Co(salen) Studies: Reactivities (TON) with different axial ligands (Co(salen):axial ligand = 1:1): 1, triethylamine; 2, methylimidazole; 3, pyridine; 4, acetonitrile; 5, water; cat:sub:oxidant ratio= 1:80:800; cat=0.8 mM; $CHCl_3$ solvent; V/V_0, 2; T, 40 °C; P, 55-63 bars; V, 3.5 mL; t, 9 h. (Reproduced with permission from reference 32. Copyright Elsevier.)

To better define the substrates appropriate for O_2 oxidations using the Co(salen)(AL) catalyst, selected substrates of varying Bond Dissociation Energies (BDEs) were studied. Specifically these experiments indicate the ability of the catalyst system to perform 1-electron oxidations in which an electron and proton are removed from the substrate.

Table I shows that substrates whose most easily abstracted hydrogens have BDEs $\leq$ 91 kcal/mol are suitable for this catalyst system.

TPA Process $Co/O_2/HOAc$ oxidation systems

Over 30 million metric tons of terephthalic acid (TPA) (40,41) are produced each year with a healthy growth rate apparent in the construction of new manufacturing plants, a fact that opens the possibility for the incorporation of new sustainable processes in the industry. Most of the existing plants have existed for decades and use 40 year old, but improved, technology. The reaction

Table I. Reactivity of Co(salen)(MeIM) catalyst system.

Substrate	(C-H)BDE (kcal/mol)	X (%)	Products identified with GC/MS
1,4-cyclohexadiene	73	85	Benzene
Cyclohexene	82	76	Cyclohexene-ol, Cyclohexene-one
Toluene	91	20	Benzaldehyde, Benzoic acid
Cyclohexane	100	0	-

SOURCE: Reproduced with permission from reference 32. Copyright Elsevier.

conditions are severe for a homogeneous catalytic process, most of them using corrosive bromide in acidic media, producing much carbon dioxide, mostly from the destruction of acetic acid solvent. Several versions of the process are used, exemplified by the MidCentury (MC) Process which uses a cobalt acetate, manganese acetate, alkali or hydrogen bromide catalyst in acetic acid, operating in the temperature range between 180 and 220°C under a pressure ~30 bar. Despite the challenges, the MidCentury and related processes remain in profitable use and may be characterized as marvels of expert process fine tuning, partly necessary because of a number of factors, several affecting the purity of the final product, TPA: (a) many intermediates have transient existence in the course of transforming p-xylene into terephthalic acid (Scheme 2), (b) the product of interest (TPA) and the most stable intermediate (p-toluic acid) have limited solubilities in the reaction medium, (c) another contaminating intermediate (4-CBA) is readily occluded in the crystallizing TPA product, (d) O_2, the primary oxidant, has very limited solubility under the reaction conditions, (e) the solubilities of catalyst components are influenced by reaction variables.

In a program dedicated to providing alternative technologies for all chemical applications, O_2 oxidation reactions demand close attention for a number of reasons, including their economic advantage, the facts that O_2 from the air is the only truly green oxidant and the basic oxidation system is, in principle, applicable on all scales from the laboratory, to fine chemicals, to the commodities, the latter exemplified by TPA production. The sustainable versions of the process would be expected to vary with the economic and other constraints of a given market. In dealing with O_2 as an oxidant, flammability and explosive limits are matters of constant concern. These risks can be strongly

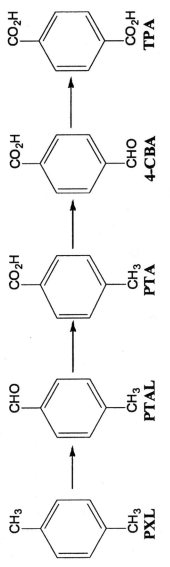

Scheme 2. Intermediates in the oxidation of p-xylene to terephthalic acid

alleviated in CXLs, providing strong motivation for their consideration as alternative media for the MidCentury and related processes.

Lessons from the Shell Process

With a $T_c = 31°C$ and $P_c = 74$ bar, the compressibility of CO_2 that facilitates the forming of CXLs in the solvent of choice, acetic acid, directs attention to the TPA processes operating under much milder temperatures than the MC process. This focused initial studies on the Shell Process which operates between 80 and 130°C, using cobalt and zirconium acetates in a molar ratio of 7/1 as the catalyst system in acetic acid (42-44). The lower temperature range and the absence of corrosive bromide are major advantages, but disadvantages include the requirement for significantly higher catalyst concentrations (~0.1M vs thousandths M), low conversion, and incomplete oxidation of intermediates in the transformation of *p*-xylene into TPA. Figure 8 very simply displays the increase in TPA yield (background bars) that accompanies replacement of N_2 as the inert gas in the system with CO_2, especially at higher gas pressures where CXLs are more likely to form. The light blue overlapping bars show the concentration of the contaminating 4-carboxybenzaldehyde (4-CBA). The yields of 4-CBA appear on the right in the graph. The 4-CBA impurity also decreases dramatically upon replacing N_2 with CO_2 in the system. However, these advantages are accompanied by certain disadvantages. Optimization in the CO_2 systems requires more zirconium than the traditional Shell system, Co/Zr moves from 7/1 to 3.5/1. Also, the solubility of the catalyst is problematic. Whereas the presence of zirconium acetate improves the solubility of cobalt acetate in acetic acid at 120°C, in the presence of reasonable concentrations of the *p*-xylene substrate, the solubility enhancement due to zirconium disappears in the presence of reasonable concentrations of the *p*-xylene substrate (Figure 9). It may be concluded that the positive advantages of CO_2 media for the Shell process are substantial, but the process is not a competitive one.

Radical Initiated Processes

Additional environmental benefits are to be expected if catalytic air or O_2 oxidation could be used to replace stoichiometric oxidation processes using large amounts of environmentally undesirable materials such as heavy metals like chromium, copper, manganese, or the halogens, especially chlorine, or bromine. This concept and/or the desire to make the MC process proceed at lower temperatures have motivated research using organic reagents as promoters or initiators in O_2 oxidation systems (45-47). In 1960, Brill reported the oxidation

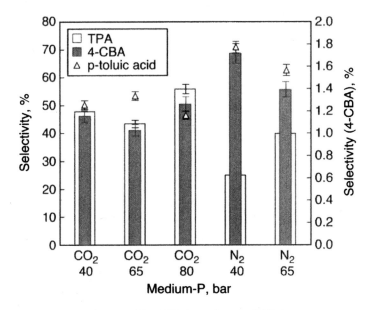

Figure 8. TPA Studies: Yields of TPA product and CBA contaminant in the presence of N_2 and CO_2 in Shell Process Experiments. Red bars (left labels) indicates yield of TPA; Blue-gray bars (right labels) indicate 4-carboxybenzaldehyde yields. Reproduced with permission from reference 33.

of p-xylene directly to terephthalic acid using methyl ethyl ketone (MEK) as the promoter (48). In 1977 the activating effect of zirconium acetate, ZrO(OAc)$_2$, was reported by Chester (49), replacing both bromide and manganese acetate and laying the ground work for the Shell patent (44). The addition of small amounts of CO_2 to the medium in the MC process (50,51) led to a correspondingly small increase in the yield of TPA from p-xylene, or, in the case of toluene, benzoic acid. In that early study, the amount of added CO_2 was not significant enough to form a CXL.

Since cobalt is the main catalyst in the many MC-related processes, the primary hydrogen atom abstraction of the process, converting toluene or p-xylene into the corresponding benzyl radical, is often attributed to a Co(III) species (52/53), but this can hardly be the sole contribution of the cobalt. The function of zirconium acetate in a radical chain reaction must be questioned because of its nature as a Lewis acid catalyst. It has been suggested that the Zr(IV) may promote the oxidation of Mn(II) to Mn(III), but the mechanism remains uncertain (54). Considering the tendency of metal acetates to form clusters in acetic acid, Chester suggested that zirconium may intercede and

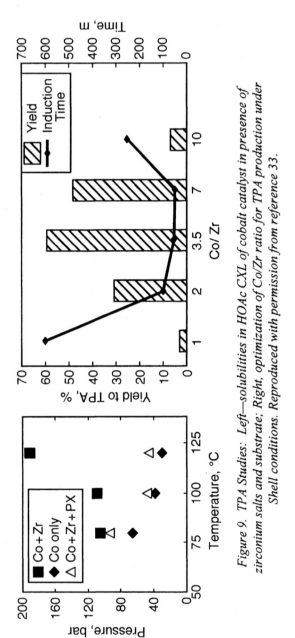

Figure 9. *TPA Studies: Left—solubilities in HOAc CXL of cobalt catalyst in presence of zirconium salts and substrate; Right, optimization of Co/Zr ratio for TPA production under Shell conditions. Reproduced with permission from reference 33.*

produce monomeric cobalt(III) in the reactor by forming weakly bound cobalt/zirconium clusters (49). In contrast, Partenheimer focused on the strong Lewis acid character of zirconium(IV) and pointed out the likelihood that it might catalyze the dehydration of the intermediate benzyl hydroperoxide to yield a second intermediate benzaldehyde, both of which are highly probable reactants in these systems (55). The initiator ketone may shorten the induction time in the MC process by accelerating cobalt oxidation (56). Ketones have been credited with promoting the oxidation of cobalt(II) by forming hydroperoxides which readily oxidize Co(II) to cobalt(III) under some circumstances (48). In turn, this could contribute to sustaining a catalytically competent cobalt(III) concentration in the reaction.

The work reported here has brought together all of the aforementioned beneficial factors. Cobalt acetate is the main catalyst, ketones having secondary hydrogens are used as initiators producing organic radicals, zirconium is present as the Lewis acid component of this complex catalyst system and studies have been performed using dense phase CO_2, providing both safety and enhanced O_2 solubility, along with adjustable solvent parameters and improved transport properties. The toluene and p-toluic acid are the test substrates and the reaction conditions are mild. The ketones chosen to characterize the dependence of productivity on initiator are shown in Scheme 3. The data in Table II (57) show the relative reactivities as a function of the ketone initiator in systems using all three synergistic components, cobalt and zirconium acetates and a ketone. Reaction conditions were chosen to minimize the reactivity of the weakest initiator; i.e., only negligible reactivity was observed with MEK in this case. These conditions clearly show that the relative effects of the several initiators decrease as the reactivities of their hydrogen atoms decrease. Catalytic activities are: triacetyl methane (TAM) > diacetyl methane ("acac", alias acetylacetone) > 2,3-pentanedione (2,3-PD) > MEK. Significantly, entry # 5 shows that, when present as a ligand directly bonded to the zirconium(IV) ion, diacetylmethane is more effective than its more highly substituted sibling, triacetylmethane. This suggests mechanistic relationships between the zirconium and the ketone, but their nature remains unclear. One can speculate that the binding of the radical initiator (ketone) to the Lewis acid (zirconium(IV)) makes a multibody/multistep process, such as successive formation and destruction of a peroxide, into a unimolecular process. This perspective receives substantial support in a second example of the coupling of two components of a cobalt/radical initiator catalyst system. The catalytic system, Co/Zr(acac)$_4$, has also been used to demonstrate the advantages of CO_2 enriched media. In these experiments, the Zr/Co molar ratio was kept at 1/6 and, accordingly, the acac/Co molar ratio was 4/6, and, unless otherwise stated, the CO_2 pressure was applied at 26 bar. Experiments involving the oxidation of toluene, with either N_2 or CO_2 as the inert component, were run at 80°, 60° and 50°C and the results are shown below.

Benzoic acid yields at 80°C: N_2/O_2, 93.3% (5 hrs); CO_2/O_2, 91.3% (5 hrs).

Benzoic acid yields at 50°C: N_2/O_2, 41.1% (48 hrs); CO_2/O_2, 80.5% (48 hrs)

Benzoic acid yields at 60°C: N_2/O_2, 87.3% (28 hrs); CO_2/O_2, 89.0% (24 hrs)

These preliminary studies show a CO_2 effect that is far more evident at lower reaction temperatures. It is plausible that the burning of acetic acid solvent will be significantly curtailed at the lower temperatures, although this could not be verified easily due to the abundance of the added CO_2 in the gas phase.

As indicated in studies related to the Shell version of the TPA process, lower temperature reactions generally require high catalyst concentrations, commonly with $[Co^{2+}]$ ~0.1 M and with a substrate/catalyst ratio in the vicinity of 10. With an effective catalyst, increased O_2 solubility and more favorable transport conditions associated with CO_2 as a solvent component should facilitate operations at lower, more sustainable catalyst concentrations.

To test these relationships, a cobalt concentration was sought at which the CO_2 reaction system performs well but the system based on N_2 is relatively unproductive. Figure 10 uses the rate at which the pressure of O_2 in contact with the reacting solution decreases during reaction to zero under such a set of conditions. At a Co/toluene ratio of 3/80 and 80°C the reaction under the N_2/O_2 gaseous mixture is severely retarded, while an 83% yield of benzoic acid is produced under CO_2/O_2. Clearly the $Co(OAc)_2/Zr(acac)_4$ catalyst system shows much promise under moderate reaction conditions, especially in CXLs.

$Co(OAc)_2$ with N-Hydroxysuccinimide, a Highly Versatile Catalyst

Having explored the common and most studied *N*-hydroxyimides, including phthalimide, for their effectiveness as promoters in O_2 oxidations with only cobalt acetate as a catalyst, *N*-hydroxysuccinimide has been found to be exceptionally versatile in the range of reactions it will promote, at characteristically low temperatures, as Table III shows (58). It should be emphasized that the times given are contact times and, in most cases, are much longer than the period of time during which the reaction actually occurs in these batch reactions. The latter has been monitored in many cases by the fall in pressure due to the use of O_2. Also, these data are the results of exploratory experiments and do not represent optimization. Having qualified these preliminary studies, it must be reiterated that this catalyst oxidizes a huge range of substrates and deserves consideration for replacement of many stoichiometric oxidation systems. As examples in the patent application (58) show, the huge

Table II. Oxidation of toluene with Co/Zr/ketone catalyst system.[a]

Entry	Catalyst	Conversion of toluene (%)	Yield of benzaldehyde[b] (%)	Yield of benzoic acid (%)
1	Co(OAc)$_2$/ZrO(OAc)$_2$/MEK	<1	trace	0
2	Co(OAc)$_2$/ZrO(OAc)$_2$/2,3-PD	26.6	2.6	19.2
3	Co(OAc)$_2$/ZrO(OAc)$_2$/Hacac	85.1	0.5	83.9
4	Co(OAc)$_2$/ZrO(OAc)$_2$/TAM	88.2	0.3	87.3
5	Co(OAc)$_2$/Zr(acac)$_4$	94.4	0.2	93.3

[a] Reaction conditions: Solvent = acetic acid; [toluene] = 1.2 M, Co/toluene = 1/10 (molar ratio), Zr/Co = 1/6 (molar ratio), ketone/Co = 2/3 (molar ratio), P_{N2} = 26 bar, P_{O2} = 34 bar, T = 80oC, t = 5 hr.
[b] In addition to benzaldehyde, small amounts of a few other by-products were detected but not identified.

SOURCE: Reproduced with permission from reference 57. Copyright 2008 American Chemical Society.

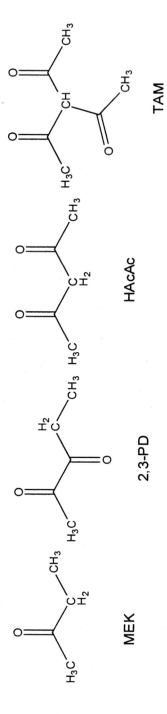

Scheme 3. Ketones studied as promoters for $Co(OAc)_2/Zr(OAc)_4$/ketone catalyst system

range of substrates includes hydrogen abstraction from allylic and benzylic groups, olefin epoxidations, cyclic olefin cleavages and saturated ring cleavages. Because of the relatively high operating temperature for the MC process, ~200°C, the ability of these Co(OAc)$_2$/NHSI catalyst systems to generate high yields of white TPA with mild CXLs is displayed. More impressive is the ability of this catalyst system to take cyclohexane directly to adipic acid in substantial yields at 80°C. Similar experiments with other cycloalkanes give very interesting products under similarly mild conditions; e.g., such difunctional major products as 2,6-diheptanedione from 1,2-dimethyl cyclohexane, octanedioic acid from cyclooctene, and 5-ketohexanoic acid from 1-methylpentene are all candidate polymer intermediates. p-Nitrotoluene is an intermediate for the synthesis of Novocain. Scheme 4 indicates the relative amounts of the multiple products from some of these systems. Recognizing that the results are preliminary in nature, the yields of some of these materials are interesting and deserve further exploration aimed at optimization. Further it is apparent that selectivity can be adjusted with temperature in many cases. For example, the reaction with cyclohexene is limited to epoxidation at 25-40°C, but ring cleavage occurs in the range from 80 to 100°C, producing the diacid and other products. It is not clear why NHSI is much better than, especially, the much studied N-hydroxyphthalimide. On the basis of the BDE of the OH linkage in NHSI, this promoter is intermediate in character when compared to other compounds with the same functional group. Investigation of the coordination chemistry of the Co^{2+}/NHSI system in the solvent (HOAc) has suggested a similarity in the behavior of this system and the Co/Zr(acac)$_4$ catalyst system described above. Isothermal calorimetry has shown that two moles of the N-hydroxysuccinimidate anion can be coordinated to a single cobalt(II) ion (Figures 11 and 12). The binding of the first mole of ligand is highly exothermic (ΔH = -56 kcal/mol) while the second is endothermic (ΔH = +13 kcal/mol). Further, the diaqua adduct of the 2:1 complex has been crystallized from an acetic acid solution prepared under reaction conditions and the structure has been determined (Figure 11).

Benchmark MC Process

The MC Process was benchmarked (33) as a semi-continuous process with continuous air flow, pulsed p-xylene addition, and retention of the precipitated TPA product, using a 50 mL automated stirred titanium Paar reactor (Figure 13). Having considered several sources, the MC Process parameters used in studies at CEBC were from U.S. Patent 6,153,790 (44). The catalyst composition was Co:Mn:Br = 1:1:1 with the catalyst present in the reaction at 2.0 mM

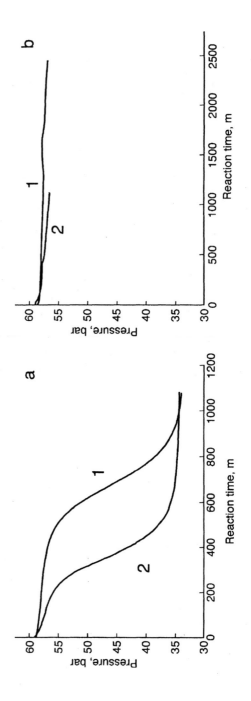

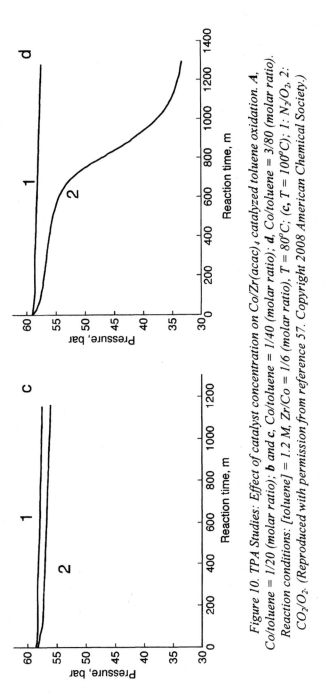

Figure 10. TPA Studies: Effect of catalyst concentration on Co/Zr(acac)₄ catalyzed toluene oxidation. A, Co/toluene = 1/20 (molar ratio); b and c, Co/toluene = 1/40 (molar ratio); d, Co/toluene = 3/80 (molar ratio). Reaction conditions: [toluene] = 1.2 M, Zr/Co = 1/6 (molar ratio), T = 80°C; (c, T = 100°C); 1: N₂/O₂, 2: CO₂/O₂. (Reproduced with permission from reference 57. Copyright 2008 American Chemical Society.)

Table III. Oxidation of various substrates using Co(OAc)$_2$/NHSI catalyst in acetic acid CXL.

Substrate	Product	Yield (%)	S/NHSI	P O$_2$ (psi)	P CO$_2$ (psi)	Time (h)	Temp (°C)
p-xylene	TPA	89	5/1	350	506	1	60
p-xylene	TPA	76	5/1	340	537	12	30
p-xylene	TPA	68	5/1	300	302	0.5	21
cyclohexane	adipic acid	56	9/1	120	745	6	80
1,2-dimethyl cyclohexane	2,6- heptanedione	60	7/1	120	745	6	60
cyclooctene	octanedioic acid	60	18/1	120	745	1	100
1-methyl pentene	5-ketohexanoic acid	60	24/1	120	745	1	100
p-nitrotoluene	p-nitrotoluic acid	90	4/1	120	745	6	50

SOURCE: Reference 58.

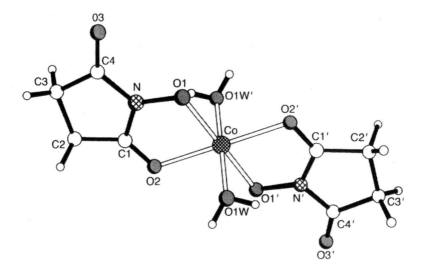

*Figure 11. TPA Studies: Crystal structure of
diaquabis(N-hydroxysuccinimidato)cobalt(II).*

concentration. The solvent was acetic acid containing 6-7% water, and the reactions were conducted at 200±4°C, with a stirring rate of 1200 rpm. Typical benchmarking data are shown in Table IV.

The products listed under S (for selectivity) reflect the dominant products encountered in the air oxidation of p-xylene. In addition to the terephthalic acid product, TPA, p-toluic acid, PTA, or P, is often a product because it is more difficult to oxidize than the other intermediates and the starting material.

4-Carboxybenzaldehyde, CBA or C, tends to be occluded in the precipitating TPA as the process proceeds and it is an undesirable contaminant, both leading to discoloration of the white TPA and interfering with its use in polymer formation. In this challenging system, these and most but not all of the by-products are intermediates in the conversion of the substrate to TPA, so that their elimination is essential to obtain a product of the necessary purity. As is evident in the data of Table IV, the color of the solid product is a crude indicator of its purity, decreasing in intensity as the purity of the product improves. This is a simple example of the tuning necessary to conduct the MC Process. The rate of stirring, which influences the gas/liquid mixing, is very important and initial experiments with stirring rates of several hundred rpm yielded unacceptable product quality. Further, air must be sparged at high flow rates into the liquid phase and the flow rate of p-xylene must be mediated in order to prevent oxygen starvation and its adverse consequences on product selectivity. In this reactor and under these conditions, product quality improves up to 1200 rpm, but higher

174

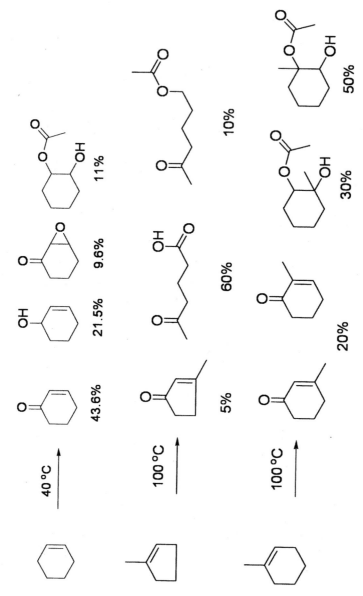

Scheme 4. Products from oxidations of cycloalkanes by Co(OAc)₂/NHSI catalyst system

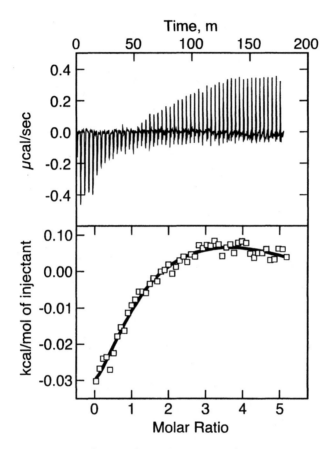

Figure 12. TPA Studies. Isothermal titration calorimetry results confirm existence of $Co^{II}(NHSI)^{+}$ and $Co(NHSI)_2^{0}$ complexes in acetic acid solution.

**Table IV. Benchmark data for the MidCentury Process,
based on U.S. Patent 6,153,790.** [a]

S#	(V_L) mL	$P_{reactor}$ (bars)	v (slpm)	X ($\pm 0.3\%$)	S ($\pm 0.5\%$) (T/C/P)	Color of solids (OD_{400})
1	31	30	5	77	80/3/14	Yellow (8.5)
2	31	30	25	95	95/1/4	Yellow (5.2)
3	35	30	25	99.6	98/0.1/2	Yellow (2.8)
4	35	33	25	99.5	99+	White(0.05)

[a] V_L = liquid volume; v = air flow rate; p-xylene addition rate = 0.1 ml/min; X = p-xylene conversion; S = selectivity , for T = terephthalic acid; C = 4-carboxybenzaldehyde; P = p-toluic acid; OD_{400} = optical density of crude TPA measured at 400 nm.
SOURCE: Reproduced with permission from reference 33.

stirring rates are without further effect, signifying the maximum extent to which the gas/liquid mass transfer rates can be improved with this system. The vigorous stirring and air sparging at high rates requires the maintenance of a free overhead reactor volume of approximately 50-67% of the total reactor volume (~50 mL) to accommodate the froth and assure return of splashing solids to the bulk of the liquid system. These researchers have been assured by experienced BP personnel that MC Process reactor optimization to maximize gas-liquid contact in the MC process is challenging and depends upon both the reactor configuration and scale. Indeed, the operating parameters for a desired product quality change if such minor changes as the placement of the O_2 sparger and p-xylene input rates are altered. Under the operating conditions of Table IV, a reactor pressure of 33 bar (air + solvent vapor) is necessary, whereas the literature reliably reports 30 bar. Under the conditions stated for the bottom line of Table IV, TPA selectivity is > 99% and the color intensity is very low (OD_{400}=0.05). The results of our benchmarking studies and of others (59) strongly suggest that even at lower temperatures (as low as 120°C) compared to the MC process (200°C), the oxidation rates are so high that the O_2 is essentially consumed at the gas-liquid interface. This result is significant and clearly reinforces the need to overcome O_2 starvation in the MC process to both prevent incomplete oxidation that results in unacceptable product quality and ensure full utilization of the substrate to maximize productivity. The design and operation (and any future improvement) of the industrial scale MC process are aimed at eliminating these potential process drawbacks.

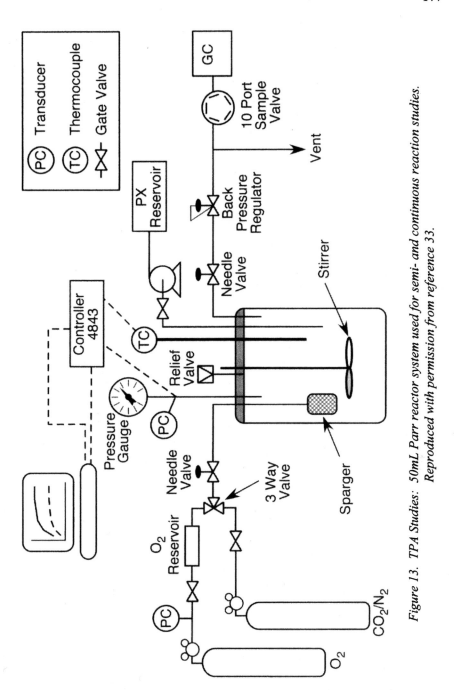

Figure 13. TPA Studies: 50mL Parr reactor system used for semi- and continuous reaction studies. Reproduced with permission from reference 33.

MC-like Processes in Dense CO_2

With the benchmark as guidance, the exploration of TPA systems that might enjoy competitive advantages over MCP has moved on to evaluating the benefits of replacing the inert N_2 of the air with CO_2.   In the simplest version, these systems simply replace the inert N_2 with CO_2 while retaining other reaction variables constant. Most notably, this retains the 200°C operating temperature, a condition under which CXLs are not likely to form. The results summarized in Table V indicate that performance with CO_2 as the inert gas is essentially as good as the benchmark experiments using air as the oxidant plus inert.

**Table V. p-Xylene oxidation under MCP conditions,
but replacing N_2 with CO_2. [a]**

S#	PX Addition rate	X (±0.3%)	S (±0.5%) (T/C/P)	Color of solids (OD$_{400}$)
1	1 ml at 0.1 ml/min	99.4	99+%	White (0.08)
2	1 ml at 0.2 ml/min	99.6	99+%	White (0.12)

[a] Reaction conditions: Co = 600 ppm; Cl:Mn:Br = 1:1.1:1; Stirring rate = 1220 rpm; T = 200 °C; V_L = 35 mL; p-xylene flow rate = 0.08 ml/min for 20 min; other definitions same as in Table IV.
SOURCE: Reproduced with permission from reference 33.

As reference 33 indicates, operating the TPA processes under CO_2 at reduced temperatures to produce CXLs offers additional challenges. These are under study as we seek to evaluate the promise of CXLs for the various versions of the MC process.

The New PO process

The Process and Its Origins

Inspired by the facile and successful epoxidation of higher (C_{5+}) olefins, this research focused on the epoxidation of light olefins for very sound reasons

(60,61). Over six million metric tons of propylene oxide are produced each year and industry is planning or constructing additional facilities. The chlorohydrin process and the peroxidation process (62,63) are the most widely practiced technologies and they represent the fact that current propylene to propylene oxide processes are not sustainable in terms of today's environmental perspective. The chlorohydrin process begins with propylene, chlorine, and lime, producing the epoxide, and large amounts of calcium chloride as waste. The peroxidation process begins with isobutane and O_2, which it converts to t-butyl hydroperoxide, and then uses that oxidant to epoxidize the propylene. Major limitations are the elevated reaction temperature and the production of twice as much by-product, t-butanol, as desired product, propylene oxide. Major companies are currently building or planning new manufacturing facilities to exploit hydrogen peroxide propylene oxidation (HPPO) technologies for the conversion of propylene to propylene oxide (64). BASF and Dow are planning a 300,000 metric ton PO plant with a 230,000 ton H_2O_2 plant on site. Degussa and Uhde, in Germany, and SKC, a Korean Company, plan a 100,000 ton/year PO manufacturing facility, to be supplied by an H_2O_2 plant to be operated by Degussa and Headwaters. These initiatives reflect the need for new sustainable technologies for the conversion of light olefins into their many useful oxidation products. The catalysts to be used in the planned facilities are proprietary, although literature reports suggest that they may be some version of titanium(IV) on silica.

An exceptionally active organometallic catalyst, methyltrioxorhenium (MTO), has been proposed in two alternative processes for propylene oxide production. Herrmann (65) proposed the H_2O_2 oxidation at low temperatures (-10°C) in media from which the water had been removed, ostensibly to prevent hydrolysis of the propylene oxide. Crocco patented an MTO process using H_2O_2, produced by O_2 oxidation of a ternary benzylic alcohol solvent, and also operating at low water concentrations. Advantages of this process are a ~20°C higher operating temperature and the introduction of pyridine as a preferred accelerating ligand (66). A third MTO/PO process, that is being developed in the laboratories of the authors, the pressure intensified MTO process (67), is illustrated in Figure 14. This diagram emphasizes the biphasic (gas/liquid) nature of a propylene to propylene oxide process that uses an oxidant (H_2O_2) and catalyst in solution. The gaseous propylene substrate has relatively little solubility in the wet methanol solution at ambient conditions so that this kind of process can easily become both solubility and transport limited.

The CO_2-based catalytic epoxidation research began in the authors' laboratories with the background described above, beginning with studies of O_2 oxidations in various solvents including CXLs. That work was followed by the demonstration, also above, that CO_2 accelerates H_2O_2 epoxidation reactions by converting biphasic phase transfer systems, involving pairs of immiscible liquids such as water and an organic solvent, into homogeneous reaction

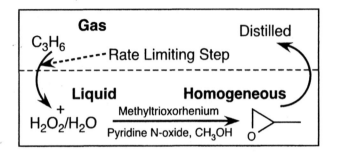

Figure 14. Pressure Intensified MTO process for PO production, a mixed gas/liquid phase process. Reproduced with permission from reference 60.

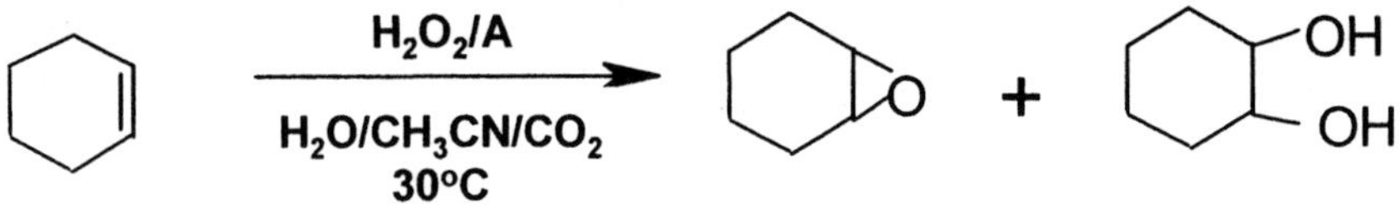

$$\text{cyclohexene} \xrightarrow[\substack{H_2O/CH_3CN/CO_2 \\ 30^\circ C}]{H_2O_2/A} \text{cyclohexene oxide} + \text{trans-cyclohexanediol}$$

Scheme 5. First monophasic epoxidation reaction using CXL media.

systems within a single CXL phase Scheme 5. Building on these beginnings, the first light olefin experiments used a methanol CXL as the solvent for the MTO catalyzed PO process. The bar graph that is Figure 15 shows at least three issues worthy of comment. First of all, the conversions in all cases were quite good. Second, the *least conversion* is that for the CXL. In fact, by applying only the gas pressure due to the volatile substrate, propylene, and the vapor in equilibrium with the liquid phase consisting of methanol and water, a substantial improvement in yield was observed. The reason is simple. In forming the CXL, the volume of the reaction medium was increased, thereby diluting the concentrations of reactants and catalyst. Apparently, the increased solubility of the propylene in the CXL did not compensate for the dilution.

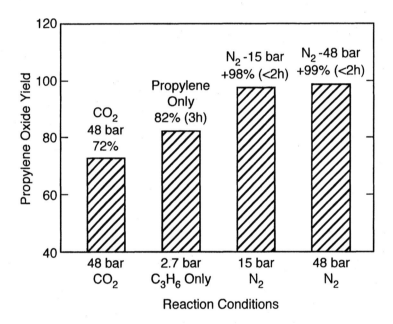

Figure 15. Influence of pressurized gas on yield of propylene oxide. Reactions at 40°C; CH₃OH, 50/50 H₂O₂ in H₂O; MTO catalyst. (Reproduced with permission from reference 60. Copyright 2007 American Chemical Society.)

Continuing the important results summarized in Figure 15, upon replacing the gas pressure exerted by CO_2 in the case of the CXL with the same pressure of the inert, sparingly soluble, gas N_2, the propylene to PO reaction goes to completion with high selectivity. Figure 16 shows how the absorbance due to C-H stretching modes of the propylene molecule increases in the infrared spectrum as the propylene solubility increases under greater and greater N_2 pressure. This

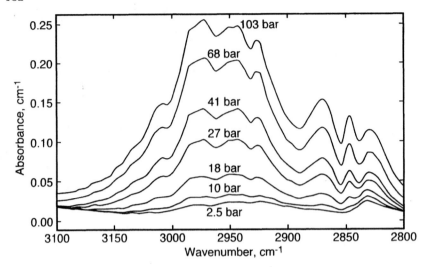

Figure 16. C-H Stretching vibrations in infrared spectrum show increasing propylene solubility with increasing N_2 pressure. (Reproduced with permission from reference 60. Copyright 2007 American Chemical Society.)

is a dramatic result and, is easily understood in terms of the system phase behavior. The increase in N_2 pressure sharply increases the solubility of the substrate in the wet methanol solvent, because the propylene at 25°C is a dense gas below its critical temperature (91.9 °C). Hence, increasing the pressure beyond 10 bars causes substantial amounts of gaseous propylene to condense into the liquid phase and that liquid is miscible with the methanol solvent. Thus, in the pressure intensified P to PO process, the substrate is the expanding gas and hence the gas expanded liquid is a *substrate expanded liquid*, which might be denoted as an SXL.

This simple MTO catalyzed process epitomizes green (68,69): it operates at high atom economy, with only water as a significant by-product, using safe aqueous H_2O_2 as the oxidant, in methanol as the solvent, with pyridine-N-oxide as the accelerating ligand, and operates under mild conditions (30 °C, ~20 bar of N_2 pressure), using a facile separation scheme for product, solvent, catalyst and water. As Table VI shows, the high throughput and selectivity of this process augment the advantages it enjoys from its green footprint.

This pressure-intensified PO process, and the previously proposed MTO processes remain paradoxical in that this long known, exceptionally reactive catalyst, methyltrioxorhenium, has not been exploited. MTO was discovered by Beattie and Jones (70) in 1978, and studied in great detail by Hermann (71,72), Espenson (73,74), Sharpless (75) and others. MTO is a unique and superb H_2O_2 oxidation catalyst (76,77) and it has the endearing properties of being soluble in

Table VI. Comparison of the pressure intensified MTO process with major industrial processes.

System		Conditions	Yield PO
Chlorohydrin	Cl$_2$, Caustic, Lime	45-90°C, 1.1-1.9 bar	Selectivity = 88-95%
Peroxidation	W, V, Mo	100-130°C, 15-35.2 bar	95% (S=97-98%)/ 2 h
Reference 60	MTO	30°C, 17 bar N$_2$, pyNO	+98%(S=99%) / < 2 h

SOURCE: Reproduced with permission from reference 60. Copyright 2007 American Chemical Society

most solvents and stable in the presence of the ambient atmosphere. MTO stability is a fundamental issue that may account for its not having been implemented. Despite the appearance of fragility that seems implicit in an organometallic compound, MTO is recognized as robust. According to Espenson, whose group has studied MTO in greatest detail, "MTO is a remarkably stable substance both as the pure solid and in dilute solutions, in which neither oxygen nor acid nor water has deleterious effects."(73) However, this statement in support of the inherent robustness of MTO does not address its stability under catalytic H$_2$O$_2$ reaction conditions. In fact, to avoid the base hydrolysis of the MTO, its catalytic reactions are almost always carried out in acidic media. Since the epoxide products are hydrolyzed to the corresponding glycols in acidic media, this widely recognized base sensitivity is a source of problems. As will be discussed in more detail in the following section, the durability of the catalyst is important to its possible commercialization and batch experiments provided the first indications of good catalyst life. Figure 17 is an example of many experiments in which multiple successive runs have been made with the same catalyst with little loss of activity. This Figure shows the cumulative yields in two sets of five successive measurements using a single catalyst sample, with fresh substrate and aqueous H$_2$O$_2$ added before each run. Two alcohols, MeOH and t-BuOH, were used as solvents and 2-fluoropyridine as an added axial ligand. The linearity of the data indicates the absence of deactivation. In this case, five successive 1 hr batch runs were made at 25°C. The conditions for these experiments were: 2.38 mmol propylene, 4.68 mmol H$_2$O$_2$ (50 wt% in H$_2$O), 0.049 mmol MTO, and 0.056 mL 2-fluoropyridine were dissolved in 4.1 mL methanol or tBuOH, under a total pressure of 20 bar.

Theoretical In-service Lifetime of MTO

The extensive and highly detailed studies of Espenson et al. revealed two sets of reactions that are especially important to the possibility of implementing

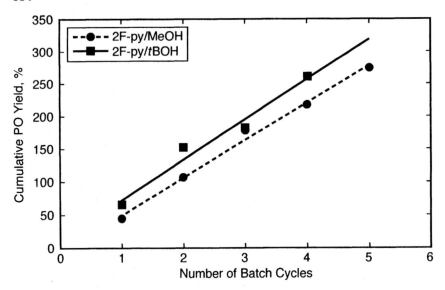

Figure 17. Catalyst stability in 5 successive runs using 2F-pyridine as axial ligand at 25°C. (Reproduced with permission from reference 60. Copyright 2007 American Chemical Society.)

this catalyst in, for example, commercial light olefin epoxidation processes. The first of these is an economic issue and involves the destruction of H_2O_2 by the MTO catalyst. The second issue is the durability of the MTO catalyst under reaction conditions and involves identification of the event in which the cleavage of the Re-C bond occurs destroying the Re-CH$_3$ group. Both of these chemical processes are closely linked to the formation of the activated catalyst which exists in two forms, the 1:1 and 2:1 H_2O_2:MTO adducts. The pertinent equations follow:

$$CH_3Re(O)_3 \;+\; H_2O_2 \;\leftrightarrow\; CH_3Re(O)_2(O_2) \;+\; H_2O$$

$$CH_3Re(O)_2(O_2) \;+\; H_2O_2 \;\leftrightarrow\; CH_3Re(O_2)_2(O) \;+\; H_2O$$

For simplicity, we use Espenson's labels for the peroxide adducts: the 1:1 peroxide adduct, $CH_3Re(O)_2(O_2)$, will be called **A** and the 2:1 adduct, $CH_3Re(O_2)_2(O)$, will be called **B**. The reactions forming **A** and **B** from MTO and H_2O_2 have been studied by Espenson and coworkers in methanol (78), acetonitrile(79), 1:1 aqueous acetonitrile (80), nitromethane (81), and aqueous solutions.(82,83).

The catalytic destruction of H_2O_2 is attributed by Espenson to the decomposition of **B**, according to the following equation:

$$CH_3Re(O_2)_2(O) \rightarrow CH_3Re(O)_3 + O_2$$

In the course of this reaction, the bis(peroxo) complex **B** is converted back to the original catalyst MTO by evolving a mole of O_2. This reaction is reported to be very slow under the conditions promoted by either Hermann or Crocco in their early patents of MTO oxidation processes. However, at higher temperatures, this catalase-like reaction could destroy sufficient oxidant to impair economic use of the system.

The remaining reaction pathway is even more critical to the utility of MTO for commercial light olefin epoxidation. Complex **A** has a limited catalyst life and that is the matter of concern. **A** decomposes into methanol and perrhenate anion by a pathway that is dependent on both the hydrogen peroxide concentration and the hydroxide concentration and this relationship could arise from either of two alternative decomposition pathways, 1 or 2. Unfortunately, experiments available to the Espenson group were unable to distinguish between these alternatives (84).

$$CH_3Re(O)_2(O_2) + OH^- \rightarrow CH_3OH + ReO_4^- \tag{1}$$

$$CH_3ReO_3 + HO_2^- \rightarrow CH_3OH + ReO_4^- \tag{2}$$

Alternative 1 says that **A** is unstable in base and decomposes according to an obvious second order equation. Alternative 2 does not include **A**, suggesting that **A** itself is stable in base. Instead of **A** being unstable, the reaction between the hydroperoxide ion and MTO is responsible for the destruction of the catalyst. Presumably, when HO_2^- attacks MTO, some fraction of the events involve nucleophilic attack at the rhenium center of $CH_3Re(O)_3$, simply producing stable **A**, and in the remaining fraction of these events, in some way, HO_2^- attacks both the carbon atom and the rhenium atom of $CH_3Re(O)_3$, producing CH_3OH and ReO_4^-. Absent any related information, such a process could either occur concerted or stepwise.

The durability of MTO as a catalyst is almost certainly going to be very different depending on whether catalyst destruction occurs by equation 1 or equation 2. If equation 1 describes the true reaction pathway, then some catalyst will be destroyed anytime **A** exists in the solution. Since **A** and **B** are the active forms of the catalyst and since when **B** acts as a catalyst it gives an oxygen atom to a substrate and is converted into **A**, any catalytic event is accompanied with some probability of catalyst decomposition.

$$CH_3Re(O)(O_2)_2 \; + \; X \; \rightarrow \; CH_3Re(O)_2(O_2) \; + \; XO$$

In contrast, if equation 2 describes the decomposition, then the MTO is safe so long as it is in the form of a peroxo complex, either **A** or **B**, because the decomposition occurs while MTO is being transformed into **A**. This suggests that conducting catalysis in the presence of a large enough excess of H_2O_2 to minimize the concentration of free MTO offers some promise of long catalyst in-service lifetime. Recent experiments in these laboratories (85) have shown that the oxygen in the methanol formed by MTO decomposition during catalysis comes exclusively from H_2O_2 and not from the water in the system. This is definitive experimental evidence that the "decomposition of **A**" occurs during its formation from MTO and suggests that **A** is stable, since there is no identified pathway to its destruction. It follows that any reaction system that involves only **A** and **B** need not lead to catalyst destruction. It must be noted that this result was anticipated in the elegant theoretical work of Goddard et al. (86).

Acknowledgements

We thank our many colleagues and collaborators who are identified in the references cited above that describe their excellent work. We acknowledge the sponsors of this research, especially the National Science Foundation and the Environmental Protection Agency. Much of this work has been made possible by NSF Grant CHE-0328185, NSF ERC Grant EEC-0310689, the support of the University of Kansas (through the Roy A. Roberts Distinguished Professorship to DHB and the Dan F. Servey Distinguished Professorship to BS) and the Kansas Technology Enterprise Corporation.

References

1. Morgenstern, D.A.; LeLacheur, R.M.; Morita, D.K.; Borkowsky, S.L.; Feng, S.; Brown, G.H.; Luan, L.; Gross, M.F.; Burk, M.J.; Tumas, W. In *Green Chemistry: Designing Chemistry for the Environment*; Anastas, P.T.; Williamson, T.C., Eds.; ACS Symposium Series; American Chemical Society: Washington, DC, 1996; 626, pp 132-51.
2. Pesiri, D. R.; Morita, D. K.; Glaz, W.; Tumas, W. *Chem. Commun.*, **1998**, 1015.
3. Birnbaum, E. R.; Lacheur, R. M.; Horton, A. C.; Tumas, W. *J. Mol. Catal.* **1998**, *139*, A-11.
4. Hass, G. R.; Kolis, J. W. *Organometallics* **1998**, 4454.
5. Kreher, U.; Schebesta, S.; Walther, D. *Z. Anorg. Allg. Chem.* **1998**, *624*, 602.

6. Wu, X.-W.; Oshima, Y.; Koda, S. *Chem. Lett.* **1997**, 1045.
7. Musie, G.; Wei, M.; Subramaniam, B.; Busch D. H. *Coord. Chem. Rev.* **2001**, *219-221*, 789-820.
8. Jessop, P. G.; Subramaniam, B. *Chem. Rev.* **2007**, *107 (6)*, 2666-94.
9. Lee, H-J; Shi, T-P; Busch, D. H.; Subramaniam, B. *Chem. Eng. Sci.* **2007**, *62*, 7282-89.
10. Subramaniam, B.; Busch, D. H.; Danby, A.; Binder, T. P. Ozonolysis Reactions in Liquid CO_2 and CO_2-Expanded Solvents. U.S. Provisional Patent Application Filed Nov 5, 2007.
11. Busch, D. H.; Yin, G.; Lee, H.-J. In *Mechanisms in Homogeneous and Heterogeneous Catalytic Epoxidation*; Oyama, S.T.; Ed.; Elsevier: New York, NY, **in press** (scheduled April 2008 publication).
12. Yin, G.; Danby, A. M.; Kitko, D.; Carter, J. D.; Scheper, W. M.; Busch, D. H. *Inorg. Chem.* **2007**, *46 (6)*, 2173.
13. Busch, D. H.; Collinson, S. R.; Hubin, T. J. International Patent WO 98/39098, 1998.
14. Busch, D. H.; Collinson, S. R.; Hubin, T. J.; Labeque, R.; Williams, B. K.; Johnston, J. P.; Kitko, D.; Burckett-St.Laurent, J.; Perkins, C. M. International Patent WO 98/39406, 1998.
15. Hubin, T. J.; McCormick, J. M.; Collinson, S. R.; Alcock, N. W.; Busch, D. H. *Chem. Comm.* **1998**, 1675.
16. Yin, G.; Danby, A. M.; Kitko, D.; Carter, J. D.; Scheper, W. M.; Busch, D. H. *J. Am. Chem. Soc.* **2007**, *129 (6)*, 1512.
17. Yin, G.; McCormick, J. M.; Buchalova, M.; Danby, A. M.; Rodgers, K.; Smith, K.; Perkins, C. M.; Kitko, D.; Carter, J. D.; Scheper, W. M.; Busch, D. H. *Inorg. Chem.* **2006**, *45*, 8052.
18. Busch, D. H. *"Transition Metal Ions" In Supramolecular Chemistry;* Fabbrizzi, L.; Ed.; Kluwer, 1994; pp 55-79.
19. Busch, D. H. ACS Symposium Series; American Chemical Society: Washington, DC, **1994**, *Werner Centennial* Vol. 65, pp 148.
20. Hubin, T. J.; McCormick, J. M.; Collinson, S. R.; Buchalova, M.; Perkins, C. M.; Alcock, N. W.; Kahol, P. K.; Raghunathan, A.; Busch, D. H. *J. Am. Chem. Soc.* **2000**, *122*, 2512.
21. Hubin, T. J.; McCormick, J. M.; Alcock, N. W.; Busch, D. H. *Inorg. Chem.* **2001**, *40*, 435.
22. Musie, G. T.; Wei, M.; Subramaniam, B.; Busch D. H. *Inorg. Chem.* **2001**, *40*, 3336.
23. Jones, R. D.; Summerville, D. A.; Basolo, F. *Chem. Rev.* **1979**, *79*,139.
24. Smith, T. D.; Pilbrow, J. T. *Coord. Chem. Rev.* **1981**, *39*, 295.
25. Gallagher, P. M.; Coffey, M. P.; Krukonis, V. J.; Klasutis, N. ACS Symposium Series; American Chemical Society: Washington, DC, 1989; 406.
26. Eckert, C. A.; Bush, D.; Brown, J. S.; Liotta, C. L. *Ind. Eng. Chem. Res.* **2000**, *39*, 4615.

188

27. Eckert, C. A.; Liotta C. L.; Bush, D.; Brown, J. S.; Hallett, J. P. *J. Phys. Chem. B* **2004**, *108*, 18108.
28. De la Fuente Badilla, J.C.; Peters, C.J.; de Swaan Arons, J.J. *Supercrit. Fluids* **2000**, *17*, 13.
29. Houndonougbo, Y.; Jin, H.; Rajagopalan, B.; Kuczera, K.; Subramaniam, B.; Laird, B. B. *J. Phys. Chem. B* **2006**, *110(26)*, 13195-13202.
30. Wei, M.; Musie, G. T.; Busch, D. H.; Subramaniam, B. *J. Am. Chem. Soc.* **2002**, *124*, 2513.
31. Wei, M.; Musie, G. T.; Busch, D. H.; Subramaniam, B. *Green Chem.* **2004**, *6*, 387.
32. Rajagopalan, B.; Cai, H., Busch, D. H.; Subramaniam, B. *Catalysis Letters,* **in press**.
33. Rajagopalan, B. Ph.D. thesis, University of Kansas, Lawrence, KS, 2007.
34. Joseph, J. B.; Hames, B. R. *J. Org. Chem.* **1995**, *60*, 2398.
35. Poltowicz, J. A.; Pamin, K. A.; Tabor, E. A.; Haber, J.; Adamski, A. B.; Sojka, Z. *App. Cat. A*, **2006**, *299*, 235.
36. Csjernyik, G.; Ell, A. H.; Fadini, L.; Pugin, B.; Baeckvall, J. *J. Org. Chem.* **2002**, *67*, 1657.
37. Kervinen, K.; Lahtinen, P.; Repo, T.; Svahn, M.; Leskelä, M.; *Catal. Tod.* **2002**, *75*, 183.
38. Kervinen, K.; Korpi, H.; Leskelä, M.; Repo, T. *J. Mol. Cat. A* **2003**, *203*, 9.
39. Goto, M.; Koyama, M.; Usui, H.; Mouri, M.; Mori, K.; Sakai, T. *Chem. & Pharm. Bull.* **1985**, *33(3)*, 927.
40. Chang-Man P. and Sheehan, R.J., *Terephthalic acid*, in *Kirk-Othmer Encyclopedia of Chemical Technology*. 1996, John Wiley and Sons, Inc.
41. Chemtech. 2005 [cited; Available from: http://www.chemsystems.com/about/cs/news/items/PERP%200506-6%20TPA.cfm.
42. Ichikawa, Y.; Iwakuni-Shi U.S. Patent 3,299,125, 1967.
43. Chester, A.W.; Scott, E.J.; Landis, P.S. *J. Catal.* **1977**, *46(3)*, 308-319.
44. June, R. L.; Potter, M. W.; Simpson, E. J.; Edwards, C. L. U.S. Patent 6,153,790, 2000.
45. Saha, B.; Koshino, N.; Espenson, J. H. *J. Phys. Chem. A* **2004**, *108*, 425.
46. Ishii, Y.; Sakaguchi, S.; Iwahama, T. *Adv. Synth. Catal.* **2001**, *343*, 393.
47. Raghavendrachar, P.; Ramachandran, S. *Ind. Eng. Chem. Res.* **1992**, *31*, 453.
48. Brill, W. F. *Ind. Eng. Chem.* **1960**, *52*, 837.
49. Chester, A. W.; Scott, E. J. Y.; Landis, P. S. *J. Catal.* **1977**, *46*, 308.

50. Burri, D. R.; Jun, K. W.; Yoo, J. S.; Lee, C. W.; Park, S. E. *Catal. Lett.* **2002**, *81*, 169.
51. Yoo, J. S.; Jhung, S. H.; Lee, K. H.; Park, Y. S. *Appl. Catal. A-Gen.* **2002**, *223*, 239.
52. Bawn, C. E. H.; Pennington, A. A.; Tipper, C. F. H. *Disc. Faraday Soc.* **1951**, *10*, 282.

53. Ichikawa, Y.; Yamashita, G.; Tokashiki, M.; Yamaji, T. *Ind. Eng. Chem.* **1970**, *62*, 38.
54. Chavan, S. A.; Halligudi, S. B.; Srinivas, D.; Ratnasamy, P. *J. Mol. Catal. A-Chem.* **2000**, *161*, 49.
55. Partenheimer, W. *J. Mol. Catal. A-Chem.* **2003**, *206*, 105.
56. Nakaoka, K.; Miyama, Y.; Matsuhisa, S.; Wakamatsu, S. *Ind. Eng. Chem. Prod. Res. Dev.* **1973**, *12*, 150.
57. Zuo, X.; Subramaniam, B.; Busch, D. H. *Ind. Eng. Chem. Res.* **2008**, *47*, 546.
58. Givens, R.S.; Ma, C.C.; Busch, D.H.; Subramaniam, B.; Rajagopalan, B. U.S. Provisional Patent Application Serial NO. 60/856,147, filing date November 2, 2006 converted to U.S. Patent Application, 11/02/07.
59. Cincotti, A.; Orru, R.; Cao, G. *Catal. Today* **1999**, *52*, 331-47.
60. Lee, H.-J.; Shi, T.-P.; Busch, D. H.; Subramaniam, B. *Chem. Eng. Sci.* **2007**, *62*, 7282-89.
61. Lee,H.-J.; Shi, T.-P.; Subramaniam, B.; Busch, D. H. *Selective Oxidation of Propylene to Propylene Oxide in CO_2 Expanded Liquids* In *Catalysis of Organic Reactions*; Schmidt, S. R.; Ed.; CRC Press, Taylor & Francis Group: Boca Raton, FL, 2007; pp 447-51.
62. Trent, D. *Encyclopedia of Chemical Technology*; 4th ed.; John Wiley & Sons: New York, NY 1996; Vol 30, pp 271-302.
63. Gerhartz, W.; Yamamoto, Y. S.; Kaudy, L. In *Ullmann's Encyclopedia of Industrial Chemistry*; 5th Ed.; Rounsaville, J. F.; Schulz, J., Eds.; Verlag Chemie: Weinheim, Germany 1987; Vol A9, pp 531-64.
64. Tullo, H.; Short, P. L. *Chem. Eng. News* October 9, 2006, Vol 84, No 41, p. 22.
65. Herrmann, W. A.; Dieter, M.; Wagner, W.; Kuchler, J. G.; Weichselbaumer, G.; Fischer, R. U.S. Patent 5,155,247, assigned to Hoechst Aktiegesellschaft, 1992 G.L.
66. Crocco, G.L.; Shum, W. F.; Zajacek, J. G.; Kesling Jr., S. U.S. Patent, 5,166,372; assigned to Arco Chemical Technology 1992.
67. Subramaniam, B.; Busch, D. H.; Lee, H.-J.; Shi, T.-P. U.S. Provisional Patent filled 10/25/05; converted to U.S. Patent Application, 10/25/06. Assigned to University of Kansas, Lawrence, KS.
67. Anastas, P. T.; Warner J. C. *Green Chemistry: Theory and Practice;* Oxford University Press: Oxford, UK, 1998.
69. Lancaster, M. *Green Chemistry: An Introductory Text*, Royal Society of Chemistry: Cambride, UK 2002.
70. Beattie, R.; Jones, P. J. *Inorg. Chem.* **1979**, *18,* 2318.
71. Kuehn, F. E.; Santos, A. M.; Herrmann, W. A. *Dalton Trans.* **2005**, 2483.
72. Herrmann, W. A.; Kuehn, F. E. *Acc. Chem. Res.* **1997**, *30,* 169.
73. Espenson, J. B.; Abu-Omar, M. M. *Reactions Catalyzed by Methylrheniumtrioxide;* Adv. Chem. Series; American Chemical Society: Washington, DC, 1997; Vol 253, pp 99.

74. Espenson, J. H. *Chem. Commun.* **1999**, 479.
75. Rudolph, J.; Reddy, K. L.; Chiang, J. P.; Sharpless, K. B. *J. Am. Chem. Soc.* **1997**, *119*, 6189.
76. Herrmann, W. A.; Fischer, R. W.; Rauch, M. U.; Scherer, W. *J. Mol. Catal.* **1994**, *86*, 243.
77. Herrmann, W. A ; Fischer, R.W.; Marz, D.W. *Angew. Chem.; Int. Ed. Engl.* **1991**, *30*, 1638.
78. Shu, Z.; Espenson, J. H. *J. Org. Chem.* **1995**, *60*, 1326.
79. Wang, W. D.; Espenson, J. H. *Inorg. Chem.* **1997**, *36*, 5069.
80. Abu-Omar, M. M.; Espenson, J. H. *J. Chem. Soc.* **1995**, *117*, 272.
81. Wang, W. D.; Espenson, J. H. *J. Am. Chem. Soc.* **1998**, *120*, 11335.
82. Yamazaki, S.; Espenson, J. H.; Huston, P. *Inorg. Chem.* **1993**, *32*, 4683.
83. Hansen, P. J.; Espenson, J. H.; *Inorg. Chem.* **1995**, *34*, 5839.
84. Hansen, P. J.; Espenson, J. H.; *Inorg. Chem.* **1995**, *34*, 5839.
85. Yin, G.; Busch, D. H. submitted for publication, 2007.
86. Gonzales, J.M.; Distasio Jr., R.; Periana, R.A.; Goddard III, W.A.; Oxgaard, J.; J. Am. Chem. Soc., **2007**, *129*, 15794.

Chapter 9

Hydrogenation of CO_2-Expanded Liquid Terpenes: Phase Equilibrium-Controlled Kinetics

Ewa Bogel-Łukasik[1], Ana Serbanovic[2], Rafal Bogel-Łukasik[1], Anna Banet-Osuna[2], Vesna Najdanovic-Visak[1,*], and Manuel Nunes da Ponte[1]

[1]REQUIMTE, Departamento de Química, FCT Universidade Nova de Lisboa, 2829–516 Caparica, Portugal
[2]Instituto de Tecnologia Química e Biológica, Universidade Nova de Lisboa, Apartado 127, 2781–901 Oeiras, Portugal
*Corresponding author: email: vesna@itqb.unl.pt

The hydrogenation of two liquid terpenes, α-pinene and limonene, was carried out in the presence of high-pressure carbon dioxide, using carbon-supported Pt catalysts. Phase equilibrium data on terpene + CO_2 + hydrogen were used to interpret the kinetics of hydrogenation in liquid + vapor systems close to the critical lines of the mixtures.

Introduction

Supercritical fluids generally exhibit complete miscibility with gases, like hydrogen. In the cases where the reactants to be hydrogenated are also soluble, those fluids have consequently been regarded as particularly advantageous solvents for carrying out hydrogenation reactions. In these situations, all reactants will be in a single fluid phase, and the access of hydrogen to the catalyst will not be slowed down by resistance of mass transfer between phases, as in the case of biphasic, liquid–gas reaction media. The fact is, however, that at least some fast hydrogenations in carbon dioxide reported in the literature are clearly carried out in a biphasic, gas + liquid system. In these cases, the phase contacting the catalyst may be described as an expanded liquid.

Mixtures of a terpene and high pressure carbon dioxide are interesting model systems to study the factors that control heterogeneous catalysis in expanded liquids. In general, carbon dioxide is highly soluble in most terpenes. At pressures close to the critical pressure of the mixture, the liquid phase in a biphasic mixture may contain 80 mol % or more of CO_2 (1 - 4). These liquids also exhibit high total volume expansions compared to the pure terpene initial volume, and they are very sensitive to pressure in the region just below the critical pressure of the mixtures. The total pressures for mixture critical point are low, often lower than 10 MPa. By changing the pressure slightly, terpene + CO_2 systems change easily from monophasic to biphasic, and heterogeneous catalysis can be compared, for the same reacting mixture, in an expanded liquid and in supercritical conditions.

Terpenes are a class of compounds with important industrial applications, especially in the fragrance and flavor industry which consumes approximately 60,000 ton/year of pinenes and limonene (5). Hydrogenation of these substances is commonly performed for industrial purposes (6). The mechanisms of these reactions in classical conditions (hydrogen gas + liquid terpene + solid catalyst) are generally well known, establishing a good starting point for their examination in CO_2-expanded liquids. α-pinene and limonene were used as model substances for the study of phase equilibrium control of the kinetics of hydrogenation in terpene + carbon dioxide mixtures.

The hydrogenation of α-pinene over Pd (7) and Pt (8) has been performed in a high pressure CO_2 medium. E. Bogel-Łukasik et al. (9, 10) studied the hydrogenation of limonene over Pt and Pd catalysts in supercritical CO_2 (ScCO2). Figure 1 shows the chemical structure of α-pinene and limonene.

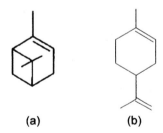

(a)   (b)

Figure 1. Chemical structure of (a) α-pinene and (b) limonene

The hydrogenation of α-pinene is the first step in a series of reactions that convert it into valuable products, such as menthol. It has been achieved using several catalysts such as Adams catalyst (PtO_2), Raney nickel and carbon-supported metals (Pt/C, Pd/C) (11). The mechanisms of these catalytic reactions are generally well known, and it is possible to control the stereoselectivity by adjusting the operating conditions. Semikolenov et al. (12) proposed a mechanism for the hydrogenation of α-pinene on Pd/C catalysts to cis- and trans-pinane. They showed that the selectivity toward cis-pinane is determined by controlling the hydrogen pressure and the reaction temperature.

Limonene has two C=C double bonds, one in the cyclohexene ring and an external one. Hydrogenation of limonene can yield different products, depending on which double bond is hydrogenated. If both C=C bonds are hydrogenated, a mixture of *cis-* and *trans-*p-menthane is produced. Grau et al. (13) and Shimazu et al. (14), have actually reported the preferential hydrogenation of the external double bond, to yield p-menthene.

Minder et al. (15) and Burgener et al. (16) studied the hydrogenation of citral in carbon dioxide. Citral has two C=C double bonds, but it also has a C=O bond. As those authors used a palladium catalyst, which is selective towards the hydrogenation of the C=C double bonds, their study is, in some aspects, similar to the one presented here. Chatterjee et al. (17) and Liu et al. (18) have also studied the hydrogenation of citral in supercritical carbon dioxide.

Pereda et al. (19) have used group contribution equation of state models to calculate the phase boundaries of the ternary α-pinene + CO_2 + H_2, in conditions similar to those used for hydrogenation by Chouchi et al. (7). They interpreted the results of these authors on the basis of their calculated ternary phase equilibrium diagram at 14 MPa and 323 K.

In this work we performed hydrogenation of α-pinene using Pt catalysts on either Norit or SKN carbon supports, in the presence of high pressure CO_2. Reaction progress was followed in either biphasic or single-phase conditions. In order to interpret the results, an experimental study of phase behavior at 323 K, along with calculations using the Peng–Robinson equation of state, was performed for (H_2 + CO_2 + α-pinene) mixtures.

Experimental and Results

Volume Expansion and Phase Equilibrium Experiments

Volume expansion and VLE measurements were performed in the same apparatus, although not simultaneously. This apparatus and the methodology used for phase equilibrium measurements are described in previous publications (20, 21). The essential part of the apparatus is a sapphire cylindrical cell (height - 15 cm; internal diameter - 1.9 cm; external diameter - 3.2 cm), allowing full visualization of the whole internal volume. It is therefore possible to measure the height of the liquid phase, as a function of pressure, and the liquid volume expansion in relation to the volume in any previously measured condition.

We measured vapor-liquid equilibrium and the volume expansion of the liquid phase for binary (α-pinene + carbon dioxide) mixtures at 323 K and at pressures from 4 MPa up to very close to the critical pressure of the mixture.

Figure 2 presents the ratio between the liquid phase total volume at the given pressure and at 4 MPa.

During the phase equilibrium measurements, the cell was connected to a volume-calibrated and manually driven screw-injector. This allowed slow addition of CO_2 to the cell. The temperature controller has been calibrated, and the precision was determined to be ± 0.03 K. Precision and accuracy of the transmitter measuring pressure inside the cell are 0.1% and 0.15%, respectively. Energetic stirring is performed using a magnetic bar driven by a magnetic stirrer.

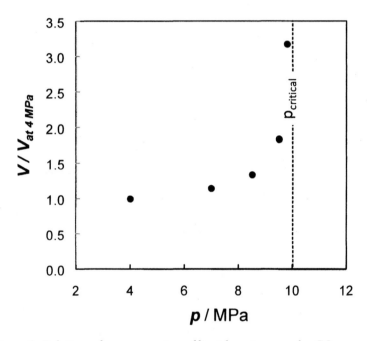

Figure 2. Relative volume expansion of liquid α-pinene under CO_2 pressure at 323 K

The cell was initially connected to a vacuum pump to facilitate loading of the liquid sample. Afterwards, CO_2 was introduced to the desired initial pressure. When the temperature in the air-bath was stable, CO_2 was added slowly to the cell. Disappearance of the phase boundary was easily observable with the naked eye.

The Peng–Robinson equation of state, using the Mathias–Klotz–Prausnitz mixing rule (22), was used to correlate the data. A calculation of phase equilibrium of the ternary mixtures was carried out using the program package PE (23).

Figure 3 presents the vapor-liquid diagram for $H_2 + CO_2 + \alpha$-pinene at 323 K, at two pressures, 11.6 MPa and 15 MPa, and in the region of high mole fraction of CO_2. These are the conditions where the liquid phase behaves as an expanded liquid.

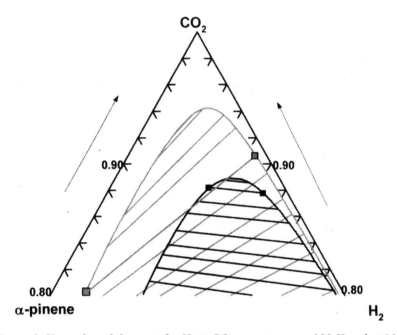

Figure 3. Vapor-liquid diagram for $H_2 + CO_2 + \alpha$-pinene at 323 K and at 11.6 MPa (gray lines) and 15.6 MPa (black lines)

If the same overall composition is considered (for instance, $x_{H2} = 0.07$ and $x_{\alpha\text{-pinene}} = 0.04$, at the intersection of the two highlighted tie-lines in the Figure 3), the most noticeable effect is that the different slopes of the tie-lines lead to radically different hydrogen-to-pinene molar ratios in the liquid phases at the two pressures; 0.1 at 11.6 MPa and 1.4 at 15.6 MPa. In Table I, these values are compared with those calculated by Bogel-Łukasik at al. (9) for $H_2 + CO_2$ +limonene, for the same overall composition (where limonene replaces α-pinene). In both cases, the H_2/terpene molar ratio in the expanded liquid, for mixtures of the same overall composition, is highly sensitive to pressure.

Table I Molar ratios of H_2 to terpene in the liquid phase at 323 K for H_2 + CO_2 + terpene (α-pinene or limonene) at selected pressures and the same overall composition (mole fractions $x_{H2} = 0.07$ and $x_{terpene} = 0.04$)

Terpene	α-pinene		limonene	
p / MPa	11.6	15.6	11	12.5
H_2 : terpene / molar ratio	0.1 : 1	1.4 : 1	0.4 : 1	1.5 : 1

Hydrogenation Experiments

Hydrogenation of α-pinene was carried out using platinum catalysts loaded on two carbon-based supports (Norit and SKN). This study followed the previous work of Milewska et al. (8) where the importance of the catalyst support to the kinetics of the reaction was revealed.

We used the same experimental apparatus as used by Milewska et al. (8). The system consisted of a 50 cm^3 sapphire-windowed cell that allowed direct visual observation of the phase behavior. The cell was connected via a circulation pump to a tubular reactor that contains a catalyst bed. Samples were taken at regular intervals through a system of two valves with a sampling loop, at the top of the tubular reactor.

Comparisons were made between reaction kinetics for α-pinene + carbon dioxide + hydrogen mixtures at supercritical, single phase conditions, slightly above the critical pressure of the mixture, and at biphasic (or expanded liquid) conditions, slightly below that pressure. The conditions used were 12 MPa CO_2 + 4 MPa H_2 for the supercritical mixtures and 8.5 MPa CO_2 + 4 MPa H_2 for the biphasic ones.

Figure 4 shows conversions obtained for the hydrogenation of α-pinene over two different catalysts with 1 wt.% Pt at 323 K and at two partial pressures of CO_2, 12 MPa and 8.5 MPa. Figure 5 shows the influence of platinum loading of the catalyst, using the same support, on the hydrogenation of α-pinene.

Discussion

The expansion of the overall volume of the liquid phase shown in Figure 2 induces changes in solution that should affect the kinetics of hydrogenation in opposite ways. On the one hand, the increase in volume must decrease the concentration (per unit volume) of the reactants, but on the other hand, higher amounts of hydrogen should be brought into the liquid phase as the CO_2 pressure increases. Brennecke and collaborators (24) have recently studied this last effect. They measured the solubility of hydrogen in acetonitrile, acetone and methanol

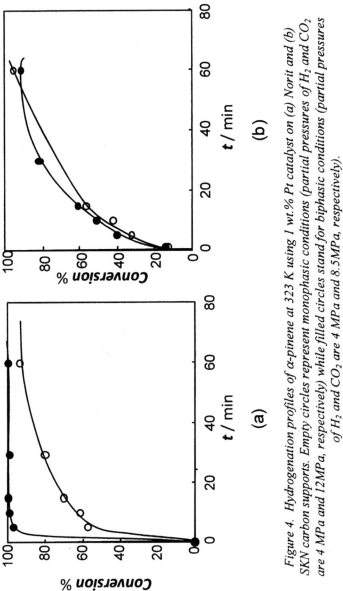

Figure 4. Hydrogenation profiles of α-pinene at 323 K using 1 wt.% Pt catalyst on (a) Norit and (b) SKN carbon supports. Empty circles represent monophasic conditions (partial pressures of H_2 and CO_2 are 4 MPa and 12MPa, respectively) while filled circles stand for biphasic conditions (partial pressures of H_2 and CO_2 are 4 MPa and 8.5MPa, respectively).

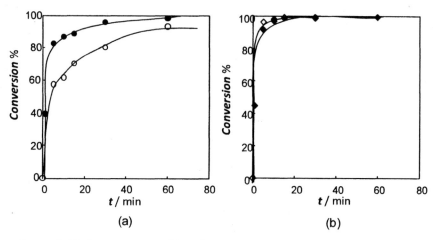

Figure 5. Hydrogenation profiles of α-pinene using 1 wt.% Pt catalyst (empty symbols) or 2 wt.% Pt catalyst (filled symbols) on Norit carbon support at 323 K (a) under monophasic conditions (partial pressures of H_2 and CO_2 are 4 MPa and 12 MPa, respectively) and (b) under biphasic conditions (partial pressures of H_2 and CO_2 are 4 MPa and 8.5 MPa, respectively)

under CO_2 pressure, and they concluded that the solubility of H_2 in expanded solvents could be improved by the presence of CO_2, but not significantly. Their measurements were, however, performed at pressures up to 9 MPa and further from the ternary mixture critical region than in this work. Our measurements show a very significant increase of the mole fraction of hydrogen in the liquid phase, at constant partial pressure of H_2, as carbon dioxide pressure is raised.

Bogel-Lukasik et al. (10) have recently concluded that, in the hydrogenation of limonene, the combined volume expansion of the liquid phase and higher amounts of hydrogen roughly compensate each other in terms of concentration per unit volume, and that the kinetics of the reaction are mainly affected by the decrease of the volumetric concentration of the terpene. In fact, higher partial pressure of hydrogen in the gas leads to lower ratios of hydrogen to limonene in the liquid. The explanation is summarized in Figure 6. At constant total pressure, when the partial pressure of hydrogen increases, the partial pressure of CO_2 decreases. This moves the system further away from the mixture critical pressure, which decreases the degree of expansion, increasing the volumetric concentration of limonene.

As the liquid-vapor equilibrium behavior of α-pinene or limonene in mixtures with carbon dioxide + hydrogen is very similar, the conclusions should also be valid for the hydrogenation of α-pinene. Another interesting conclusion of the phase equilibrium measurements is that the hydrogen/terpene molar ratios may increase to values close to or even higher than 1 in the vicinity of the mixture critical line. This eliminates the limitations imposed by phase transfer

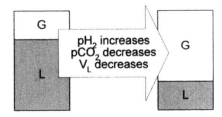

*Figure 6. Schematic representation of volume contraction of a liquid phase
as partial pressure of hydrogen increases at constant total pressure.*

resistance on the kinetics in hydrogenations in normal gas-liquid systems, as enough hydrogen can be present in the liquid phase from the start.

The results obtained in this work for the kinetics of hydrogenation are in good agreement with the findings and conclusions of Milewska et al. (8). As seen in Figures 4 and 5, hydrogenation in the expanded liquid is, in certain conditions, faster than in one supercritical phase. This means that (1) there is a higher concentration of α-pinene in the liquid phase in the vicinity of the catalyst than in supercritical conditions, and (2) mass transfer of hydrogen towards the catalyst is sufficiently fast.

Another interesting conclusion is that, in the case of monophasic conditions, the hydrogenation rate is significantly higher as the Pt loading on the catalyst increases (Figure 5), which is not the case in the expanded liquid. However, this observation might be an artificial fact appearing due to very high initial rates in the two phase system.

The hydrogenations of α-pinene or limonene show important differences when reaction rates in one and two phase mixtures are compared. In general, we found that while biphasic conditions yield faster hydrogenation of α-pinene, the opposite happened with limonene. Of course, as limonene has two double bonds that can be hydrogenated, only the hydrogenation of the first C=C bond to be hydrogenated, the external one, might be compared with the hydrogenation of α-pinene. The fact that the C=C bond in α-pinene is an internal one might explain those differences.

The interpretation of the results of hydrogenation of terpenes from other authors may follow similar lines to those explained in this work. For instance, Minder et al. (15) studied the effect of pressure on reaction rate and selectivity at 40°C, for a fixed feed composition of 100 mol CO_2 + 5 mol H_2 + 1 mol citral. The hydrogenation reaction gave good yields at homogeneous, high-pressure conditions, but the rate was very slow when at pressures below about 11 MPa where the mixtures were biphasic. Selectivity towards the fully hydrogenated product increased markedly in the one-phase region. However, when hydrogenation was studied at a fixed pressure of 13 MPa with varying compositions, the reaction rates in one and in two-phase mixtures became similar, while selectivity was mainly a function of the H_2:citral ratio.

There are no VLE data on $CO_2 + H_2 +$ citral, except for the dew point results presented by Minder et al. (15). These data do not include compositions, and they are not sufficient to produce a good correlation through an equation of state, as presented above for α-pinene. However, the vapor-liquid equilibrium behavior of binary mixtures of citral with carbon dioxide is similar to limonene (4), although the critical pressures were higher for each temperature. If this similarity is taken as a good indicator of similarity of phase equilibrium in ternary mixtures with hydrogen, the results of Minder et al. (15) could be interpreted on the same basis as the results presented in this work.

Conclusions

The hydrogenation of α-pinene and limonene was studied under various conditions. The main advantage of performing hydrogenation of terpenes in CO_2-expanded biphasic systems is the ability of CO_2 to bring the hydrogen into the liquid medium in a controlled way, which facilitates mass transfer and accessibility of hydrogen, but which can also be used for increasing selectivity.

Although various factors may affect the kinetics of hydrogenation carried out in CO_2-expanded terpenes, the decisive factor in controlling the kinetics may be the volume expansion that appears close to critical conditions.

References

1. Matos, H. A.; Gomes de Azevedo, E. J. S.; Simões, P.C.; Carrondo, M.T.; Nunes da Ponte, M. *Fluid Phase Equilibr.* **1989**, *52*, 357-364 .
2. Akgün, M.; Akgün, N. A.; Dinçer, S. *J. Supercrit. Fluids* **1999**, *15*, 117-125.
3. Reverchon, E. *J. Supercrit. Fluids* **1997**, *10*, 1-37.
4. Fonseca, J; Simões, P. C.; Nunes da Ponte, M. *J. Supercrit. Fluids* **2003**, *25*, 7-17.
5. Albert, R. M.; Traynor, S. G. In *Naval Stores-Production, Chemistry and Utilization*; Zinkel, D. F.; Russel, J., Eds.; Pulp Chemical Association: New York, 1989; Chapter 12.
6. Chapuis, C.; Jacoby, D. *Appl. Catal. A* **2001**, *221*, 93-117.
7. Chouchi, D.; Gourgouillon, D.; Courel, M.; Vital, J.; Nunes da Ponte, M. *Ind. Eng Chem. Res.* **2001**, *40*, 2551-2554.
8. Milewska, A.; Banet Osuna, A. M.; Fonseca, I. M.; Nunes da Ponte, M. *Green Chem.* **2005**, *7*, 726-732.
9. Bogel-Łukasik, E.; Fonseca, I.; Bogel-Łukasik, R.; Tarasenko, Y. A.; Nunes da Ponte, M.; Paiva, A.; Brunner, G. *Green Chem.* **2007**, *9*, 427-430.

10. Bogel-Łukasik, E.; Bogel-Łukasik, R.; Kriaa, K.; Fonseca, I.; Tarasenko, Y.; Nunes da Ponte, M. *J. Supercrit. Fluids* **2008**, *45*, 225-230.
11. Cocker, W.; Shannon, P. V. R.; Stanilan, P. A. *J. Chem. Soc. C* **1966**, *1*, 41.
12. Semikolenov, V; Ilyna, I. I.; Simakova, I. L. *Appl. Cat. A: General* **2001**, *211*, 91-107.
13. Grau, R. J.; Zgolicz, P. D.; Gutierrez, C.; Taher, H. A. *J. Mol. Catal. A: Chem.* **1999**, *148*, 203-214.
14. Shimazu, S.; Baba, N.; Ichikuni, N.; Uematsu, T. *J. Mol. Catal. A: Chem.* **2002**, *182*, 343-350.
15. Minder, B.; Mallat, T.; Pickel, K. H.; Steiner, K.; Baiker, A. *Catal. Lett.* **1995**, *34*, 1-9.
16. Burgener, M.; Furrer, R.; Mallat, T.; Baiker, A. *Appl. Catal. A* **2004**, *268*, 1-8.
17. Chatterjee, M.; Chetterjee, A.; Ikushima, Y. *Green Chem.* **2004**, *6*, 114-118.
18. Liu, R.; Zhao, F.; Fujita, S.-I.; Arai, M. *App. Catal. A* **2007**, *316*, 127-133.
19. Pereda, S.; Bottini, S. B.; Brignole E. A. *Appl. Catal. A* **2005**, *281*, 129.
20. Gourgouillon, D.; Nunes da Ponte, M. *Phys. Chem. Chem. Phys.* **1999**, *23*, 5369-5375.
21. Lopes, J. A.; Nunes da Ponte, M. *J. Supercrit. Fluids* **2005**, *34*, 189-194.
22. Mathias, P. M.; Klotz, H. C.; Prausnitz, J. M. *Fluid Phase Equilibr.* **1991**, *67*, 31-44.
23. Pfohl, O; Petkov, S.; Brunner, G. PE V2.9.9a - Software for Phase Equilibria Calculations, Technische Universität Hamburg-Harburg, Hamburg (1998).
24. Lopez-Castillo, Z. K.; Aki, S. N. V.; Stadtherr, M. A. K.; Brennecke, J. F. *Ind. Eng. Chem. Res.* **2008**, *47*, 570-576.

Chapter 10

Hydroformylation in CO_2-Expanded Media

Ruihu Wang[1], Hu Cai[1], Hong Jin[2], ZhuanZhuan Xie[2], Bala Subramaniam[2], and Jon A. Tunge[1]

Departments of [1]Chemistry and [2]Chemical and Petroleum Engineering, The University of Kansas, Lawrence, KS 66045

CO_2-expanded solvents provide a unique reaction medium for hydroformylation of 1-octene. The tunable syngas solubility in CXLs coupled with their tunable solubility properties leads to solvents that can be tailored to a particular reaction. Thus, hydroformylation of 1-octene using rhodium catalysts in CO_2-expanded 1-octene or toluene proceeds with higher selectivities and the catalysts also exhibit higher activities. For such a process to be economically viable, it is essential that the expensive rhodium catalyst is recycled. Herein we detail our intial efforts to effect efficient recycle of rhodium by supporting the catalyst on soluble polymers that can be readily precipitated and recycled.

Hydroformylation of higher olefins

Since its serendipitous discovery in 1938, the catalytic hydroformylation of olefins (i.e. oxo synthesis) has become one of the most important industrial processes, having an annual production of aldehydes approaching 10 million tons (1-2). Hydrofomylation involves the addition of syngas to an alkene to produce aldehydes in a reaction that exhibits perfect atom economy (Scheme 1). Typical catalysts for hydroformylation include cobalt or rhodium carbonyl-based catalysts and the selectivities of these catalysts can often be improved by modifying the catalyst by addition of ligands. In general, the rhodium-based catalysts are significantly more active and selective yet they operate at lower temperatures and pressures (ca. 90 °C, 15-40 bars, TOF ~600 h^{-1}) than the cobalt-based catalysts (ca 160 °C, 200 bars, TOF ~ 25 h^{-1}). Thus the hydroformylation of propylene is conducted with rhodium catalysts in either aldehyde condensate (UCC process) or in an aqueous biphasic process (Ruhrchemie/Rhone-Poulenc = RCH/RP) that allows the recycle of the expensive rhodium catalyst (3-5). However, the RCH/RP process does not extend to the hydroformylation of higher olefins (> C5) where the low solubility of the olefin in the aqueous solution obviates the use of aqueous biphasic reaction conditions (1, 6). Furthermore, the instability of catalysts at the temperatures required for distillation of product olefins (i.e. nonanal) complicates catalyst recycle (7), thus making the rhodium-catalyzed process economically unfeasible. Hence, the hydroformylation of higher olefins is commonly catalyzed by cobalt catalysts, albeit under harsher conditions than those required for rhodium catalysts. However, if one could develop simple methods for the recycle of rhodium catalysts, our preliminary environmental and economic analyses indicate that a rhodium-based process would be economically comparable and more environmentally benign than the cobalt-based processes (8).

One challenge in the development of a rhodium-catalyzed hydroformylation of 1-octene is the propensity of high-activity rhodium catalysts to form less reactive internal olefins under syngas starved conditions (Scheme 2) (2). Thus, it was our hypothesis that increasing syngas solubility and/or mass transfer would increase the selectivity toward aldehydes. Since CO$_2$-expanded liquids (CXLs) increase the solubility of H$_2$ and CO in solution (Figure 1), it appeared that CXLs may be the optimal media for performing hydroformylation with highly active rhodium catalysts. Previous work on hydroformylation in CO$_2$-based media, including supercritical CO$_2$, and the rationale for employing CO$_2$-expanded solvents have been discussed in detail elsewhere (9-11). More recently, Cole-Hamilton and coworkers reported a continuous homogeneous hydroformylation process using dense CO$_2$ to transport the reactants into and

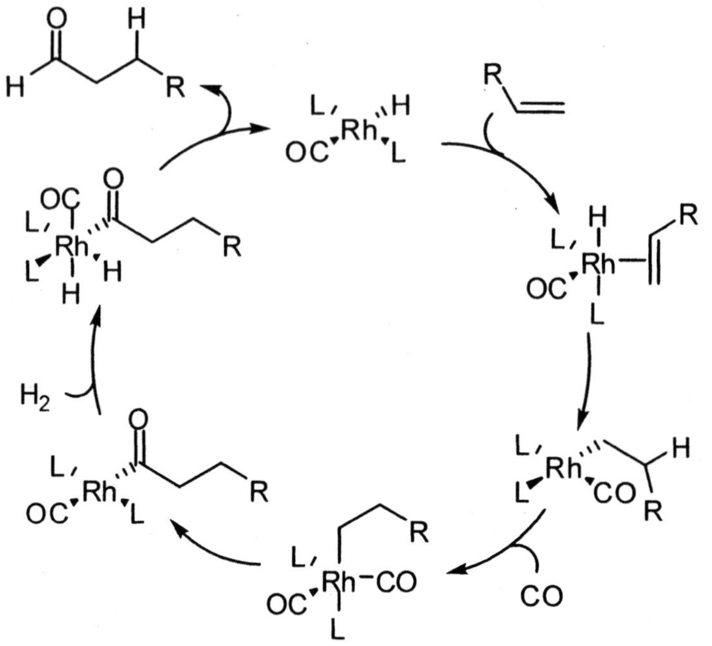

Scheme 1. General mechanism for hydroformylation.

transfer the products out of the reactor leaving behind the CO_2-insoluble Rh-based catalyst complex in the reactor solution (12). The CXL-based hydroformylation concept is a complementary approach that requires relatively low pressures on the order of a few tens of bars.

We began by comparing the hydroformylation of 1-octene in non-expanded solvent vs. CXL using a simple rhodium catalyst that was unmodified by external ligands (9). Doing so showed that indeed the activity of rhodium catalysts in CXLs is higher than in normal solvents as evidenced by the significant increase in conversion. Moreover, and more importantly, hydroformylation in the CXL had a marked positive effect on the selectivity of the reaction toward the desired aldehyde products; the mass balance is made up of internal octene isomers (Scheme 3).

While the experiments with unmodified catalyst demonstrated enhanced rates and selectivities of hydroformylation in a CXL, the use of unmodified catalyst produced low selectivities for the desired linear (n) aldehyde (n/i = 1.5). Thus we turned our attention to investigating ligand modified rhodium catalysts (10). Toward this end, the most common ligand modifier for rhodium catalysts, PPh$_3$, was chosen for initial studies. Investigating the effect of CO_2-expansion on the rate and selectivity of hydroformylation of 1-octene revealed that the n/i

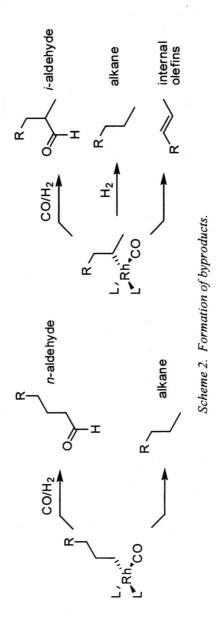

Scheme 2. Formation of byproducts.

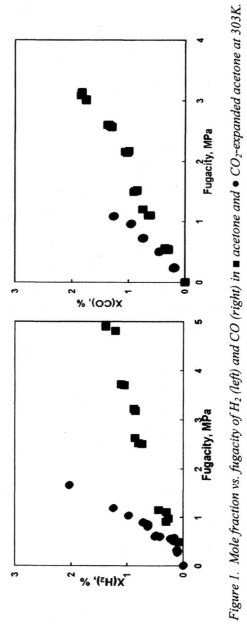

Figure 1. Mole fraction vs. fugacity of H_2 (left) and CO (right) in ■ acetone and ● CO_2-expanded acetone at 303K.

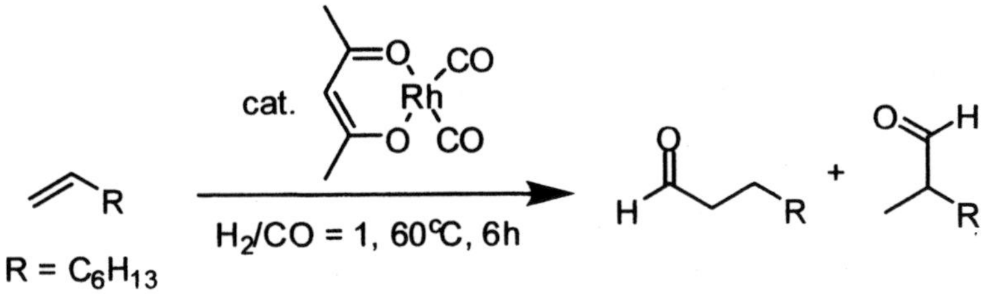

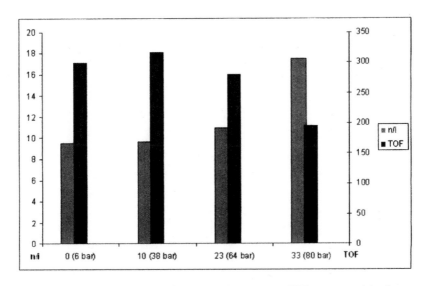

solvent	conversion	S_{ald}
acetone	36%	46%
1:1 acetone:CO_2	97%	73%

Scheme 3

Figure 2. Comparison of the percent expansion (CO_2 pressure) in the hydroformylation of 1-octene with 0.04 mol% $Rh(CO)_2(acac)$, 10 mol% PPh_3, 1:1 H_2/CO, 60°C. TOF = mol aldehyde/(mol Rh·h).

selectivity showed a marked increase at higher degrees of CO_2 expansion (Figure 2). However, the reactions showed a somewhat decreased rate at higher CO_2 expansion. The decrease in rate is readily attributed to the larger volume of the CO_2-expanded liquid and thus lower concentration of reactants.

To further examine the effect of CO_2-expansion, the hydroformylation of 1-octene was conducted at a constant volume and pressure (Scheme 4). The results conclusively show that CO_2 enhances the activity (TOF) and n/i selectivity of hydroformylation when compared to equivalent pressures of syngas or an inert gas (N_2).

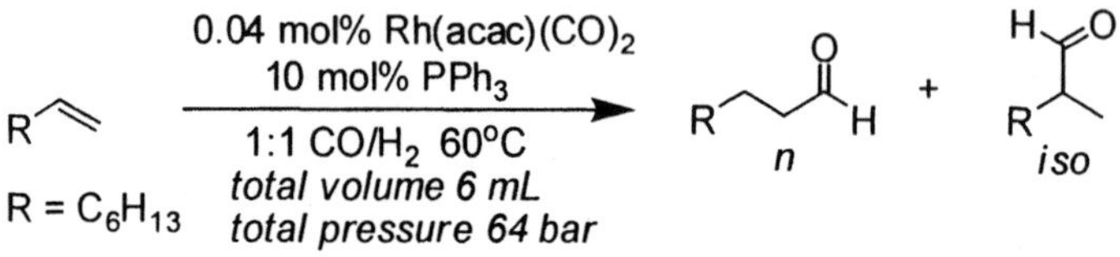

$P_{(syngas)}$	$P_{(CO2)}$	$P_{(N2)}$	n/i	TOF
64 bar	0 bar	0 bar	3.9	175
6 bar	58 bar	0 bar	10.8	316
6 bar	0 bar	58 bar	4.7	141

Scheme 4

While the selectivities and activities of the hydroformylation in CXL's are comparable to, or better than, those reported for similar reactions performed in neat 1-octene or supercritical CO_2, the issue of catalyst recycle still needs to be addressed in order to make the process economically viable (10).

Hydroformylation with soluble-polymer supported ligands

The recycle of rhodium catalysts has often been accomplished by the heterogenerization of the catalyst by immobilization of the homogenous catalysts on solid support that allows catalyst separation by filtration or facilitates use of a fixed catalyst bed for continuous operation (4-5, 13-17). Unfortunately, such methodologies often lead to a significant decrease of catalytic activity and selectivity of the reaction due to the slower diffusion of substrates into catalytic centers. From a catalytic viewpoint, an ideally engineered system should combine the advantages of homogeneous and heterogenous catalysis. An underutilized method to achieve this goal is to utilize *soluble* polymer-supported catalysts that can be readily precipitated or separated by size exclusion (18, 19). For example, low to medium molecular weight polystyrene and its analogues are usually soluble in non-polar solvents like toluene, CH_2Cl_2 and THF, but they are highly insoluble in polar solvents such as methanol and water. So, homogenous reactions can be performed in non-polar solvents, and the recovery and recycle of catalysts are realized by changing the solvent polarity to initiate precipitation of polymer-supported catalysts. The potential benefits of utilizing a homogeneous polymer-supported catalyst were recognized and outlined by van Leeuwen (18). In addition, van Leeuwen detailed the benefits of using polymer-supported phosphites as compared to the more commonly used supported phosphines. However, the polymer-supported phosphite catalysts prepared were only utilized for cyclooctene and styrene hydroformylation, where complications from olefin isomerization are avoided. Thus, we began our investigations by studying the use of polymer-supported monophosphites in the hydroformylation of 1-octene.

The synthesis of polymer-supported phosphites was accomplished by the copolymerization of a hydroxystyrene derivative with styrene as previously described (Scheme 5, 17). The resulting polymeric alcohol can be modified with a variety of chlorophosphites to produce a polystyrene supported triarylphosphite. Analysis of the resulting polymer by ^{1}H NMR spectroscopy in the presence of triphenylphosphine internal standard shows that the ligand loading on the polymer is 0.39 mmol/g; a similar, yet sterically larger polymer-supported phosphite was prepared by the same method and shown to have a ligand load of 0.8 mmol/g by ^{1}H NMR analysis.

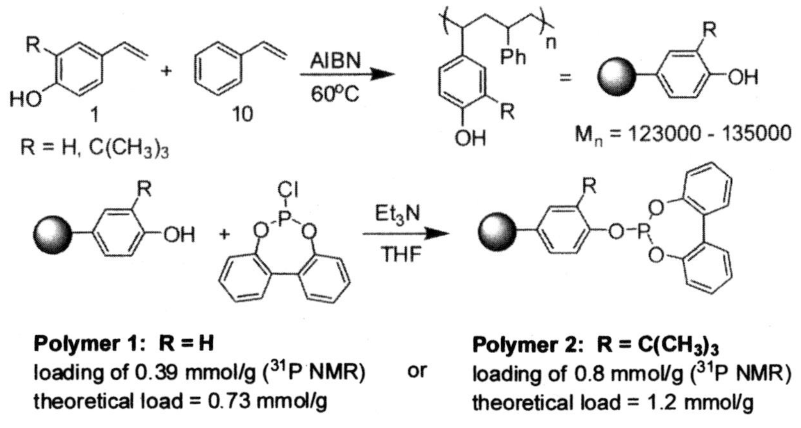

Polymer 1: R = H
loading of 0.39 mmol/g (^{31}P NMR)   or
theoretical load = 0.73 mmol/g

Polymer 2: R = C(CH$_3$)$_3$
loading of 0.8 mmol/g (^{31}P NMR)
theoretical load = 1.2 mmol/g

Scheme 5

Next, the hydroformylation of 1-octene was conducted in neat 1-octene with a ligand/Rh(CO)$_2$(acac) ratio of 2:1 at 60 °C. Both polymers 1 and 2 provided active and selective catalysts for the production of nonanals (Table I). Polymer 1-modified rhodium catalyst showed a selectivity for aldehydes of 80% while the bulkier polymer 2 exhibited a selectivity of 88%; both selectivities are higher than those for P(OPh)$_3$-modified catalyst operating under similar conditions (63%). However both polymer 1 (n/i = 4) and polymer 2 (n/i = 4.3) showed lower selectivity for the linear aldehyde than that of P(OPh)$_3$ (8.2). We hypothesized that the difference in n/i selectivity was due to the formation of mono-phosphite ligated rhodium complexes with the polymers while P(OPh)$_3$ was able to form bis-phosphite ligated rhodium catalysts that are known to be more selective for the n-aldehyde (1).

Table. I. Hydroformylation of 1-octene with polymer-supported monophosphite Polymer 2.[a]

entry	cycle	octene/Rh	% conv.	S_{ald} (%)	n/i	S_{octane} (%)	$S_{2\text{-octene}}$ (%)
1	1	100	78	86	7.0	6	7
2	2[b]	100	98	77	7.2	12	11
3	3	100	86	41	7.9	40	19
4	1	200	44	98	5.2	1	1
5	2	200	41	93	5.3	2	4
6	3	200	35	92	5.3	3	4
7	4	200	22	92	5.3	3	4
8[c]	1	3868	97	70	1.4	3	0

[a] reactions run at 60 °C for 1h at 6 bar of 1:1 H$_2$/CO [b] 94.2% catalyst recovery
[c] unpurified 1-octene

Next we turned our attention to the issue of catalyst recycle. We began by testing the polymer 2-modified catalyst in the hydroformylation of 1-octene. After each reaction, the catalyst was precipitated by the addition of methanol and the polymer-supported catalyst was isolated by filtration either under air (entries 1-3, Table I) or under an inert Ar atmosphere (entries 4-7, Table I). It is noteworthy that the n/i ratios in each of these reactions is higher than that determined in our initial reaction screen (4.3). This highlights the difficulty in producing identical batches of supported ligands, thus each batch of polymer 2 has somewhat unique characteristics. What is apparent from the reactions where recycle is conducted under air is that exposure to air produces a supported catalyst that is better at hydrogenation than hydroformylation. Thus, we turned our attention to performing the catalyst recycle in an inert atmosphere. Doing so indeed led to reproducible selectivites to aldehydes, thus the catalyst apparently remained intact upon, and after, recycle. However, the conversion of the reaction decreased with each recycle, indicating that catalyst was likely leaching from the polymer support during filtration. This was further evidenced when we performed the hydroformylation under more realistic conditions of high 1-octene/catalyst ratio (entry 8). Doing so led to nearly complete conversion of the 1-octene, however the selectivity to aldehydes and the low n/i ratio were nearly identical to those produce from the *unmodified* rhodium catalyst. Thus, we believe that as the reaction progressed, the catalyst simply leached from the polymer support.

In addition to the poor recyclability of the polymer-supported catalysts, we noted that precipitation of the polymer was difficult. Often precipitation would produce a free-flowing powder, however sometimes a polymer gel was produced that was difficult to separate. The difficulty of precipitating the linear polymer supported catalysts was also noted by van Leeuwen, who addressed the problem by grafting the polymers to an insoluble silica surface. While the recovery was facilitated by grafting, the production of a heterogeneous catalyst leads to the problems outlined in the introduction. We hypothesized than many of these problems could be rectified by (A) utilizing an inherently bidentate phosphite to create a stable catalyst, limit leaching, and ensure consistent catalyst speciation and (B) utilizing a cross-linked polystyrene support that will be more easily separated. Finally, we anticipated that the solubility of such a catalyst would be tunable based on the CXL cosolvent and/or pressure of CO_2.

While there are many potential bidentate phosphite ligands to support on a polymer, we chose to investigate the production of a polymer-supported derivative of BiPhePhos. BiPhePhos is known to modify rhodium to produce catalysts that have high activity and selectivity for n-aldehydes (19). Moreover the symmetric nature of the ligand is amenable to production of a derivative that will be an efficient cross-linker for polystyrene.

Synthesis of the supported ligand began by performing the radical co-polymerization of styrene with our biaryl styrene cross-linker under the conditions of van Leeuwen (17). Polymers where the ratio of styrene to cross-linker used was 10:1 (PBB10) and 20:1 (PBB20) were produced (Scheme 6). The reactive alcohol groups on the resulting polymer were demasked which then allowed addition of the phosphite to the polymer. Once again, ^{1}H NMR analysis of the polymers showed that the ligand incorporations in the polymers were 0.34 mmol/g for the 10:1 polymer and 0.12 mmol/g for the 20:1 polymer.

The optimal reaction conditions were briefly explored using PBB20 in toluene solvent. First, it was noted that the solublization of the polymer supported catalyst was incomplete when only small amounts of toluene solvent were used. The resulting polymer gel catalyst was still active; however, the selectivity to aldehydes was relatively low under these conditions (65%), and hydrogenation (2.9%) and isomerization (32%) were significant. However, upon proper dissolution of the catalyst, the selectivities to aldehydes were very high

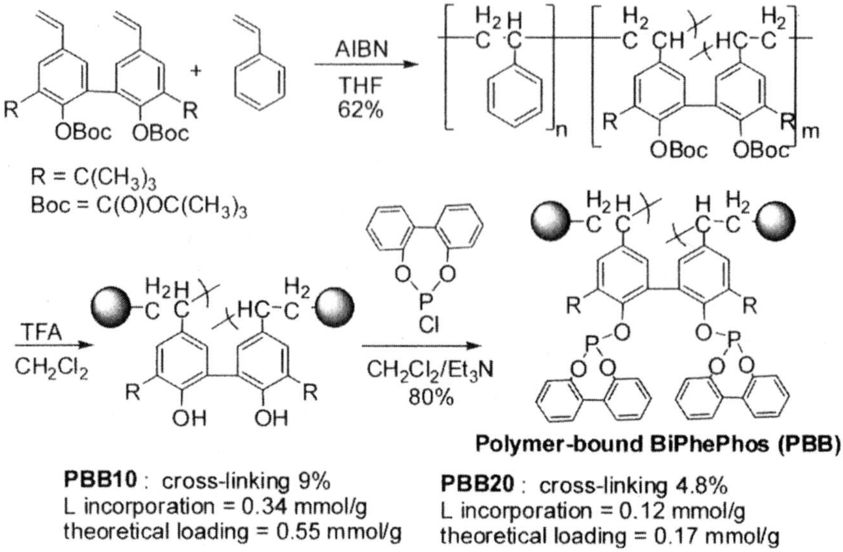
AIBN
THF
62%
R = C(CH3)3
Boc = C(O)OC(CH3)3
OBoc OBoc
TFA
CH2Cl2
OH OH
CH2Cl2/Et3N
80%
Polymer-bound BiPhePhos (PBB)
PBB10 : cross-linking 9%
L incorporation = 0.34 mmol/g
theoretical loading = 0.55 mmol/g
PBB20 : cross-linking 4.8%
L incorporation = 0.12 mmol/g
theoretical loading = 0.17 mmol/g

Scheme 6

Table II. Effect of solvent volume and temperature on hydroformylation of 1-octene.

entry	toluene (mL)	T (°C)	TOF (h^{-1})	S_{ald}	S_{octane}	*n/i*
1[b]	2	60	-	65.0	2.9	4.4
2[b]	6	60	-	86.5	1.6	2.5
3[b]	11	60	-	91.7	0.4	4.3
4[b]	20	60	-	87.3	0.4	4.7
5[c]	11	40	29	90.0	1.7	4.7
6[c]	11	60	203	89.1	0.9	6.1
7[c]	11	80	750	60.2	1.3	6.9

[a] $Rh(CO)_2(acac)$ (0.03 mmol), PBB20 (0.06 mmol), syngas (1:1; 6 bar).

[b] 1-octene/Rh = 100; [c] 1-octene:Rh = 1000.

(entries 2-4, Table II). Next, we determined the activity of the catalysts at various temperatures at a constant solvent volume. The results show that at 40 °C, the catalyst is active, but not highly so. An increase in temperature to 60 °C leads to a much higher TOF and the catalyst maintains its high selectivity toward aldehydes. Further increase in temperature to 80 °C, leads to a very active catalyst, however the selectivity to the desired aldehydes decreased significantly; isomerization to less active internal alkenes was responsible for the poorer selectivity.

With the proper conditions in hand, we then briefly investigated the recyclablility of the polymer-supported catalysts (Table III). In each case, we performed the hydroformylation, then precipitated the polymer by addition of an equal volume of methanol. The resulting polymer was isolated by filtration under an inert atmosphere, and the resulting polymer was charged into the reaction vessel to perform the next hydroformylation. Completely drying the polymer (PBB20) under vaccum prior to use allowed us to determine a 94% recovery of the catalyst. While the loss of 6% of the rhodium is not acceptable, the recovery was performed by crude filtration and higher recoveries are expected when more sophisticated methods of catalyst isolation are utilized. As can be seen from the data in Table III, the reproducibility of reaction activities and selectivities from run to run is much higher than for the linear monophosphite polymers. In each run, the recycled rhodium catalyst provides linear aldehyde quite selectively and with moderate turnover with the PBB10 modified catalyst. The selectivity toward aldehydes and octanes is 85-90% and 1-2%, respectively. The mass balance of converted 1-octene is made up by internal octenes (2-octene, 3-octene, and 4-octene). With that in mind we wanted to quantify the reactivity of our catalyst toward internal octenes. Thus, 2-octene was used as a substrate instead of 1-octene during run 5 (entry 5). The reactivity of 2-octene is an order of magnitude less than that of the terminal olefin 1-octene. Moreover, reaction of the 2-octene proceeds to give primarily alkane and branched aldehyde. Thus, any production of 2-octene by isomerization of 1-octene will also lead to the production of unwanted alkane and internal olefin byproducts. This is in contrast to several other systems, where reaction of 2-octene proceeds via rapid isomerization to 1-octene, thus it does not affect product selectivities. Finally, the data allow us to compare the hydroformylation of 1-octene with ligand PBB10 vs. PBB20. First, the selectivity for aldehydes with PBB10 is significantly higher than for PBB20. However, the activity of the PBB20 ligand is higher than that for PBB10. Both ligands provide similar selectivities for n-aldehydes. If one examines these data just in terms of production of the desired n-aldehyde, it can be seen that PBB10 provides higher selectivities to the n-aldehyde (74.5-75.5% vs. 53.6-58.2% for PBB20). However, PBB20 allows the production of n-nonanal at a higher rate ($TOF_n \sim 75$ h^{-1} for PBB10 vs 360 h^{-1} for PBB20).

Table III. Recyclability of soluble polymer-supported rhodium.

entry	L	run	TOF (h^{-1})	S_{ald} (%)	S_{octane} (%)	n/i
1[b]	PBB10	1	80	85.5	2.1	6.8
2[b]	PBB10	2	114	87.7	1.0	6.0
3[b]	PBB10	3	94	89.9	1.2	5.7
4[b]	PBB10	4	103	88.6	0.9	6.0
5[b]	PBB10	5	7	42.2	34.8	0.2
6[c]	PBB20	1	444	67.4	0.9	6.3
7[c]	PBB20	2	>974	61.6	2.5	6.7
8[c]	PBB20	3	672	63.5	2.7	6.2
9[c]	PBB20	4	517	64.8	1.1	6.0

[a] Rh(CO)$_2$(acac) (0.03 mmol), 1-octene (30 mmol); PBB (0.06 mmol), syngas (1:1; 6 bar), 60 °C. [b] toluene = 11 mL [c] toluene = 6 mL.

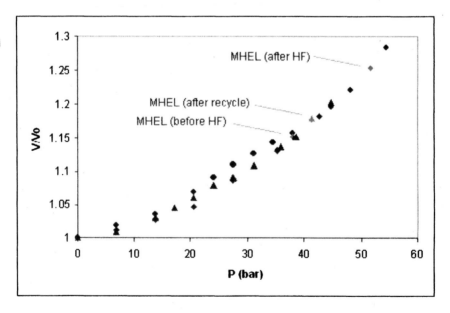

Figure 3. Effect of CO$_2$ expansion on catalyst solubility.

With the proper ligands in hand, we set out to examine their use in CXLs. Prior to their use, it was necessary to determine their solubility in CO_2-expanded toluene. A Jerguson® view cell was used to determine both the volume expansion and the timing of precipitation (10). The point at which the solution becomes cloudy, indicating catalyst precipitation, is termed the maximum homogeneous expansion limit (MHEL). Data were collected for the PBB20 (0.06 mmol) in 5 mL toluene at three points: (A) before hydroformylation, (B) after the hydroformylation while the catalyst resides in the product mixture, and (C) after recycle. The data are shown in figure 3. While the polymer precipitated at ca. 37 bar CO_2 before hydroformylation, the catalyst did not precipitate until 50 bars CO_2 pressure in the product mixture of hydroformylation. Thus, the catalyst becomes more soluble in the reaction media as the olefin is transformed into aldehydes. The solubility of the recycled catalyst is slightly higher than that of the orginal polymer; this may be due to reaction of the olefinic groups that terminate the polymer.

Finally, we performed the hydroformylation of 1-octene in CO_2-expanded toluene. As can be seen from the data (entries 2-5) pressurization with CO_2 has a beneficial effect on both the activity and selectivity of the catalyst up to a pressure of 21 bars (Table IV). When the reaction was run at 35 bar CO_2 pressure, which is at or near the MHEL, the catalyst becomes less active and selective. We attribute this loss in activity to partial precipitation of the polymer-supported catalyst. In addition the catalyst has a higher selectivity to aldehydes than the small molecule analog (BiPhePhos); the major byproducts in each case are those of olefin isomerization. However, the polymer supported catalyst is not as active or regioselective as the BiPhePhos ligated catalyst. In addition, our catalyst hydroformylates the internal olefin 2-octene to produce mainly branched aldehyde, while BiPhePhos catalyzes the isomerization-hydroformylation of internal octenes to give n-aldehydes (20).

Table IV. Effect of CO_2 pressure on the hydroformylation of 1-octene using polymer supported catalyst.[a]

entry	L	CO_2 (bar)	L/Rh	TOF (h^{-1})	S_{ald} (%)	n/i
1	BiPhePhos	0	6	>975	26.2	55
2	PBB10	0	3	192	39.6	9.8
3	PBB10	6	3	357	34.7	10.8
4	PBB10	21	3	390	56.9	13.0
5	PBB10	35	3	273	32.6	11.6
6	PBB10	21	3	64	20.8	0.4
7	PBB20	21	3	399	47.8	8.6

[a] Rh(acac)(CO)$_2$ (0.0067 mmol), 1-octene (6.7 mmol), in toluene (2.5 mL) at 60 °C for 1h.

In conclusion, expansion of hydroformylation media with CO_2 is beneficial for both selectivity to aldehydes as well as to linear aldehydes. Supporting catalysts on soluble polymers that can be readily precipitated factilitates the recycle of expensive rhodium catalysts. Future work will focus on creating polymer supports that maintain the high activity of the small molecule analogs and produce aldehydes with high chemo- and regioselectivity. In addition, quantification of rhodium recovery using the polymer supports is ongoing.

Acknowledgements

Support from the KU Center for Environmentally Beneficial Catalysis under the National Science Foundation Engineering Research Center Grant EEC-0310689 is gratefully acknowledged.

References

1. van Leeuwen, P. W. N. M. In *Rhodium Catalyzed Hydroformylation*. Claver, C; Ed.; Catal. Met. Complexes, Kluwer, Dordrecht, Neth. 2000.
2. Frohning, C. D.; Kohlpaintner, C. W. In *Carbon Monoxide and Synthesis Gas Chemistry: Hydroformylation*; Cornils, B. Herrmann, W. A. Eds.; Applied Homogeneous Catalysis with Organometallic Compounds, Wiley-VCH, Weinheim, Germany, 2002.
3. Kohlpaintner, C. W.; Fischer, R. W.; Cornils, B. *Appl. Catal., A* **2001**, *221*, 219.
4. *Catalyst separation, recovery and recycling, Chemistry and Process Design*; Cole-Hamilton, D. J.; Tooze, R. P. Eds.; Springer, Dordrecht, 2006.
5. Banister, J. A.; Harrison, G. E. U.S. Patent 6,946,580, April, 30, 2004.
6. Tijani, J.; El Ali, B. *Appl. Catal., A* **2006**, *303*, 158.
7. P. B. Webb, T. E. Kunene, D. J. Cole-Hamilton, *Green Chem.* **2005**, *7*, 373.
8. Fang, J.; Jin, H.; Ruddy, T.; Pennybaker, K.; Fahey, D.; Subramaniam, B. *Ind. Eng. Chem. Res.* **2007**, *46*, 8687.
9. Jin, H.; Subramaniam, B. *Chem. Eng. Sci.* **2004**, *59*, 4887.
10. Jin, H.; Subramaniam, B.; Ghosh, A.; Tunge, J. *AIChE J.* **2006**, *52*, 2575.
11. Jessop, P. G. and Subramaniam, B. *Chem. Rev.* **2007**, *107*, 2666.
12. Frisch, A.C.; Webb, P. B.; Zhao, G.; Muldoon, M. J.; Pogorzelec, P. J.; Cole-Hamilton, D.J. *Dalton Trans.* **2007**, 5531.
13. El Ali, B.; Tijani, J.; Fettouhi, M. *J. Mol. Catal. A.* **2006**, *250*, 153.
14. Reynhardt, J. P. K.; Yang, Y.; Sayari, A.; Alper, H. *Adv. Syn. Catal.* **2005**, *347*, 1379.
15. Riisager, A.; Eriksen, K. M.; Wasserscheid, P.; Fehrmann, R. *Catal. Lett.* **2003**, *90*, 149.

16. Chen, R. B.; Raymond P. J.; Kamer, P. C. J.; Van Leeuwen, P. W. N. M.; Reek, J. N. H. *J. Am. Chem. Soc.* **2004**, *126*, 14557.
17. van Leeuwen, P. W. N. M.; Sandee, A. J.; Reek, J. N. H.; Kamer, P. C. J. *J. Mol. Catal. A.* **2002**, *182*, 107.
18. van Leeuwen, P. W. N. M.; Jongsma, T.; Challa, G. *Macromol. Symp.* **1994**, *80*, 241.
19. Jongsma, T.; Fossen, M.; Challa, G.; van Leeuwen, P. W. N. M. *J. Mol. Catal.* **1993**, *83*, 17.
20. Vogl, C.; Paetzold, E.; Fischer, C.; Kragl, U. *J. Mol. Catal. A.* **2005**, 232, 41.

Chapter 11

Hydrogenation in Biphasic Ionic Liquid–Carbon Dioxide Systems

Azita Ahosseini, Wei Ren, and Aaron M. Scurto[*]

Department of Chemical and Petroleum Engineering and NSF–ERC Center
for Environmentally Beneficial Catalysis, University of Kansas,
Lawrence, KS 66045
[*]Corresponding author: phone: +1 785–864–4947; fax: +1 785–864–4947;
email: ascurto@ku.edu

A biphasic ionic liquid (IL) and compressed carbon dioxide
system has a number of advantages for efficient homogeneous
catalysis. The hydrogenation of 1-octene catalyzed by
rhodium-triphenylphosphine was used as a model reaction to
determine the effects of CO_2 pressure in a biphasic ionic
liquid/CO_2 system with 1-hexyl-3-methyl-imidazolium bis-
(trifluormethylsulfonyl)imide ([HMIm][Tf$_2$N]). For reactions
that were limited by mass transfer, the presence of CO_2 tended
to increase the apparent reaction rate. However in well
agitated systems and at constant moles of 1-octene, increased
pressure of CO_2 decreased the apparent reaction rate. Detailed
phase equilibrium studies were conducted to determine
volume expansion of the IL phase with CO_2 and the phase
behavior and mixture critical points between the reactant,
product and CO_2. The volume expansion decreases the
concentration (molarity) of the reactant. In addition, the
mixture critical points indicate that at higher pressures, the
reactant can partition away from the IL phase. Proper
understanding of the phase equilibria is needed to engineer
biphasic IL/CO_2 reaction systems for efficient processes.

Introduction

Homogeneous catalysis with organometallic complexes can efficiently perform a variety of chemical transformations with high chemo-, regio- and enantio-selectivity (1, 2). In most cases, the economic feasibility of a catalytic system is determined primarily by the ability to separate and recycle the often costly catalyst. Multiphase systems are constructed where one phase immobilizes or sequesters the catalyst and the other phase acts as a mobile phase to deliver reactants and to remove products. An idealized biphasic system would have the following traits as illustrated in Figure 1: complete immiscibility (no cross contamination); no catalyst partitioning; reactants partition into the catalytic phase; and products partition out of the catalytic phase. One of the largest examples of multiphase homogeneous catalysis is the Ruhrchemie/Rhône-Poulenc (2) process for short-chain olefin hydroformylation. This process uses water to sequester and recycle a rhodium catalyst with modified triphenylphosphine ligands. However, its application is limited by the solubility of the olefin, which makes it practical for only the short chain terminal alkenes ($<C_4$). Many multiphase methods may suffer from thermodynamic (solubility) issues, mass transfer limitations, catalyst separation, and cross-contamination problems. Innovative and generic methodologies for catalyst immobilization for continuous reactions are fundamental for process development. Any new process should also be based upon the principles of modern "green"/sustainable chemistry and engineering (3). A biphasic ionic liquid and compressed carbon dioxide system may solve these important issues.

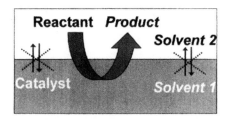

Figure 1. Idealized multiphase catalysis system with no solvent miscibility and complete catalyst immobilization.

This contribution will illustrate the necessary thermodynamic, kinetic and mass transport properties needed for reaction engineering a biphasic ionic liquid/CO$_2$ system. The hydrogenation of 1-octene with rhodium and a simple triphenylphosphine ligand was chosen as a model reaction system for illustration.

Background

Compressed, dense-phase or supercritical (SCF) carbon dioxide provides environmentally benign characteristics with tunable physico-chemical properties (4). CO_2 can be made miscible with liquids and, importantly, "permanent" and reaction gases (viz. H_2, CO, O_2, etc.), thus eliminating phase boundaries and mass transfer limitations. However for homogeneous catalysis, often elaborate ligand systems are required to render enough catalyst soluble in the supercritical CO_2 for high-throughput production (5, 6).

Ionic liquids (ILs) are organic salts that are liquid at or near room temperature ($T_m < 100°C$). There are a myriad of cation/anion combinations that can yield an organic salt. **Error! Reference source not found.** illustrates a few of the common classes of ionic liquids: imidazolium, pyridinium, quaternary ammonium, and phosphonium. Ionic liquids can be molecularly engineered for specific physico-chemical properties through various "R-"groups and cation/anion selection, *e.g.* viscosity, solubility properties, density, acidity/basicity, co-ordination properties, stereochemistry, etc. It has been estimated that $\sim 10^{14}$ unique cation/anion combinations are possible (7). Current toxicology studies of ILs (8-11) illustrate that ILs can have low to high toxicity depending on the cation, anion, and especially the organism studied, but most ILs have no measurable vapor pressure, thus eliminating air pollution. Negligible vapor pressure also results in highly-elevated flash points, usually by the flammability of the decomposition products at temperatures greater than 300°C. These properties have led researchers to claim ionic liquids as potential "green" solvents. Novel ILs and processes are continually being developed for extractions, reactions (8, 9), and material processing (10-14).

Upon closer inspection, ionic liquids can present a number of challenges for their use in catalysis. Most ionic liquids are more viscous than their organic-solvent counterparts (15). This leads to lower diffusivity and lower mass transfer rates (lower reaction rates, extraction efficiency, etc.). ILs often have a lower solubility of reaction gases (16) (H_2, CO, O_2, etc.) which, with the lower diffusivity, can hinder reaction rates. Separating substances from ionic liquids can often be difficult, as common separation and fractionation techniques, such as distillation, are no longer possible for the non-volatile IL solvent. While, the theoretical number of ionic liquids is astonishing (e.g. $\sim 10^{14}$) (7), the vast majority are *actually* <u>solids</u> with melting points, T_m, above 100°C.

Biphasic Ionic Liquids/ CO_2 Systems

However, coupling ionic liquids with a CO_2 solvent may overcome many of these challenges. Biphasic IL/CO_2 systems are a unique sub-set of gas or CO_2 expanded liquids (GXL or CXL); see Chapter 1. While, the solubility of CO_2 in

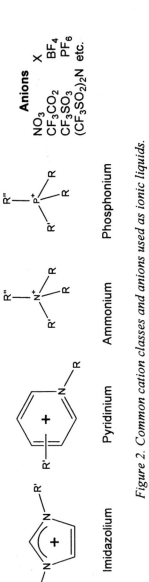

Figure 2. Common cation classes and anions used as ionic liquids.

ILs is high, the volume expands relatively little compared to CO_2 with most organic solvents (17). A biphasic system exists even at hyperbaric conditions (see below), whereas organic solvents become miscible/critical with high pressure CO_2. In addition, the polarity of ILs changes very little with even high concentrations of CO_2 (18, 19).

CO_2 can dramatically decrease the viscosity of ILs to that of common solvents, which increases the diffusivity of compounds in IL. In concurrent studies in our laboratory, the viscosity of the IL, 1-hexyl-3-methyl-imidazolium bis(trifyl)imide ([HMIm][Tf$_2$N]), decreases approximately 80+% with up to 60 bar of CO_2 at 40°C. This decrease in viscosity has been used to predict the diffusivity of the ionic liquid with approximately a 3 fold improvement over the same pressure range. Brennecke and coworkers found that presence of CO_2 can even increase the solubility of reaction gases into the IL-phase (20, 21). Scurto *et al.* (22, 23) have demonstrated the ability to separate ILs from organics and water using CO_2 pressure which induces immiscibility for extraction. Recently, Scurto and Leitner (24, 25) have demonstrated that CO_2 can dramatically decrease the melting point of many ionic solids, even inducing melting over 100°C below their normal melting points. This new technique with CO_2 now allows much more of the estimated 10^{14} ILs to be usable in a process as they have shown for a variety of catalytic reactions.

A biphasic ionic liquid/CO_2 system draws on the advantages of each of the respective technologies and helps overcome their challenges. Ionic liquids have

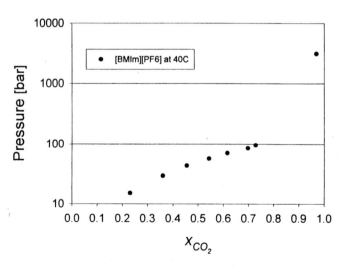

Figure 3. Phase behavior of the IL, [BMIm][PF$_6$] with CO$_2$ at 40 °C to ultra-high pressures from Blanchard et al. (17).

unique phase behavior with CO_2 as illustrated in Figure 3. CO_2 is very soluble in the ionic liquid, but the ionic liquid is immeasurably insoluble in the CO_2 phase and does not become miscible even at hyperbaric pressures, e.g. 3.1 kilobar (17). This is unlike CO_2 with most organic solvents, which become miscible (critical; one phase) at moderate pressures (100-200bar). However, CO_2 expanded liquids (CXLs) (5, 26-32), as pioneered by the groups of Subramaniam, Jessop, Eckert, Foster and others, have been used as a tunable reaction media (see Chapter 1). Few biphasic systems, except for those involving solids, can boast such lack of cross-contamination as with ILs and CO_2. Solid supported metal complexes have their own inherent limitations which include: poorer activity and selectivity than their homogeneous counterparts, deactivation, and the need to perform complete shutdown of reactors to replace the catalyst/support.

The combination of compressed CO_2 and ionic liquids offers a number of benefits for catalyzed reactions. Here, the ionic liquid phase sequesters the organometallic catalyst, and the CO_2 phase becomes the mobile phase for reactants and products. The role of CO_2 and its effect on the reaction is often conflicting among various reports especially in terms of reaction rate/activity.

Literature Survey of Hydrogenation in Biphasic IL/CO$_2$ Systems

One of the first examples of an IL/CO_2 biphasic system was olefin hydrogenation by Liu *et al.* (33). They compared the results in a biphasic IL/CO_2 and IL/hexane system and found little difference in the reaction rate with pressurized CO_2 or hexane. Jessop, Eckert and co-workers performed asymmetric Ru-catalyzed hydrogenations in an ionic liquid/CO_2 biphasic system. Substrates were chosen whose enantio-selectivity (%ee) increased or decreased depending on high or low H_2 concentration (34). They found a large increase in %*ee* with CO_2 pressure for substrates optimal under high H_2-concentration, which they attributed to an increase in H_2 solubility induced by CO_2. However, only a slight decrease in %ee was observed for the substrates optimal under dilute H_2-conditions. Leitner and coworkers (35) demonstrated that the anion of the IL greatly influences the selectivity for iridium catalyzed hydrogenation of aromatic imines. They also observed an increase in reaction rate with the presence of CO_2 compared with just H_2 pressure. The activity increased with CO_2 pressure and improved the stability of the catalyst as shown through recycling studies. From high-pressure NMR studies, the solubility of H_2 in the IL (as indicated by the size of the H_2 peak) increases with increasing CO_2 pressure while maintaining constant H_2 loading. Scurto and Leitner (24) illustrated the Rh-catalyzed hydrogenation of vinyl-naphthalene in a biphasic IL/CO_2 system, where CO_2 was used to induce melting of a quaternary ammonium salt over 100°C below its normal melting point.

Experimental

Safety

The experiments described here were performed under elevated pressures and should be handled with care. All equipment should have the proper pressure ratings and standard operating procedures as determined by trained professionals.

Reaction Catalysis

The biphasic IL/CO_2 hydrogenation reactions were carried out in a 10-cm^3 316 stainless steel high-pressure autoclave reactor. For the reaction experiments, catalyst (acetylacetonatodicarbonylrhodium, $Rh(CO)_2Acac$), and ligand (triphenylphosphine, TPP) were dissolved in dichloromethane. Dichloromethane solutions were used to obtain dilutions to achieve the small desired catalyst loading to high precision. The solution was added to the ionic liquid inside the reactor and then slowly heated under vacuum to remove the dichloromethane without entraining the catalyst or TPP. Argon was then added as an inert blanket and the catalyst solution was stirred for 30 minutes to allow complexation of the rhodium precursor with the ligand; this time allowed the complete complexation of TPP with the Rh precursor as determined by ^{31}P NMR studies. The reactant (1-octene) was added to the above prepared catalyst mixture and loaded into the reactor. The reaction mixture was heated to 70°C, and then the desired amount of H_2 and CO_2 pressures were injected to the system, respectively. The gas phase is vented in hexane during depressurizing to trap the reactants and products, as significant quantities are soluble in the CO_2 phase at the higher pressures. The liquid phase is extracted by ~3mL of hexane three times for quantitative extraction. Both solutions are combined and analyzed by GC.

Phase Equilibrium

The phase equilibrium data were measured by a static vapor-liquid equilibrium apparatus. The details of the apparatus setup are described by Ren and Scurto (36). This apparatus consists of a high-pressure equilibrium cell, high-pressure and precision syringe pump filled with CO_2, a water bath and accessories to measure the temperature and pressure. The solubility, density of the liquid solution, molar volume, volume expansion and molarity are computed from the apparatus. These calculations are based on the mass balance by determining the amount of CO_2 delivered from the pump, the moles/mass of CO_2 in the headspace above the liquid and in the tubing/lines to the equilibrium cell. This apparatus yields high resolution of solubility data (often better than ±0.001 mole fraction), pressure with accuracy of ±0.2 bar, temperature of ±0.01°C, density up to ±0.4 %, and volume expansion to ±0.05%. In addition, there is no need for analytical methods, nor calibration for each specific system of interest.

Synthesis of Ionic Liquid

1-Hexyl-3-methylimidazolium bis(trifluoromethylsulfonyl)imide ([HMIm][Tf₂N]), was prepared by anion exchange from the corresponding bromide salt of the imidazolium cation ([HMIm][Br]) with lithium bis(trifluoromethylsulfonyl)imide (Li[Tf₂N]) in deionized water similar to other literature techniques (37, 38). The bromide salt of the imidazolium cation was prepared from a quaternization reaction of 1-methylimidazole with a slight excess amount of the corresponding 1-bromohexane in acetonitrile at 40°C under an argon atmosphere with stirring for three days. *Caution: This reaction can be highly exothermic and adequate solvent volumes and/or cooling must be provided during the reaction.* The bromide salt of 1-Hexyl-3-methyl-imidazolium was purified with activated charcoal (10%) stirring for 24 hours. Acetonitrile was added to reduce the viscosity of the ionic liquid, and the mixture was filtered. The mixture was then passed through a plug of celite (depth=7 cm, diameter=3cm) and through a short column (height=20 cm, diameter=2.5 cm) of acidic alumina. The solvent was removed on a rotary evaporator under reduced pressure at 40°C and then connected to a high vacuum (<10^{-4} torr) at 50°C for at least 48 hours. [HMIm][Tf₂N] was synthesized by anion exchange between the [HMIm][Br] and Li[Tf₂N]. The denser hydrophobic IL phase was decanted and washed with water for 6-8 times. The IL was then dried under vacuum. ^{1}H NMR chemical shifts (relative to TMS internal standard) and coupling constants J/Hz: δ=8.65 (s, 1H), 7.39 (d, 2H, J=4.19), 4.17 (q, 2H, J=7.4), 3.93(s, 3H), 1.87(m, 4H), 1.32(m, 6H) 0.87(t, 3H, J=6.53). Elemental analysis calculated for $C_{12}H_{19}N_3F_6S_2O_4$: C, 32.2; H, 4.28; N, 9.39; S, 14.33. Found: C, 32.21; H, 4.27; N, 9.25; S, 14.19, the water content is less than 100 ppm and the Br content is below 8 ppm.

Analysis

Elemental analysis was performed by Desert Analytics Transwest Geochem. ^{1}H-NMR spectra were recorded on a Bruker 400 NMR Spectrometer using TMS as a reference for ^{1}H chemical shifts. Gas Chromatography analyses of the product mixtures were carried out on a Varian CP-3800 Series gas chromatograph equipped with a Flame Ionization Detector (GC-FID for quantitative analyses) and a chrompack capillary column CP-Sil 5 CB (25m, .32mm, 1.2 μm). The water content was determined by a Mettler Toledo DL32 Karl Fisher Coulometer and the Br content was measured by a Cole Parmer Bromide Electrode (27502-05) read with an Oakton Ion 510 series meter.

Materials

Coleman Instrument grade CO_2, argon (extra dry grade 99.998%), H_2 (high purity grade 99.995%) were obtained from Airgas, Inc. 1-Octene (CAS 111-66-0) 98%, acetylacetonato dicarbonylrhodium (CAS 14874-82-9) 98%, triphenylphosphine (CAS 603-35-0) 99%, hexane (CAS 110-54-3) HPLC grade≥ 95% , 1-methylimidazole (CAS 616-47-7) 99+%, lithium bis(trifluoromethylsulfonyl)imide (CAS 90076-65-6) 99.95%, and acetonitrile (CAS 75-05-8) ≥99.9% were purchased from Sigma-Aldrich. 1-Bromohexane (CAS 111-25-1) 99+%, was obtained from Acros. The $Rh(CO)_2Acac$ was placed in a Schlenk tube and stored under dry argon. 1-Methyl-imidazole and 1-bromohexane were vacuum-distilled and used immediately.

Results and Discussion

The hydrogenation of 1-octene was chosen as a model reaction in the biphasic ionic liquid [HMIm][Tf_2N] and CO_2 system. Rhodium with triphenylphosphine ligands was chosen as a simple catalyst system. Focus was placed on the effect of mass transport and phase equilibria on the reaction rate.

Mass Transfer Effects

CO_2 pressure was found to enhance significantly the reaction rate as measured in turn-over frequency (TOF) of the hydrogenation reaction operated in a mass-transfer limited regime. As previously discussed, the presence of CO_2 dramatically lowers the viscosity of the ionic liquid and increases the diffusivity. The hydrogenation of 1-octene was performed without agitation and maintained with a small gas-liquid surface area. As shown in Figure 4, the TOF increased by 25+% with pressure of gaseous CO_2 to 60 bar (5 bar H_2) at a constant concentration (molarity) of 1-octene. This rapid increase came to an asymptotic maximum after approximately 50 bar of CO_2 pressure, most probably due to

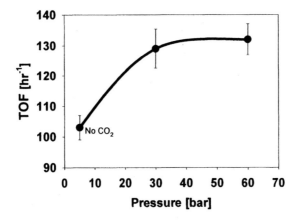

Figure 4. Reaction rate for the hydrogenation of 1-octene with Rh- TPP (1:25) in [HMIm][Tf₂N] with and without CO₂ without mechanical stirring, i.e. under initially mass-transfer limited conditions. Reaction conditions: 70 ℃, P_{H2}=5 Bar; Rh - TPP (1:25); 1-octene concentration = 156mM; 1 hour; line is smoothed data.

transition to the kinetically-controlled regime. Thus, CO_2 increases the apparent reaction rate by increasing intra- and inter-phase mass transport. This may explain why some researchers observed an increase in the reaction rate with CO_2 pressure (see above).

Effect of Pressure on Reaction Rates

From the literature, the effect of pressure on reaction rates in biphasic IL/CO_2 reaction systems is somewhat inconsistent: some report that CO_2 pressure increases the reaction rate; others say that it decreases the rate. Here, reactions were performed under well-mixed conditions and at various levels of pressure of CO_2 with 30 bar of hydrogen initially (>100% excess of H_2 loading over 1-octene) and Rh:TPP = 1:4 for three hours. As shown in Figure 5, the reaction rate seems to decrease with increasing CO_2 pressure. These results can be contrasted with the observed increase in reaction rate in the mass-transport controlled regime as discussed above. *What are the dominant phenomena with increases of CO_2 pressure? As reaction rates are direct functions of concentration, how does the presence of CO_2 affect concentrations of the reactants?*

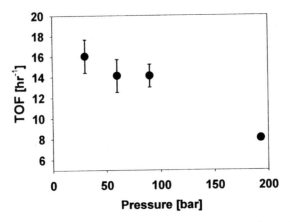

Figure 5. Reaction rate of the hydrogenation of 1-octene with total pressure by adding CO_2. Reaction conditions: P_{H2} = 30 bar, 70°C; 3 hours catalyzed by Rh-TPP (1:4)

Phase Equilibria

Volume Expansion of the Ionic Liquid

As discussed, CO_2 is highly soluble in ionic liquids. As CO_2 begins to dissolve into the IL phase, the total volume (and often molar volume) expands. However, this expansion is often much smaller than the expansion in organic liquids at a similar solubility of CO_2. For example, CO_2 can expand the volume of methanol to 200% at at 69.73 bar and 35°C, and 686.15% at 72.63 bar (39). The volume expansion of the [HMIm][Tf$_2$N] with CO_2 and with a mixture of CO_2 and hydrogen have been measured at 70°C. As seen in Figure 6, the IL phase expands by approximately 25% over the volume of the IL without CO_2 for a pressure range of 120 bar. With an initial pressure of approximately 6 bar of H_2, the volume expansion is less at a given pressure over that of just pure CO_2. The pressure where the volume expansion is equal with and without H_2 is, not just the initial 6 bar of H_2 initially loaded, but approximately 8-10 bar higher than the pure CO_2 case.

Phase Behavior of the Reactant/Product

The phase behavior of the reactants and products with CO_2 (without the IL) also yields valuable insight into the reaction behavior. Organic liquids, such as 1-octene and octane, can become miscible with CO_2 at conditions (temperature,

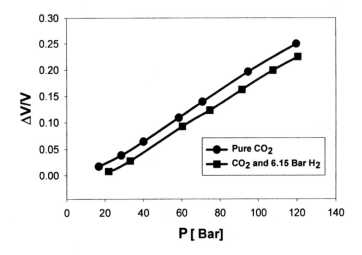

Figure 6. Volume expansion of [HMIm][Tf₂N] with CO₂ pressure with/without H₂ at 70 °C; lines are smoothed data.

Table I. Volume expansion[a] of [HMIm][Tf₂N] with CO_2 and CO_2 with H_2

Pure CO_2		CO_2 with initially 6.15 bar H_2	
P [bar]	$\Delta V/V^0$	P [bar]	$\Delta V/V^0$
16.86	0.0174	21.87	0.0078
28.44	0.0386	33.06	0.0280
39.97	0.0644	60.29	0.0928
58.44	0.1091	74.66	0.1230
70.8	0.1392	91.43	0.1625
94.65	0.1964	107.58	0.1992
119.78	0.2495	120.62	0.2247

[a] $\Delta V/V^0 = [V(P_{CO2}) - V(P_{CO2}=1 \text{ bar})]/ V(P_{CO2}=1 \text{ bar})$

pressure, and composition) above the bubble point (solubility of the CO_2 in the liquid phase) or at conditions beyond the mixture critical point. The mixture critical point is the highest pressure and composition at a given temperature in which a vapor and liquid phase can coexist. For liquid mixtures with two or more components, the mixture critical points can change with composition. Figure 7 and Table II illustrate the mixture critical pressure at 70°C as the proportion of 1-octene to octane is varied between pure 1-octene and pure octane (on a CO_2-free basis); above the line, a one phase mixture exists and below the line, 2-phase vapor-liquid equilibrium exists. This change in concentration can also be considered the conversion for the hydrogenation reaction. As seen in the figure at pure 1-octene (0% conversion), pure octane (100% conversion), or intermediate concentrations, the mixture critical pressure is very similar due to similar physical and critical point properties of 1-octene and octane. Thus, reaction conditions above the mixture critical point will remain in one phase regardless of the level of conversion. The figure also demonstrates that the increases in initial pressures of H_2 increase the mixture critical pressure beyond the pure CO_2 case. With an initial loading of 10 bar of H_2, the mixture critical pressure increases by approximately 12 bar over the pure CO_2 conditions.

Effects of Phase Equilibrium on the Catalytic Reaction Rate

The pressure of the mixture critical point, indicates a region where the solute/reactant, 1-octene, is normally completely soluble/miscible in the CO_2 phase. These data indicate that in the reaction mixture, the driving force for partitioning of the 1-octene from the catalytic IL-phase to the CO_2 phase increases as the CO_2 pressure increases. As 1-octene partitions away from the catalytic phase, the apparent reaction rate will decrease as the corresponding molarity of the reactant decreases. Thus, the more dramatic drop in reaction rate at the higher pressures can be ascribed to the decrease in solubility or concentration of the 1-octene reactant as seen in Figure 5. However, at lower pressures of CO_2, partitioning will not be the dominant factor.

The concentration of a reactant in the IL phase, at any pressure, is affected by the volume expansion. The data in Figure 6 indicate that the volumes of the reaction mixtures illustrated in Figure 5 will increase with the increasing pressure. While keeping mass or moles of the reactant in the initial IL constant, the addition of CO_2 and the accompanying volume expansion will *decrease* the molarity of the reactant. For instance, if the concentration of 1-octene in the IL is 1M, then at 90 bar, the volume increases by 15% and the concentration reduces to 0.87 M. The actual reduction to the reaction rate will be determined by the order of the kinetic rate expression. Schrock and Osborn (40) and Grubbs et al. (41) have determined that Rh-catalyzed hydrogenation of olefins was 1st

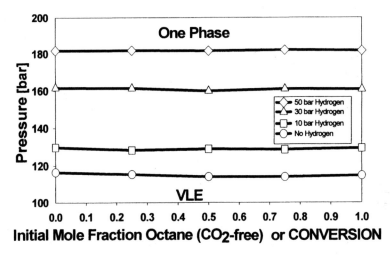

Figure 7. Phase behavior of 1-Octene and Octane and CO_2 as percent of initial amount of octane (CO_2-free basis); of can be read as phase behavior as a function of the conversion; line is smoothed data.

Table II. Mixture critical points of 1-octene, octane and CO_2 with H_2

Mole Fraction Octane (CO_2-free Basis)	Mixture Critical Points CO_2 with Initial Pressures of H_2			
	Pure CO_2	10 bar	30 bar	50 bar
0.00	116.45	129.79	161.62	181.74
0.25	115.29	128.27	161.36	181.76
0.50	114.03	128.90	159.90	181.55
0.75	113.95	128.56	160.98	182.00
1.00	114.69	129.28	160.77	181.65

order in the olefin. Thus at similar catalyst and H_2 concentrations and constant moles of 1-octene, a CO_2 pressure that increases the volume by 15% will reduce the concentration by 13% and, thus, the reaction rate by 13%.

Thus, the combination of volume expansion of the IL and the partitioning of the reactant will affect the reaction rate. With thorough characterization of the phase equilibrium, the actual concentrations of the reactants, including H_2, can be determined. The actual concentrations will allow intrinsic kinetic constants for a given rate mechanism to be determined. The effect of CO_2 concentration on the intrinsic kinetics is unknown. The kinetic constant can often be correlated to measures of "polarity" of the solvent such as Kamlet-Taft parameters (42). Brennecke and coworkers (43) have determined the Kamlet-Taft parameters for several ILs with CO_2 and have found that the polarity of the IL changes little with even large amounts (>50%mole) of CO_2 dissolved. Thus, one may assume that the intrinsic kinetics may change very little with the presence of CO_2. However, this proposed phenomenon is currently under investigation. Ultimately, both phase equilibrium and kinetics will allow reaction systems to be properly designed for reactant/catalyst loading, pressure, reactor configuration, etc. to achieve high throughput from a biphasic IL/CO_2 system.

Conclusion

The hydrogenation of 1-octene was used as a model reaction to determine the effects of CO_2 pressure in biphasic ionic liquid/CO_2 systems. Rhodium with triphenylphosphine was chosen as a simple catalyst system. For reactions that were limited by mass transfer, the presence of CO_2 tended to increase the apparent reaction rate. However, in well agitated conditions and at constant loading of 1-octene, increased pressure of CO_2 decreased the apparent reaction rate. Detailed phase equilibrium studies were conducted to determine the volume expansion of the IL phase with CO_2 and the phase behavior between the reactants/product and CO_2. Both volume expansion and the increased solvent power of CO_2 with pressure affect the concentration of the reactant which affects the apparent kinetics. A biphasic IL/CO_2 system represents a highly tunable and flexible platform for conducting homogeneously catalyzed reactions. Detailed phase equilibrium is needed to properly understand and engineer these reactions.

Acknowledgements

This work was supported by the NSF-ERC Center of Environmentally Benign Catalysis (CEBC EEC-0310689).

References

1. Cornils, B.; Herrmann, W. A., *Applied homogeneous catalysis with organometallic compounds.* Wiley-VCH: Weinheim, 1996.
2. Cornils, B.; Herrmann, W. A., *Aqueous phase organometallic catalysis.* Wiley-VCH: Weinheim, 1998.
3. Anastas, P.; Warner, J. C., *Green chemistry: Theory and practice.* Oxford University Press: New York, 1998.
4. Jessop, P. G.; Leitner, W., *Chemical synthesis using supercritical fluids.* Wiley-VCH: Weinheim, 1999.
5. Jessop, P. G., *J Supercrit Fluid* **2006**, *38*, 211-231.
6. Leitner, W., *Acc. Chem. Res.* **2002**, 35, 746-756.
7. Holbrey, J. D.; Seddon, K. R., *Clean Prod. Proc.* **1999**, *1*, 223-236.
8. Wasserscheid, P.; Keim, W., *Angew. Chemie Int. Ed.* **2000**, *39*, 3773-3789.
9. Welton, T., *Chem. Rev.* **1999**, *99*, 2071-2083.
10. Barisci, J. N.; Wallace, G. G.; MacFarlane, D. R.; Baughman, R. H., *Electrochem. Commun.* **2004**, *6*, 22-27.
11. Murugesan, S.; Linhardt, R. J., *Curr. Org. Synth.* **2005**, *2*, 437-451.
12. Zhao, D. C.; Xu, H. T.; Xu, P.; Liu, F. Q.; Gao, G., *Prog. Chem.* **2005**, *17*, 700-705.
13. Lu, W.; Fadeev, A. G.; Qi, B. H.; Mattes, B. R., *J. Electrochem. Soc.* **2004**, *151*, H33-H39.
14. Cao, J. M.; Fang, B. Q.; Wang, J.; Zheng, M. B.; Deng, S. G.; Ma, X. J., *Prog. Chem.* **2005**, *17*, 1028-1033.
15. Dzyuba, S. V.; Bartsch, R. A., *ChemPhysChem* **2002**, *3*, 161-166.
16. Anthony, J. L.; Maginn, E. J.; Brennecke, J. F., *J. Phys. Chem. B* **2002**, *106*, 7315-7320.
17. Blanchard, L.; Gu, Z.; Brennecke, J., *J. Phys. Chem. B* **2001**, *105*, 2437-2444.
18. Fredlake, C. P.; Muldoon, M. J.; Aki, S. N. V. K.; Welton, T.; Brennecke, J. F., *Phys Chem Chem Phys* **2004**, *6*, 3280-3285.
19. Lu, J.; Liotta, C. L.; Eckert, C. A., *J. Phys. Chem.* **2003**, *107*, 3995-4000.
20. Hert, D. G.; Anderson, J. L.; Aki, S. N. V. K.; Brennecke, J. F., *Chem. Comm.* **2005**, 2603-2605.
21. Lopez-Castillo, Z. K.; Aki, S. N. V. K.; Stadtherr, M. A.; Brennecke, J. F., *Ind. Eng. Chem. Res.* **2006**, 45.
22. Scurto, A. M.; Aki, S. N. V. K.; Brennecke, J. F., *J. Am. Chem. Soc.* **2002**, *124*, 10276-10277.
23. Scurto, A. M.; Aki, S. N. V. K.; Brennecke, J. F., *Chem Comm* **2003**, 572-573.
24. Scurto, A. M.; Leitner, W., *Chem. Comm.* **2006**, 3681-3683.
25. Scurto, A. M.; Newton, E.; Weikel, R. R.; Draucker, L.; Hallett, J.; Liotta, C. L.; Leitner, W.; Eckert, C. A., *Ind. Eng. Chem. Res.* **2008**, *47*, 493-501.

234

26. Chamblee, T. S.; Weikel, R. R.; Nolen, S. A.; Liotta, C. L.; Eckert, C. A., *Green Chem.* **2004**, *6*, 382-386.
27. Combes, G.; Coen, E.; Dehghani, F.; Foster, N., *J Supercrit Fluid* **2005**, *36*, 127-136.
28. Heldebrant, D. J.; Jessop, P. G., *J. Am. Chem. Soc.* **2003**, *125*, 5600-5601.
29. Jessop, P. G.; Subramaniam, B., *Chem. Rev.* **2007**, *107*, 2666-2694.
30. Jin, H.; Subramaniam, B., *Chem. Eng. Sci.* **2004**, *59*, 4887-4893.
31. Jin, H.; Subramaniam, B.; Ghosh, A.; Tunge, J., *AIChE J.* **2006**, *52*, 2575-2581.
32. Wei, M.; Musie, G. T.; Busch, D. H.; Subramaniam, B., *Green Chem.* **2004**, 6, 387-393.
33. Liu, F.; Abrams, M. B.; Baker, R. T.; Tumas, W., *Chem. Comm.* **2001**, 433-434.
34. Jessop, P. G.; Stanley, R.; Brown, R. A.; Eckert, C. A.; Liotta, C. L.; Ngo, T. T.; Pollet, P., *Green Chem.* **2003**, *5*, 123-128.
35. Solinas, M.; Pfaltz, A.; Cozzi, P. G.; Leitner, W., *J. Am. Chem. Soc.* **2004**, *126*, 16142-16147.
36. Ren, W.; Scurto, A. M., *Rev. Sci. Instrum.* **2007**, *78*, 125104-125107.
37. Bonhôte, P.; Dias, A. P.; Papageorgiou, N.; Kalyanasundaram, K.; Gratzel, M., *Inorg. Chem.* **1996**, *35*, 1168-1178.
38. Nockemann, P.; Binnemans, K.; Driesen, K., *Chem. Phys. Lett.* **2005**, *415*, 131-136.
39. Bezanehtak, K.; Combes, G. B.; Dehghani, F.; Foster, N. R.; Tomasko, D. L., *J. Chem. Eng. Data* **2002**, *47*, 161-168.
40. Schrock, R. R.; Osborn, J. A., *J. Am. Chem. Soc* **1976**, *98*, 2134-2143.
41. Grubbs, R. H.; Kroll, L. C.; Sweet, E. M., *J. Macromol. Sci. A* **1973**, *A7*, 1047-1063.
42. Kamlet, M. J.; Abboud, J. M.; Abraham, M. H.; Taft, R. W., *J. Org. Chem.* **1983**, 2877-2887.
43. Fredlake, C. P.; Muldoon, M. J.; Aki, S. N. V. K.; Welton, T.; Brennecke, J. F., *Phys. Chem. Chem. Phys.* **2004**, *6*, 3280 - 3285.

Chapter 12

Sulfur Trioxide Containing Caprolactamium Hydrogen Sulfate: An Expanded Ionic Liquid for Large-Scale Production of ε- Caprolactam

I. T. Horváth[1], V. Fábos[1], D. Lantos[1], A. Bodor[1], A.-M. Bálint[1], L. T. Mika[1], O. E. Sielcken[2], and A. D. Cuiper[2]

[1]Institute of Chemistry, Eötvös University, Pázmány Péter sétány 1/A, H–1117 Budapest, Hungary
[2]DSM Research, Geleen, The Netherlands

The Beckmann rearrangement of cyclohexanone oxime to ε-caprolactam in the presence of oleum proceeds in an ionic liquid, the caprolactamium hydrogen sulfate. We report its remarkable capability to keep the vapor pressure of dissolved sulfur trioxide below 10 kPa even at 140°C, allowing its safe use in large scale processes for a long-time.

Controlling the vapor pressure of hazardous chemicals (*1*) is perhaps the most important safety and environmental challenge during their use in storage, transportation, chemical reactions, and processes. While their replacement would be the best solution to mitigate risk, it is frequently difficult and sometimes not possible to find an alternative chemical with a similar performance profile. One of the possible approaches to lower the risks of toxic gases is the use of appropriate solvents, which lower the vapor pressure by participating in reversible chemical reactions. For example, the vapor pressure of hydrogen fluoride and pyridine mixtures is very low, but it is high enough to release HF to catalyze chemical reactions (*2*).

Ionic liquids (*3*) are among the solvents (*4*) which can be used by molecular designers (*5*) to fine tune the performance of chemical systems. In particular, they can provide molecular level control of the vapor pressure of the medium (*6*), a key issue of process safety and environmental protection (*7*). Ionic liquids are starting to leave academic labs and find their way into a wide variety of industrial applications (*8*). We have been investigating the mechanism of the Beckmann rearrangement of cyclohexanone oxime (**1**) to ε-caprolactam (**2**) in oleum as described in literature (*9*) and concluded that the caprolactam process is in fact the largest scale industrial technology that has been using an ionic liquid, the caprolactamium hydrogensulfate (**3**) (*10*), as the reaction medium for decades. We report here that the vapor pressure of 12% dissolved sulfur trioxide in **3** is below 10 kPa even at 140°C, allowing its safe use in large scale processes for decades.

The commercial processes of the Beckmann rearrangement of **1** to **2** are performed in the so-called *"rearrangement mixture"* (*11*) in the presence of oleum. As the reaction proceeds to complete conversion of **1** in 99.5% selectivity to **2**, an appropriate amount of the *rearrangement mixture* is removed and neutralized with ammonia to produce **2** and $(NH_4)_2SO_4$. The molar ratio of the oleum to **1**, calculated as $([H_2SO_4]+[SO_3])/[1]$, is generally higher than one because the dissolved sulfur trioxide significantly increases the rate of the Beckmann rearrangement (*12*). Since the reaction is very exothermic (*11*), the control of the vapor pressure of SO_3 has been a key safety issue.

The reaction medium, traditionally called *"rearrangement mixture"*, can be prepared by dissolving **2** in oleum. It should be noted that the addition of one equivalent of **2** to 100% sulfuric acid results in the formation of the *caprolactamium hydrogen sulfate (3)*, a colorless solid at room temperature, which becomes a viscous liquid around 60°C. The viscosity of **3** decreases by the addition of SO_3 and becomes easy flowing at 12% SO_3, even at room temperature. While the density of **3** decreases by increasing the temperature, its ion conductivity increases, as expected (*10*). In comparison to other ionic liquids (*13-19*), the density of **2**-based ionic liquids are similar, but their conductivity is noticeably lower. ^1H- and ^{13}C-NMR measurements have revealed that the carbonyl group of ε-caprolactam (**2**) is protonated to form a *caprolactamium hydrogensulfate (3)* as the main species:

It is important to emphasize that ionic liquids are considered as greener solvents because their vapor pressure is extremely low (*6*). The temperature-dependent vapor pressures of 2/100% H_2SO_4 (1/1), 2/100%H_2SO_4/SO_3 (3.3/4.7/1), and 100% H_2SO_4/SO_3(12/1) were measured (Figure 1.) (*10*).

The vapor pressure measurements were performed in a 25 mL Hasteloy-C Parr reactor connected to a Rosemount® Hasteloy-C-276 digital pressure gauge. The system was validated by measuring the vapor pressure of methanol at different temperatures and comparing the experimental values to literature data. Surprisingly, the vapor pressures of ε-caprolactam (**2**) based ionic liquids are much lower above 70°C than that of oleum, which explains why the *rearrangement mixture* has been used without any serious problems for decades. While the observed low vapor pressure of dissolved SO_3 could be the result of

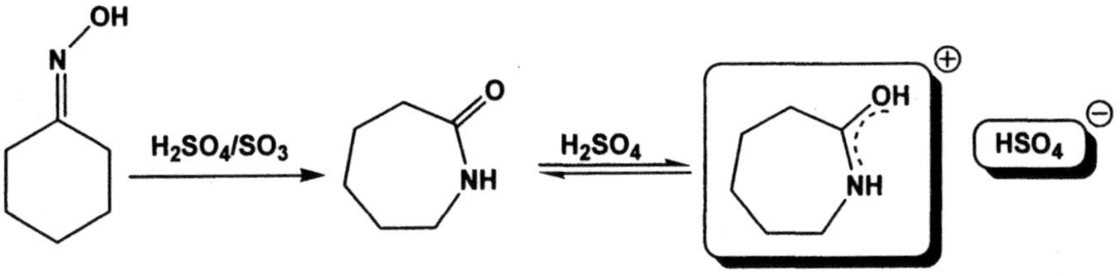

Scheme 1.

Scheme 2.

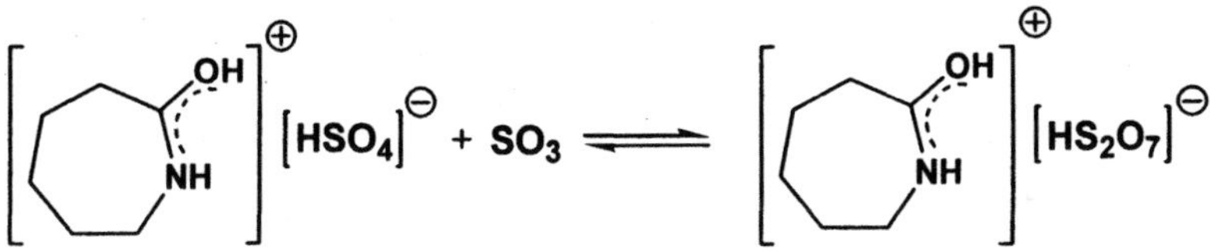

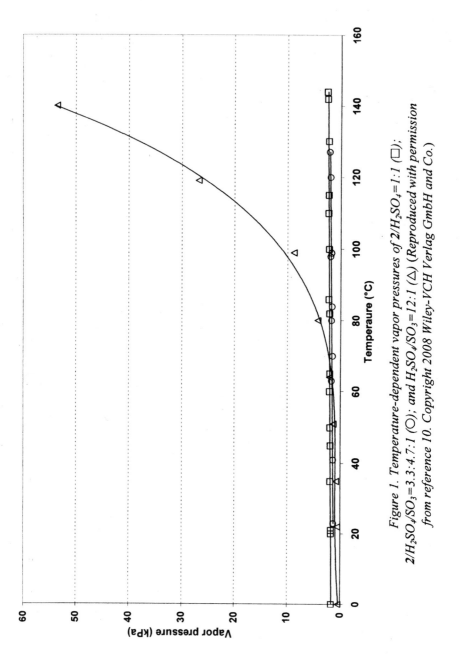

Figure 1. Temperature-dependent vapor pressures of 2/H₂SO₄=1:1 (□);
2/H₂SO₄/SO₃=3.3:4.7:1 (○); and H₂SO₄/SO₃=12:1 (△) (Reproduced with permission
from reference 10. Copyright 2008 Wiley-VCH Verlag GmbH and Co.)

strong interactions between **3** and sulfur trioxide, the formation of another anion such as [HS$_2$O$_7$]$^-$ in **4** (*20*) seems also plausible:

In conclusion, various mixtures of **2** and sulfuric acid or oleum must be considered as ionic liquids, which could hold SO$_3$ very strongly up to 140°C and therefore provide a safe environment for a very exothermic reaction.

References

1. Bender, H. F.; Eisenbarth, P.; *Hazardous Chemicals,* Wiley VCH, **2007**.
2. Olah,G. A. U.S. Patent 5,073,674, **1991**.
3. Welton, T. *Chem. Rev.* **1999**, *99*, 2071-2083.
4. Reichardt, C. *Org. Process Res. Dev.* **2007**, *111*, 105-113.
5. Horváth, I. T. *Acc. Chem. Res.* **2002**, *35*, 685.
6. Earle, M. J.; Esperanca, M. S. S.; Gilea, M. A.; Canongia Lopes, J. N.; Rebelo, L. P. N.; Magee, J. W.; Seddon, K. R.; Wildegren, J. A. *Nature* **2006**, *439*, 831-834.
7. Anastas, P. T.; Warner, J. C. *Green Chemistry:Theory and Practice*, Oxford University Press, Oxford, **1998**.
8. Short, P. L. *Chem. Eng. News* **2004**, *84*, 15.
9. Beckmann, E. *Ber. Dtsch. Chem. Ges.* **1886**, *89*, 988.
10. Fábos, V.; Lantos, D.; Bodor, A.; Bálint, A-M.; Mika, L. T.; Sielcken, O. E.; Cuiper, A.; Horváth, I. T. *Chem. Sus. Chem.* **2008**, *1*, 189-192.
11. Ritz, J.; Fuchs, H.; Kieczka, H.; Moran, W. C. "Caprolacatam" in *Ullmann's Encyclopedia of Industrial Chemistry*, Wiley VCH, Weinheim, Germany, **2005**.
12. Wichterle, O.; Roèek, J. *Collect Czech.Chem. Commun.* **1951**, *16*, 591-598.
13. Du, Z.; Li, Z. ; Guo, S. ; Zhang, J.; Zhu, L. ; Deng, Y. *J. Phys. Chem. B* **2005**, *109*, 19542.
14. Bonhote, P.; Dias, A. P.; Papageorgoiu, N.; Kalyanaundaram, K.; Gratzel, M. *Inorg, Chem.* **1996**, *35*, 1168-1178.
15. Earle, M. J.; Katdare, S. P. U.S 7,009,077, **2006**.
16. Jacquemin, J. ; Husson, P. ; Padua, A. A. H. ; Majer, V. *Green. Chem.* **2006**, *8*, 172-180.
17. Tokuda, H. ; Tsuzuki, S.; Md. Abu Bin Hasan Susan, Hayamizu, K.; Watanabe, M. *J. Phys. Chem. B* **2006**, *110*, 19593-19600.
18. Guo, S.; Du, Z.; Zhang, S.; Li, D.; Li, Z.; Deng, Y. *Green Chem.*, **2006**, *8*, 296.
19. Du, Z.; Li, Z.; Gu,Y.; Zhang, J.; Deng, Y. *Journal of Molecular Catalysis A: Chemical*, **2005**, *237*, 80.
20. Gillespie, R. J. *J. Chem. Soc.* **1950**, 2493-2503.

Materials Processing

Chapter 13

Emulsion-Templated Porous Materials Using Concentrated Carbon Dioxide-in-Water Emulsions and Inexpensive Hydrocarbon Surfactants

Colin D. Wood, Bien Tan, Jun-Young Lee, and Andrew I. Cooper[*]

Department of Chemistry, University of Liverpool, Crown Street, Liverpool L69 3BX, United Kingdom
[*]Corresponding author: email: aicooper@liv.ac.uk

Supercritical carbon dioxide is the most extensively studied supercritical fluid (SCF) medium for polymerization reactions and organic transformations. This can be attributed to a list of advantages ranging from solvent properties to practical environmental as well as economic considerations. In addition, when CO_2 is used for the synthesis and processing of porous materials additional advantages are also realized. The subject of this review will be the synthesis of porous materials and in particular emulsion-templated polymers. The development of inexpensive hydrocarbon surfactants will also be discussed.

Introduction

Supercritical CO_2 ($scCO_2$) is a sustainable solvent because it is nontoxic, nonflammable, and readily available in high purity from a number of sources.[1] CO_2 has been shown to be a versatile solvent for polymer synthesis and processing,[2-5] and it has been exploited quite widely for the preparation of porous materials:[6] for example, $scCO_2$ has been used for the production of microcellular polymer foams,[7,8] biodegradable composite materials,[9] macroporous polyacrylates,[10-13] and fluorinated microcellular materials.[14]

A significant technical barrier with using CO_2 is that it is a relatively weak solvent: important classes of materials which tend to exhibit low solubility in $scCO_2$ include polar biomolecules, pharmaceutical actives, and high-molecular weight polymers.[1,2,6,15-18] Until recently, the only polymers found to have significant solubility in CO_2 under moderate conditions (<100°C, <400 bar) were amorphous fluoropolymers[1] and, to a lesser extent, polysiloxanes.[17] Therefore, the discovery of inexpensive CO_2-soluble materials or "CO_2-philes" has been an important challenge.[19-21] This chapter will discuss the development of inexpensive hydrocarbon surfactants for use in CO_2 as well as the synthesis of emulsion-templated polymers.

Porous Materials and Supercritical Fluids

Porous materials are used in a wide range of applications, including catalysis, chemical separations, and tissue engineering.[6] However, the synthesis of these materials is often solvent intensive. Supercritical carbon dioxide can circumvent this issue as an alternative solvent for the synthesis of functional porous materials as well as affording a number of specific physical, chemical, and toxicological advantages. For example, energy intensive drying steps are required in order to dry porous materials whereas the transient "dry" nature of CO_2 overcomes these issues. Pore collapse can occur in certain materials when removing conventional liquid solvents; this can be avoided using supercritical fluids (SCFs) because they do not possess a liquid-vapour interface. Porous structures are important in biomedical applications (*e.g.*, tissue engineering) where the low toxicity of CO_2 offers specific advantages in terms of minimizing the use of toxic organic solvents. In addition, the wetting properties and low viscosity of CO_2 offers specific benefits in terms of surface modification.

A number of new approaches have been developed in the past few years for the preparation of porous materials using supercritical fluids (SCF).[6,22-,25] Current routes include foaming,[9,26-29] CO_2-induced phase separation,[30,31] reactive-*(10-13)* and nonreactive*(14,32)* gelation of CO_2 solutions, nanoscale casting using supercritical CO_2,*(33,34)* and CO_2-in-water (C/W) emulsion templating.*(22-25,35,36)* Each of these methods uses a different mechanism to generate the porosity in the material.

$scCO_2$ has been used for the formation of permanently porous crosslinked poly(acrylate) and poly(methacrylates) monoliths and beads using $scCO_2$ as the porogenic solvent.*(10-13)* Materials of this type*(37,38)* are useful in applications such as high-performance liquid chromatography, high-performance membrane chromatography, capillary electrochromatography, microfluidics,*(39)* and high-throughput bioreactors.*(40)* In the CO_2-based process, no organic solvents are used in either the synthesis or purification. The variable density associated with SCF solvents was exploited in order to "fine-tune" the polymer morphology. As such, the apparent Brunauer-Emmett-Teller (BET) surface area and the average pore size of the materials varied substantially for a series of crosslinked polymers synthesized using $scCO_2$ as the porogen over a range of reaction pressures.*(10-13)*

Recently, an entirely new approach to preparing porous materials was developed by templating the structure of *solid* CO_2 by directional freezing.*(41)* In this process, a liquid CO_2 solution was frozen in liquid nitrogen unidirectionally. The solid CO_2 was subsequently removed by direct sublimation to yield a porous, solvent-free structure with no additional purification steps (Figure 1). Other CO_2-soluble actives could be incorporated into the porous structure uniformly. This was demonstrated by dispersing oil red uniformly in the aligned porous sugar acetates.*(41)* This method differs fundamentally from the other CO_2-based techniques*(6)* and offers the unique advantage of generating materials with aligned pore structures. Materials with aligned microstructures and nanostructures are of interest in a wide range of applications such as organic electronics,*(42)* microfluidics,*(43)* molecular filtration,*(44)* nanowires,*(45)* and tissue engineering.*(46)*

In addition to producing materials with aligned porosity, there are a number of additional advantages associated with this new technique. The method avoids the use of any organic solvents, thus eliminating toxic residues in the resulting material. The CO_2 can be removed by simple sublimation, unlike aqueous-based processes where the water must be removed by freeze-drying.*(47-49)* Moreover, the method can be applied to relatively nonpolar, water-insoluble materials. These aligned porous structures may find numerous applications, for example, as biomaterials. Aligned porous materials with micrometer-sized pores are of importance in tissue engineering where modification with biological cells is required. We are particularly interested in the use of such porous materials as scaffolds for aligned nerve cell growth. This latter application will be greatly facilitated by the recent development of biodegradable CO_2-soluble hydrocarbon polymers as potential scaffold materials.*(50)*

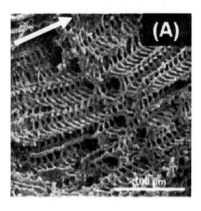

Figure 1. Porous sugar acetate with an aligned structure prepared by directional freezing of a liquid CO_2 solution. Reproduced from reference 41. Copyright 2005 American Chemical Society

Polymer solubility in CO_2

One of the fundamental issues that one must consider when implementing CO_2 for polymer synthesis or processing is polymer solubility. As mentioned, CO_2 is a weak solvent and there has been considerable research effort focused on discovering inexpensive biodegradable CO_2-soluble polymers from which inexpensive CO_2-soluble surfactants, ligands, and phase transfer agents could be developed. This problem is not unique to CO_2. Indeed, an enormous variety of surfactants and phase transfer agents have been developed to disperse poorly soluble molecules in water. A technical barrier to the use of scCO_2 is the lack of an equivalent range of inexpensive CO_2-soluble (and preferably biodegradable) surfactants, ligands, and phase transfer agents. However, it is very difficult to predict which polymer structures would be CO_2-soluble, despite attempts to rationalize specific solvent-solute interactions by using *ab initio* calculations.(51) Only a few examples of CO_2-soluble polymers currently exist and, as such, there are a limited number of "design motifs" to draw upon. Moreover, it is clear that polymer solubility in CO_2 is influenced by a large number of interrelated factors(17) such as specific solvent-solute interactions,(20,51-53) backbone flexibility,(20,52,54) topology,(54) and the nature of the end-groups.(54) Given the current limits of predictive understanding, the discovery of new CO_2-soluble polymers might be accelerated using a parallel or 'high-throughput' methodology. The synthetic approaches for such a strategy are already well in place; for example, a growing number of methods exist whereby one may synthesize and characterize polymer libraries.(55) By contrast, there are no examples of techniques for the rapid, parallel determination of solubility for libraries of materials in $scCO_2$ or other SCFs. The conventional method for evaluating polymer solubility in SCFs is cloud point measurement,[17,20,52,54] which involves the use of a variable-volume view cell. This technique is not suitable for rapid solubility measurements and would be impractical for large libraries of materials.

A number of research groups have synthesised 'CO$_2$-philic' fluoropolymers or silicone-based materials for use as steric stabilisers in dispersion polymerization[5,56-59] as phase transfer agents for liquid–liquid extraction,[60] as supports for homogeneous catalysis,[61,62] and as surfactants for the formation of water / CO_2 emulsions and microemulsions.[63,64] Unfortunately, the high cost of fluorinated polymers may prohibit their use on an industrial scale for some applications. Fluoropolymers also tend to have poor environmental degradability, and this could negate the environmental advantages associated with the use of $scCO_2$. The lack of inexpensive CO_2-soluble polymers and surfactants is a significant barrier to the future implementation of this solvent technology.[65]

Inexpensive poly(ether carbonate) (PEC) copolymers have been reported to be soluble in CO_2 under moderate conditions.[20] Similarly, sugar acetates are highly soluble and have been proposed as renewable CO_2-philes.[51,66] Such materials could, in principle, function as CO_2-philic building blocks for inexpensive ligands and surfactants, but this potential has not yet been realized and numerous practical difficulties remain. For example, CO_2 solubility does not in itself guarantee performance in the various applications of interest. Effective surfactants, in particular, tend to require specific asymmetric topologies such as diblock copolymers.[63,67] This in turn necessitates a flexible and robust synthetic methodology to produce well-defined architectures for specific applications.

High throughput solubility measurements in CO_2

We reported a new method which allows for the rapid parallel solubility measurements for libraries of materials in supercritical fluids.[68] The method is based on parallel gravimetric extraction. In a typical experiment, polymer samples (ca. 100 mg) are accurately weighed into borosilicate glass sample tubes and loaded into a specially-designed sample holder which will accept up to 60 tubes on this scale. This holder is then placed into a custom-built (Thar Designs) SCF extractor consisting of a vertically-mounted 500 ml extraction vessel and computer-controlled syringe pump/back pressure regulator. CO_2 is then passed through the vessel at a controlled pressure, temperature, and flow rate for a predetermined period of time. Thus, all 60 samples are subjected, in parallel, to precisely the same extraction conditions. The CO_2 is then slowly vented, the sample holder removed from the vessel, and the samples reweighed individually to determine the sample mass loss (if any) under those extraction conditions. The procedure is then repeated (typically at progressively higher CO_2 pressures) in order to build a cumulative extraction profile for the library of samples. The technique was used to evaluate the solubility of a mixed library of 100 synthetic polymers including polyesters, polycarbonates, and vinyl polymers. It was found that poly(vinyl acetate) (PVAc) showed the highest solubility in CO_2, the anamolously high solubility of PVAc has been shown previously.[65] This method is at least 50 times faster than other techniques in terms of the rate of useful information that is obtained and has broad utility in the discovery of novel SCF-soluble ligands, catalysts, biomolecules, dyes, or pharmaceuticals for a wide range of materials applications.

Inexpensive and Biodegradable CO_2-Philes

Poly(vinyl acetate) (PVAc) is an inexpensive, high-tonnage bulk commodity polymer which, unlike most vinyl polymers, is moderately biodegradable and has been used in pharmaceutical excipient formulations. PVAc has also been shown to exhibit anomalously high solubility in CO_2 with respect to other vinyl hydrocarbon polymers,[65] although the polymer is soluble only at relatively low molecular weights under conditions of practical relevance (P < 300 bar, T < 100 °C).

We presented a simple and generic method for producing inexpensive and biodegradable polymer surfactants for $scCO_2$ for solubilization, emulsification, and related applications.[50] In this method, the terminal hydroxyl group of a poly(vinyl acetate) (PVAc) oligomer is transformed into an imidazole ester by reaction with carbonyl diimidazole (CDI). This route has a number of advantages. First, the OVAc imidazolide intermediate can be isolated, purified, and then coupled with a wide range of alcohols (or amines) to produce a variety of structures. Second, the route introduces a carbonate linkage that may further enhance CO_2 solubility[52,69] and could also improve the biodegradability of the resulting materials. To illustrate the use of OVAc as a solubilizing group, an organic dye, Disperse Red 19 (DR19), was functionalized with OVAc (M_n) 1070 g/mol, M_w) 1430 g/mol) to produce **1** (Figure 2). The stoichiometry of the reaction was controlled such that one OVAc chain was attached to each DR19 molecule, as confirmed by GPC and ^1H NMR. DR19 itself had negligible solubility in CO_2 up to pressures of 300 bar/25 °C (no color was observed in the CO_2 phase). By contrast, the functionalized dye, **1**, was found to be soluble in CO_2 (100-200 bar) at least up to concentrations of around 1 wt %. This suggested that OVAc has potential as a less expensive and more biodegradable replacement for the highly fluorinated materials used previously to solubilize species such as dyes, catalysts, proteins, and nanoparticles in CO_2.[1,4,5,59-61,63,64,70,71] Another important area in $scCO_2$ technology is the formation of water-in-CO_2 (W/C) and CO_2-in-water (C/W) emulsions and microemulsions.*(64,72-75)* The same CDI route was used to couple OVAc with poly(ethylene glycol) monomethyl ethers (HO-PEG-OMe) and poly(ethylene glycol) diols (PEG) to produce diblock (Figure 3, **2a**) and triblock (**2b**) copolymers, respectively (Figure 3).*(50)*

These polymers were found to be useful surfactants. For example, an OVAc-b-PEG-OVAC triblock surfactant was found to emulsify up to 97 v/v % C/W emulsion which was stable for at least 48 h. An OVAc-b-PEG diblock copolymer was used to form a 90 v/v % C/W emulsion. Materials of this type were used to template SCF emulsions and will be discussed in detail in the following section.

Templating of supercritical fluid emulsions

Emulsion templating is useful for the synthesis of highly porous inorganic,*(76-79)* and organic materials.*(80-82)* In general, the technique involves forming a high internal phase emulsion (HIPE) (>74.05% v/v internal droplet phase) and locking in the structure of the continuous phase, usually by reaction-induced phase separation (e.g., free-radical polymerization, sol-gel chemistry). Subsequent removal of the internal phase gives rise to a porous replica of the emulsion. In principle, it is possible to access a wide range of porous hydrophilic materials by reaction-induced phase separation of concentrated oil-in-water (O/W) emulsions. This can lead to novel, porous hydrophilic materials for applications such as separation media, catalyst supports, biological tissue scaffolds, and controlled release devices. A significant drawback to this process is that large quantities of a water-immiscible oil or organic solvent are required as the internal phase (usually >80 vol.%). In addition, it is often difficult to remove this oil phase after the reaction.

Based on studies concerning SCF emulsion stability and formation,*(73)* we have developed methods for templating high internal phase CO_2-in-water (C/W) emulsions (HIPE) to generate highly porous materials in the absence of any organic solvents – only water and CO_2 are used.*(35)* If the emulsions are stable one can generate low-density materials ($\sim$ 0.1 g/cm^3) with relatively large pore volumes (up to 6 cm^3/g) from water-soluble vinyl monomers such as acrylamide and hydroxyethyl acrylate. Figure 4 shows a crosslinked polyacrylamide material synthesized from a high internal phase C/W emulsion, as characterized by SEM and confocal microscopy (scale = 230 μm x 230 μm). Comparison of the two images illustrates quite clearly how the porous structure shown in the SEM image is templated from the C/W emulsion (as represented by the confocal microscopy image of the pores). In general, the confocal image gives a better measure of the CO_2 emulsion droplet size and size distribution immediately before gelation of the aqueous phase. Initially we used low molecular weight

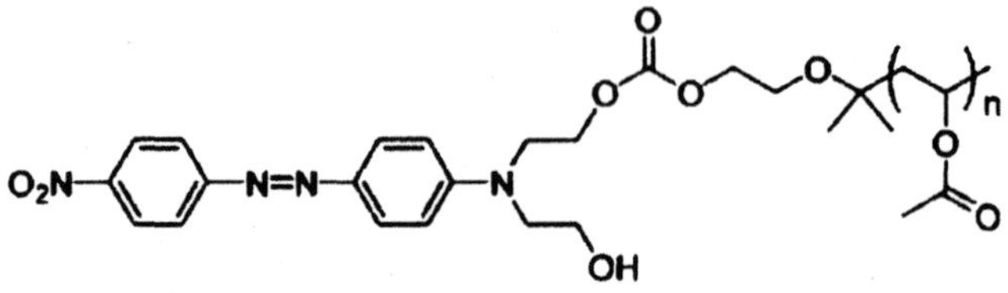

Figure 2. OVAc-functionalized dye, 1, dissolves in CO_2 (200 bar, 20 °C, 0.77 wt %). Reproduced from reference 50. Copyright 2005 American Chemical Society

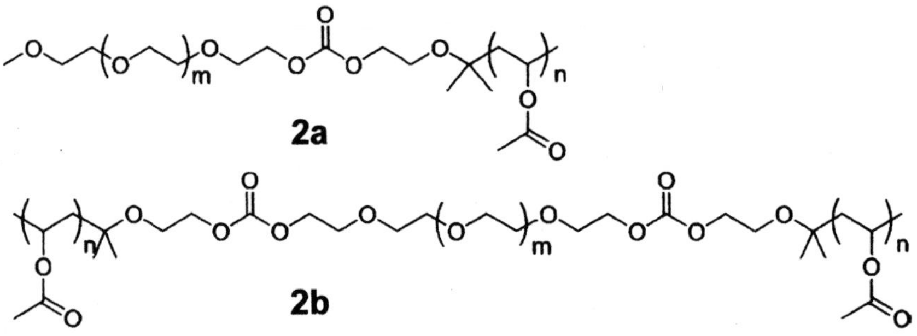

Figure 3. Structures of CO_2-philic surfactants for C/W emulsion formation: OVAc-b-PEG diblock polymer (2a) and OVAc-b-PEG-b-OVAc triblock polymer (2b). Reproduced from reference 50. Copyright 2005 American Chemical Society

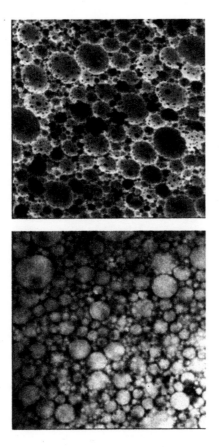

Figure 4. Emulsion-templated crosslinked polyacrylamide materials synthesized by polymerization of a high-internal phase CO_2-in-water emulsion (C/W HIPE). (a) SEM image of sectioned material. (b) Confocal image of same material, obtained by filling the pore structure with a solution of fluorescent dye. As such (a) shows the "walls" of the material while (b) show the "holes" formed by templating the $scCO_2$ emulsion droplets. Both images = 230 μm x 230 μm. Ratio of CO_2/aqueous phase = 80:20 v/v.d Pore volume = 3.9 cm³/g. Average pore diameter = 3.9 μm. (Reproduced with permission from reference 6. Copyright 2003 Wiley VCH Verlag GmbH & Co.)

($M_w \sim 550$ g/mol) perfluoropolyether ammonium carboxylate surfactants to stabilize the C/W emulsions(35) but as discussed there are some practical disadvantages of using these surfactants in this particular process such as cost and the surfactant is non-degradable. It was subsequently shown that it is possible to use inexpensive hydrocarbon surfactants to stabilize C/W emulsions and that these emulsions can also be templated to yield low-density porous materials.(36) In this study it was shown that all of the problems associated with the initial approach could be overcome and it was possible to synthesize C/W emulsion-templated polymers at relatively modest pressures (60 – 70 bar) and low temperatures (20 °C) using inexpensive and readily available hydrocarbon surfactants. Moreover, we demonstrated that this technique can in principle be extended to the synthesis of emulsion-templated HEA and HEMA hydrogels that may be useful, for example, in biomedical applications.(83-85)

As mentioned, we demonstrated a simple and generic method for producing inexpensive, functional hydrocarbon CO_2-philes for solubilization, emulsification, and related applications.(50) This approach was extended and water-soluble diblock and triblock surfactant architectures were accessed and it was found that both types of structure could stabilize highly concentrated C/W emulsions. A detailed investigation into the factors affecting the C/W emulsion stability was carried out in order to utilize these optimized emulsions to generate materials with significantly increased levels of porosity.(22) This new method is a simple and generic method for producing inexpensive and biodegradable polymer surfactants for use in supercritical CO_2. Low molecular weight ($M_w <$ 7000 g/mol) hydroxyl-terminated poly(vinyl acetate) (PVAc-OH) was synthesized using optimized reaction conditions and isopropyl ethanol (IPE) as the chain transfer agent. Oligomeric PVAc-OH (OVAc-OH, $M_w < 3000$ g/mol) was then obtained by supercritical fluid fractionation. The OVAc-OH species was converted to the imidazole ester by reaction with carbonyl diimidazole (CDI) and CO_2-soluble surfactants were produced by coupling these reactive blocks with poly(ethylene glycol) methyl ethers or poly(ethylene glycol) diols. The surfactants were found to be extremely effective in the production of stable CO_2-in-water (C/W) emulsions, which were then used as templates to produce emulsion-templated materials with unprecedentedly high levels of porosity for materials produced by this route. It was shown that these hydrocarbon surfactants can out-perform perfluorinated species in applications of this type. The synthetic methodology also allows fine-tuning of the hydrophilic-CO_2-philic balance to suit different applications. Surfactants of this type may find a range of additional uses in emulsion technology, particularly where biodegradability of the hydrophobic segment is required.

We recently presented a new methodology to produce highly porous cross-linked polymeric hydrogel materials by templating concentrated CO_2-in-water (C/W) emulsions.(23) Polymeric hydrogels have been studied for applications in a variety of fields such as in chemical engineering, pharmaceuticals, food, and agriculture.*(86,87)* Hydrogels can change their volume and shape reversibly as a result of external physicochemical factors such as temperature, solvent composition, pH, and ionic concentration.*(88-90)* These large volume or shape changes, which can be induced by supplying thermal, chemical, or electrical stimuli, offer a range of possibilities for advanced functional polymers. Hydrogels are appealing as scaffold materials because they are structurally similar to the extracellular matrix of many tissues, may often be processed under relatively mild conditions, and may be delivered in a minimally invasive manner. A variety of synthetic and naturally derived materials have been used to form hydrogels for tissue engineering scaffolds. Poly(vinyl alcohol) (PVA) hydrogels have been proposed as promising biomaterials to replace diseased or damaged articular cartilage. PVA has been widely explored for use in space filling and drug delivery applications. It can be physically cross-linked by repeated freeze-thawing cycles of aqueous polymer solutions*(91)* or chemically cross-linked with glutaraldehyde*(92)* or acid chlorides*(93)* to form hydrogels. It can also be blended with other water-soluble polymers and again cross-linked either physically or chemically.*(94-96)* It is not easy to control the pore size and porosity in such materials, partly as a result of the solid-liquid interface which exists between the matrix polymer and porogen (usually water). Current challenges in this field include the provision of appropriate preparation methods for controlled macroporous structures.

In our new approach poly(vinyl alcohol) (PVA), blended PVA/PEG, and naturally derived chitosan materials were produced by templating concentrated CO_2-in-water (C/W) emulsions.*(23)* Using the PVAc-based surfactants discussed above the C/W emulsions were sufficiently stable for templating to occur and for open-cell porous materials to be produced, as shown in Figure 5. It was observed that the internal structure was uniformly porous and consisted of a skeletal replica of the original C/W HIPE. The pore structure was highly interconnected and there were open pores on the surface that were connected to the interior (Figure 5a). The diameter of the macropores was found to be in the range 3–15 μm. The technique can be carried out at moderate temperatures and pressures (25 degrees C, < 120 bar) using inexpensive hydrocarbon surfactants such as PVAc-based block copolymers which are composed of biodegradable blocks. This methodology opens up a new solvent-free route for the preparation of porous biopolymers, hydrogels, and composites, including materials that cannot readily be produced by foaming.

A number of limitations were apparent in our initial C/W emulsion templating approach; namely that the PFPE surfactant was expensive and nondegradable, reaction pressures were high (250-290 bar), and reaction temperatures were elevated (50-60 °C).[36] We have since extended our methodology to produce highly porous cross-linked PVA materials, blended PVA/PEG, and naturally derived chitosan by the gelation of C/W HIPEs. Moreover, we have shown that this technique can be carried out at much lower

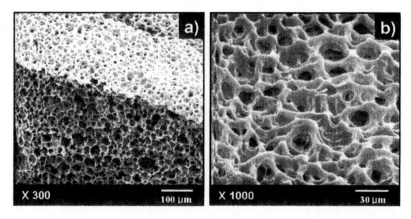

Figure 5. Electron micrographs of open-cell porous PVA hydrogel produced from C/W emulsions in the presence of PVAc-based surfactant. a) internal and surface pore structures; b) showing surface morphology with higher magnification. Reproduced from reference 23. Copyright 2007 American Chemical Society

temperatures and pressures (25 °C, < 120 bar) using inexpensive, biodegradable hydrocarbon surfactants such as PVAc-based block copolymers. Our methodology opens up a new solvent-free route for the preparation of porous biopolymers, hydrogels, and composites, including materials which cannot readily be produced by foaming. We plan to use this knowledge in future studies to develop highly porous materials, and to achieve fine control over porous structure by tuning the CO_2 density for a number of applications, particularly those in which organic solvent residues pose a problem.

Conclusions

In general, CO_2 is an attractive solvent alternative for the synthesis of polymers because it is 'environmentally friendly', non-toxic, non-flammable, inexpensive, and readily available in high-purity from a number of sources. It offers the potential of reducing organic solvent usage in the production of a range of materials. Product isolation is straightforward because CO_2 is a gas under ambient conditions, removing the need for energy intensive drying steps. This is particularly advantageous in processes where large volumes of organic solvents are used such as the production of porous materials. Moreover, the discovery of inexpensive, functional hydrocarbon CO_2-philes has opened up new opportunities for solubilization, emulsification, and related applications. It has been recently shown that these structures can outperform perfluorinated analogues in specific applications.

Acknowledgements

We thank EPSRC for financial support (EP/C511794/1) and the Royal Society for a Royal Society University Research Fellowship (to A.I.C).

References

1. DeSimone, J.M. *Science.* **2002**, *297*, 799.
2. Cooper, A. I. *J. Mater. Chem.* **2000**, *10*, 207-234.
3. Cooper, A. I. *Adv. Mater.* **2001**, *13*, 1111.
4. DeSimone, J. M.; Guan, Z.; Elsbernd, C.S. *Science* **1992**, *257*, 945.
5. DeSimone, J.M.; Maury, E.E.; Menceloglu, Y.Z.; McClain, J.B.; Romack, T.J.; Combes, J.R. *Science* **1994**, *265*, 356.
6. Cooper, A.I. *Adv. Mater.* **2003**, *15*, 1049.
7. Parks, K.L.; Beckman, E.J.; *Polym. Eng. Sci.* **1996**, *36*, 2404.
8. Parks, K.L.; Beckman, E.J. *J. Polym. Eng. Sci.* **1996**, *36*, 2417.
9. Howdle, S.M.; Watson, M.S.; Whitaker, M.J.; Popov, V.K.; Davies, M.C.; Mandel, F.S.; Wang, J.D.; Shakesheff, K.M. *Chem. Commun.* **2001**, 109.
10. Cooper, A.I.; Holmes, A.B. *Adv. Mater.* **1999**, *11*, 1270.
11. Cooper, A.I.; Wood, C.D.; Holmes, A.B. *Ind. Eng. Chem. Res.* **2000**, *39*, 4741.
12. Hebb, A.K.; Senoo, K. Bhat, R.; Cooper, A.I. *Chem. Mater.* **2003**, *15*, 2061.
13. Wood, C.D.; Cooper, A.I. *Macromolecules* **2001**, *34*, 5.
14. Shi, C.; Huang, Z.; Kilic, S.; Xu, J.; Enick, R.M.; Beckman, E.J.; Carr, A.J.; Melendez, R.E.; Hamilton, A.D.; *Science* **1999**, *286*, 1540.
15. Kendall, J.L.; Canelas, D.A.; Young, J.L.; DeSimone, J.M. *Chem. Rev.* **1999**, *99*, 543.
16. Woods, H.M.; Silva, M.; Nouvel, C.; Shakesheff, K.M.; Howdle, S.M. *J. Mater. Chem.* **2004**, *14*, 1663.
17. Kirby, C.F.; McHugh, M.A. *Chem. Rev.* **1999**, *99*, 565.
18. Jessop, P.G.; Leitner, W. *Chemical Synthesis Using Supercritical Fluids;* Wiley-VCH: Weinheim **1999**.
19. Beckman, E.J. Chem. Comm. **2004**, *17*, 1885.
20. Sarbu, T.; Styranec, T.J.; Beckman E.J. *Ind. Eng. Chem. Res.* **2004**, *39*, 4678.
21. Eastoe, J.; Paul, A.; Nave, S.; Steytler, D.C.; Robinson, B.H.; Rumsey, E.; Thorpe, M.; Heenan, R.K.; *J. Am. Chem. Soc.* **2001**, *123*, 988.

256

22. Tan, B.; Lee, J.Y.; Cooper A.I. *Macromolecules* **2007**, *40*, 1945.
23. Lee, J.Y.; Tan, B.; Cooper, A.I. Macromolecules **2007**, *40*, 1955.
24. Palocci, C.; Barbetta, A.; Grotta, A.L.; Dentini, M.; *Langmuir* **2007**, *23*, 8243.
25. Partap, S.; Rehman, I.; Jones, J.R.; Darr, J.A. *Adv. Mater.* **2006**, *18*, 501.
26. Goel, S.K.; Beckman, E.J. *Polymer* **1993**, *34*, 1410.
27. Siripurapu, S.; Gay, Y.J.; Royer, J.R. DeSimone, J.M.; Spontak, R.J.; Khan, S.A. *Polymer,* **2002**, *43*, 5511.
28. Krause, B.; Koops, G.H.; van der Vegt, N.F.A.; Wessling, M.; Wubbenhorst, M.; van Turnhout, J. *Adv. Mater.* **2002**, *14*, 1041,
29. Siripurapu, S.; DeSimone, J.M.; Khan, S.A.; Spontak, R.J. *Adv. Mater.* **2004**, *16*, 989.
30. Matsuyama, H.; Yano, H.; Maki, T.; Teramoto, M.; Mishima, K.; Matsuyama, K. *J. Membr. Sci.* **2001**, *194*, 157.
31. Matsuyama, H.; Yamamoto, A.; Yano, H.; Maki, T.; Teramoto, M.; Mishima, K.; Matsuyama, K.J. *Membr. Sci.* **2002**, *204*, 81.
32. Placin, F.; Desvergne, J.P.; Cansell, F.J. *Mater. Chem.* **2000**, *10*, 2147.
33. Wakayama, H.; Itahara, H.; Tatsuda, N.; Inagaki, S.; Fukushima, Y. *Chem. Mater.* **2001**, *13*, 2392.
34. Fukushima, Y.; Wakayama, H. *J. Phys. Chem. B* **1999**, *103*, 3062.
35. Butler, R.; Davies, C.M.; Cooper, A.I. *Adv. Mater.* **2001**, *13*, 1459.
36. Butler, R.; Hopkinson, I.; Cooper, A.I. *J. Am. Chem. Soc.* **2003**, *125*, 14473.
37. Svec, F. ; Fréchet, J.M.J. *Science.* **1996**, *273*, 205.
38. Svec, F. ; Fréchet, J.M.J. *Ind. Eng. Chem. Res.* **1998**, *38*, 34.
39. Yu, C. ; Xu, M.C. Svec, F. ; Fréchet, J.M.J. *J. Polym. Sci. A, Polym. Chem.* **2002**, *40*, 755.
40. Petro, M.; Svec F.; Fréchet, J.M.J. *Biotechnol. Bioeng.* **1996**, *49*, 355.
41. Zhang, H.; Long, J.; Cooper, A.I. *J. Am. Chem. Soc.* **2005**, *127*, 13482.
42. Gu, H.; Zheng, R.; Zhang, X.; Xu, B. *Adv. Mater.* **2004**, *16*, 1356.
43. Quake, S.R.; Scherer, A. *Science* **2000**, *290*, 1536.
44. Yamaguchi, A.; Uejo, F.; Yoda, T.; Uchida, T.; Tanamura, Y.; Yamashita, T.; Teramae, N. *Nat. Mater.* **2004**, *3*, 337.
45. Adelung, R.; Aktas, O.C.; Franc, J.; Biswas, A.; Kunz, R.; Elbahri, M.; Kanzow, J.; Schuramann, U.; Faupel, F. *Nat. Mater.* **2004**, *3*, 375.
46. Xu, C.Y.; Inai, R.; Kotaki, M.; Ramakrishna, S. *Biomaterials* **2004**, *25*, 877.
47. Mahler, W.; Bechtold, M.F. *Nature* **1980**, *285*, 27.
48. Mukai, S.R.; Nishihara, H.; Tamon, H. *Chem. Comm.* **2004**, 874.
49. Nishihara, H.; Mukai, S.R.; Yamashita, D.; Tamon, H.; *Chem. Mater.* **2005**, *17*, 683.
50. Tan, B.; Cooper, A.I. *J. Am. Chem. Soc.* **2005**, *127*, 8938.
51. Raveendran, P.; Wallen, S.L. *J. Am. Chem. Soc* **2002**, 124, 12590.
52. Sarbu, T. ; Styranec, T. ; Beckman, E.J. ; *Nature* **2000**, *405*, 165.

53. Kazarian, S.G. ; Vincent, M.F.; Bright, F.V.; Liotta, C.L. Eckert, C.A. *J. Am. Chem. Soc* **1996**, *118*, 1729.

54. Drohmann, C. ; Beckman, E.J. *J. Supercritical Fluids* **2002**, *22*, 103.

55. Hoogenboom, R.; Meier, M.A.R.; Schubert, U.S. *Macromolecular Rapid Comm.* **2003**, *24*, 16.

56. Shaffer, K.A.; Jones, T.A.; Canelas, D.A.; DeSimone, J.M.; Wilkinson, S.P. *Macromolecules* **1996**, *29*, 2704.

57. Lepilleur, C.; Beckman, E.J.; *Macromolecules* **1997**, *30*, 745.

58. Cooper, A.I.; Hems, W.P.; Holmes, A.B. *Macromolecules* **1999**, *32*, 2156.

59. Christian, P.; Howdle, S.M.; Irvine, D.J. *Macromolecules* **2000**, *33*, 237.

60. Cooper, A.I.; Londono, J.D.; Wignall, G.; McClain, J.B.; Samulski, E.T.; Lin, J.S.; Dobrynin, A.; Rubinstein, M.; Burke, A.L.C.; Fréchet, J.M.J.; DeSimone, J.M. *Nature* **1997**, *389*, 368.

61. Chen, W.P.; Xu, L.J.; Hu, Y.L.; Osuna, A.M.B.; Xiao, J.L.; *Tetrahedron* **2002**, *58*, 3889.

62. Kani, I.; Omary, M.A.; Rawashdeh-Omary, M.A. Lopez-Castillo, Z.K.; Flores, R.; Akgerman, A.; Fackler, J.P. *Tetrahedron* **2002**, *58*, 3923.

63. Johnston, K.P. *Current Opinion in Colloid & Interface Science* **2000**, *5*, 351.

64. Johnston, K.P.; Harrison, K.L.; Clarke, M.J.; Howdle, S.M.; Heitz, M.P.; Bright, F.V.; Carlier, C.; Randolph, T.W. *Science* **1996**, *271*, 624.

65. Shen, Z.; McHugh, M.A.; Xu, J.; Belardi, J.; Kilic, S.; Mesiano, A.; Bane, S.; Karnikas, C.; Beckman, E.J.; Enick, R. *Polymer* **2003**, *44*, 1491.

66. Raveendran, P.; Wallen, S.L. *J. Am. Chem. Soc.* **2002**, *124*, 7274.

67. Canelas, D.A.; Betts, D.E. DeSimone, J.M. *Macromolecules* **1996**, *29*, 2818.

68. Bray, C.L.; Tan, B.; Wood, C.D.; Cooper, A.I. *J. Mater. Chem.* **2005**, *15*, 456.

69. Tan, B.; Woods, H.M.; Licence, P.; Howdle, S.M.; Cooper, A.I. *Macromolecules* **2005**, *38*, 1691.

70. Carroll, M.A.; Holmes, A.B. *Chem. Comm.* **1998**, 1395 (1998)

71. Shah P.G.; Holmes, J.D.; Doty, R.C.; Johnston, K.P.; Korgel, B.A. *J. Am. Chem. Soc.* **2000**, *122*, 4245.

72. da Rocha, S.R.P.; Psathas, P.A.; Klein, E.; Johnston, K.P.; *J. Coll.Int. Sci.* **2001**, *239*, 241.

73. Lee, C.T.; Psathas, A.; Johnston, K.P.; deGrazia, J.; Randolph, T.W. *Langmuir,* **1999**, *15*, 6781.

74. Dickson J.L.; Binks, B.P.; Johnston, K.P. *Langmuir* **2004**, *20*, 7976.

75. Dickson, J.L.; Smith, P.G.; Dhanuka, V.V.; Srinivasan, V.; Stone, M.T.; Rossky, P.J.; Behles, J.A.; Keiper, J.S.; Xu, B.; Johnson, C.; DeSimone, J.M.; Johnston, K.P. *Ind. Eng. Chem. Res.* **2005**, *44*, 1370.

76. Imhof, A.; Pine, D.J. *Adv. Mater.* **1998**, *10*, 697.

77. Schmidt-Winkel, P.; Lukens, W.W.; Yang, P.D.; Margolese, D.I.; Letlow, J.S.; Ying, J.Y.; Stucky, G.D.; *Chem. Mater.* **2000**, *12*, 686.
78. Manoharan, V.N.; Imhof, A.; Thorne, J.D.; Pine. D.J. Adv. *Mater.* **2001**, *13*, 447.
79. Zhang, H.; Hardy, G.C.; Rosseinsky, M.J.; Cooper, A.I. *Adv. Mater.* **2003**, *15*, 78.
80. Cameron, N.R.; Sherrington. D.C. *Adv. Polym. Sci.* **1996**, *126*, 163.
81. Busby, W.; Cameron, N.R.; Jahoda, C.A.B. *Biomacromolecules.* **2001**, *2*, 154.
82. Zhang, H.; Cooper, A.I. *Chem. Mater.* **2002**, *14*, 4017.
83. Song J.; Saiz, E.; Bertozzi, C.R. *J. Am. Chem. Soc.* **2003**, *125*, 1236.
84. Luo, Y.; Dalton, P.D.; Shoichet, M.S. *Chem. Mater.* **2001**, *13*, 4087.
85. Chiellini, F.; Bizzari, R.; Ober, C,K.; Schmaljohann, D.; Yu, T.Y.; Solaro, R.; Chiellini, E. *Macromol. Rapid. Comm.* **2001**, *22*, 1284.
86. Aharoni, S. M. *Synthesis, Characterization and Theory of Polymeric Networks and Gel*; Plenum: New York, 1992.
87. Klempner, O.; Utrachi, L. A.; Sperling, L. H. *Interpenetrating Polymer Networks*; American Chemical Society: Washington DC, 1991; Vol. 239.
88. Feil, H.; Bae, Y. H.; Feijen, J.; Kim, S. W. *Macromolecules* **1992**, *25*, 5528.
89. Nozawa, I.; Suzuki, Y.; Sato, S.; Sugibayashi, K.; Morimoto, Y. *J. Biomed. Mater. Res.* **1991**, *25*, 243.
90. Yoshida, M.; Yang, J. S.; Kumakura, M.; Hagiwara, M.; Katakai, R. *Eur. Polym. J.* **1991**, *27*, 997.
91. Cascone, M. G.; Laus, M.; Ricci, D.; Delguerra, R. S. *J. Mater. Sci.Mater. Med.* **1995**, *6* (2), 71.
92. Nuttelman, C. R.; Mortisen, D. J.; Henry, S. M.; Anseth, K. S. *J. Biomed. Mater. Res.* **2001**, *57*, 217.
93. Orienti, I.; Trere, R.; Zecchi, V. *Drug Dev. Ind. Pharm.* **2001**, *27*, 877.
94. Cascone, M. G.; Barbani, N.; Cristallini, C.; Giusti, P.; Ciardelli, G.; Lazzeri, L. *J. Biomater. Sci.-Polym. Ed.* **2001**, *12*, 267.
95. CauichRodriguez, J. V.; Deb, S.; Smith, R. *Biomaterials* **1996**, *17*, 2259.
96. Cauich-Rodriguez, J. V.; Deb, S.; Smith, R. *J. Appl. Polym. Sci.* **2001**, *82*, 3578.

Chapter 14

Fluoropolymer Synthesis in Carbon Dioxide-Expanded Liquids: A Practical Approach to Avoid the Use of Perfluorooctanoic Acid

Libin Du[1], Joseph M. DeSimone[1,2], and George W. Roberts[2]

[1]Department of Chemistry, University of North Carolina at Chapel Hill, Chapel Hill, NC 27599
[2]Department of Chemical and Biomolecular Engineering, North Carolina State University, Raleigh, NC 27695

Successful polymerization of tetrafluoroethylene (TFE) in CO_2-expanded liquids to synthesize polytetrafluoroethylene (PTFE) was demonstrated using both hydrocarbon and hydrofluorocarbon solvents. Yields were satisfactory and samples with thermal properties comparable to commercial polymer were produced. The chain transfer reaction that resulted from the use of hydrogen-containing solvents was not as severe as expected. Acetic acid was identified as a particularly promising expandable solvent. Use of this solvent lowers the pressure required for polymerization of TFE, and acetic acid is less environmentally detrimental than halogenated solvents. A wide range of solvent ratios and pressures were found to be suitable for TFE polymerization in CO_2-expanded acetic acid. Colloidal PTFE samples were obtained in an emulsion-like polymerization run in CO_2-expanded acetic acid. A commercially-available, conventional hydrocarbon surfactant was used in this research; the process did not require fluorinated surfactants such as perfluorooctanoic acid (PFOA) or other the specially designed surfactants for CO_2 systems.

Introduction

Fluoropolymers have a wide range of applications (1). Their uses in certain areas are often irreplaceable due to the unique combination of properties such as high thermal stability, good chemical resistance, low surface energy, and low friction coefficient. The current primary processes in fluoropolymer synthesis include polymerization in organic solvents, mostly halocarbon solvents, and heterogeneous processes such as aqueous emulsion and suspension polymerization. Halocarbon solvents generally are not environmentally friendly, with many of them being currently under strict regulation (2). Moreover, the fluorocarbon surfactants, mostly perfluorooctanoic acid and derivatives (PFOA), used in aqueous processes will be abandoned as persistent pollutants and bio-accumulating agents (3). Earlier, in the 1990's, we developed processes using supercritical (sc) CO_2 as a polymerization medium to synthesize fluoropolymers (4,5). Since then, successful syntheses of many commercial fluoropolymers in $scCO_2$ have been demonstrated (2). This process has environmental benefits compared with other processes, and the polymers prepared in this manner have significantly fewer ionic end groups compared with the same polymers prepared via conventional aqueous processes which require a multi-step post-treatment to remove (5). Polymerization in $scCO_2$ represents a more environmentally friendly alternative to current processes and a potential solution to the above problems, and is being used in a DuPont Fluoropolymer Plant for making certain grades of fluoropolymers (6). However, a relatively high pressure is required for running this process in order to maintain a sufficient density of the CO_2 to facilitate efficient heat transfer. This is especially important in highly exothermic polymerizations such as those involving TFE. When supercritical CO_2 is used as the medium for the precipitation polymerization of fluoroolefins, there are far fewer unit operations needed compared with conventional aqueous processes since operations like flocculation, drying, surfactant stripping and recovery are not needed. However, despite a simplified process, there is still a desire to lower the pressure of such CO_2-based processes to reduce the overall costs of the process. (7).

Gas-expanded solvents (GXLs) are another class of environmentally benign solvents which may offer advantages over regular organic solvents, as well as supercritical fluids, in some applications. Formed through expansion of a traditional organic solvent by dissolution of a gas, GXLs usually have reduced viscosity, improved diffusivity and gas solubility compared with conventional organic solvents, while possessing better solvating power than supercritical fluids (8,9). The most frequently used GXL is CO_2-expanded liquids (CXLs) (9). A great advantage of CXLs in comparison with $scCO_2$ is the substantially lower pressures that may be required (10). Mainly used in gas-antisolvent crystallization processes previously, CXLs are currently finding applications in separation, nanoparticle formation, enhanced oil recovery, photoresist removal and various organic reactions, especially in catalysis involving gases and viscous solutions (9,10). However, very little work in polymer synthesis has been reported so far using CXLs as medium. A synthesis of polycyclo-hexylethylene through catalytic hydrogenation of polystyrene was reported recently (11-13). The CO_2-expanded decahydronaphthalene solution of poly-styrene was used as the reaction medium and a higher reaction rate was reported. The polymerization of methyl methacrylate (MMA) in CO_2-expanded MMA was also reported recently (14). No additional solvent was used besides the monomer, MMA.

Herein, we report for the first time a synthesis of fluoropolymer in CO_2-expanded organic solvents. A number of organic solvents could be used as CXLs with CO_2 to make PTFE. Acetic acid was chosen to assess the effect of solvent ratio and pressure on polymerization in CXLs. Advantages were found for using acetic acid as a co-solvent with CO_2 as the reaction medium. The success of this process is three-fold. First, we showed that high yields of PTFE could be obtained at relatively low pressures (5.9 to 8.5 MPa) in CO_2-expanded acetic acid. Control reactions in pure CO_2 at the same pressures did not give satisfactory results. Second, polymerization of fluoroolefins can be carried out in the presence of high concentrations of a hydrocarbon solvent (up to 60 vol% of acetic acid), which is surprising given the dogma of the high chain transfer rates of such electrophilic radicals. Last, a common hydrocarbon surfactant could be used in the above reaction medium to produce colloidal PTFE, suggesting that regular surfactants could be effective in CO_2-expanded polar organic solvents.

Experimental

Materials

TFE in CO_2 (50/50 by weight) was kindly provided by DuPont. SFC grade CO_2 was purchased from National Welders. Sodium dodecyl sulfate (SDS) was purchased from Sigma-Aldrich, Inc. All the hydrocarbon solvents: acetone, acetonitrile and acetic acid, were obtained from Fisher Scientific, Inc. Solkane® 365 mfc was purchased from Micro Care Corporation, Vertrel® XF was purchased from DuPont, and HFE-449 and HFE-356 were purchased from SynQuest Laboratories, Inc. All chemicals were used as received. Bis(perfluoro-2-propoxypropionyl) peroxide (or 3P) was prepared according to DuPont's procedure.

Measurements

Thermogravimetric Analysis (TGA)

TGA measurements of the PTFE were performed on a Perkin Elmer Pyris 1 Thermogravimetric Analyzer using nitrogen purge gas. Samples were first put in an aluminum pan and confined in a special platinum dish during the measurement. The sample weights were within 5 – 15 mg. The sample temperature was increased from 25 to 600 °C with a heating rate of 10 °C/min.

Differential Scanning Calorimetry (DSC)

Calorimetric measurements of the PTFEs were preformed using a Seiko DSC 220C Calorimeter. Samples, usually within the range of 5 – 10 mg, were sealed in a pair of Aluminum pans and referenced against another pair of empty pans. They were heated from 25 to 350 °C, then cooled to 25 °C with a rate of 10 °C/min. Molecular weights of PTFEs prepared in this study were determined using a reported method based on the heats of crystallization obtained in DSC measurements (15).

Scanning Electron Microscopy (SEM)

A Hitachi model S-4700 FE-SEM was used to inspect the morphologies of polymer particles. Samples were prepared by directly casting the diluted polymer suspension on a glass slide followed by drying the slide under vacuum for 12 hrs.

Dynamic Light Scattering (DLS)

Particles in a polymer/acetic acid mixture were analyzed using a Brookhaven 90Plus Particle Analyzer. Samples were prepared by directly diluting the mixture with water, followed by filtering with 5.00 μm Millipore Millex® Syringe Filter before measurement. The effective diameters of the particles were reported as the average particle sizes in this study.

Synthesis

A 25 ml stainless reactor equipped with a magnetic stir bar and a controller was used in this study. Solvent expansion was monitored through a sapphire window. In all the polymerizations in this study, the solvents mixed well and formed only one phase at the beginning of the polymerization. Polymer started to precipitate out shortly after CO_2 and initiator were added. In the polymerizations with surfactant, the reaction mixture started out clear and turned to milky under agitation after some time.

Polymerization of TFE in scCO₂

A 25 ml stainless steel cell was purged with carbon dioxide three times to remove air. A mixture of TFE and CO_2 (50/50 wt%, 2.6 g) was charged into the reactor cell through a high-pressure syringe pump. The reactor was then slowly heated to 35 °C. A solution of bis(perfluoro-2-propoxypropionyl) peroxide in 1,1,2-trichlorotrifluoro-ethane (0.223 M, 0.0056 ml) was injected into an addition tube under Argon flow, and the tube was resealed. Carbon dioxide was charged into the cell slowly through the addition tube, carrying the initiator solution into the cell. The initial pressure was set at approximately 17 MPa. The reaction was held for approximately 4 hrs before being cooled down to room temperature. The end pressure was approximately 12 MPa. Carbon dioxide, excess TFE and other volatiles were then slowly vented. A fluffy white powdered polymer was collected. The product was washed with a mixture of 1,1,2-trichlorotrifluoroethane and methanol (1/1 by volume) three times and dried in a vacuum oven at elevated temperature. White powdered PTFE (0.89 g, 67.9% yield) was recovered. Thermogravimetric analysis gave a 5 % weight loss at 514 °C. Differential scanning calorimetry showed a melting temperature of 328.4 °C with a 10 °C/min. heating rate. Integration of the melting peak gave a heat of crystallization of 43.8 J/g, corresponding to an estimated molecular weight of 110,000 g/mol.

Polymerization of TFE in CO₂-Expanded Acetic Acid.

A 25 ml high pressure reactor cell was purged with liquid carbon dioxide three times to remove air. Acetic acid (5 ml), pre-purged with Argon, was added into the cell under Argon flow, and the reactor was resealed. A mixture of TFE and carbon dioxide (50/50 by weight, 2.6 g) was charged into the reactor cell through a high-pressure syringe pump. The reactor was then slowly heated to 35 °C. A solution of bis(perfluoro-2-propoxypropionyl) peroxide in 1,1,2-trichlorotrifluoroethane (0.204 M, 0.006 ml) was injected into an addition tube under Argon flow, and the tube was quickly resealed. Carbon dioxide was charged into the cell slowly through the addition tube and carried the initiator into the cell. The acetic acid was quickly expanded by carbon dioxide until the reactor was fully filled with the mixture, at a pressure of approximately 7 MPa.

Carbon dioxide was further added until the initial pressure reached approximately 17 MPa. The reaction was held for 4 hours before being cooled down to room temperature. The final pressure was approximately 13 MPa. Carbon dioxide, excess TFE and other volatiles were slowly vented. The reactor was opened and the product, a white solid wetted with acetic acid, was collected. The product was washed with a mixture of 1,1,2-trichlorotrifluoroethane and methanol (1/1 by volume) three times and dried in a vacuum oven at elevated temperature. A white powdered PTFE (0.71 g, 53.7% yield) was recovered. Thermogravimetric analysis gave a 5 % weight loss at 492 °C. Differential scanning calorimetry showed a melting temperature at 328.3 °C with a 10 °C/min. heating rate. The integration of the melting peak gave a heat of crystallization of 50.1 J/g, corresponding to an estimated molecular weight of 57,000 g/mol.

Polymerization of TFE in CO$_2$-Expanded Acetic Acid Using SDS.

A 25 ml high pressure reactor cell was purged with liquid carbon dioxide three times to remove air. Sodium dodecyl sulphate (0.0144 g) was dissolved into acetic acid (5.0 ml) and then purged with Argon for 30 min. The solution was added into the cell under Argon flow, and the reactor was resealed. A mixture of TFE and carbon dioxide (50/50 by weight, 2.6 g) was charged into the reactor cell through a high-pressure syringe pump. The reactor was then slowly heated to 50 °C. Di(2-ethylhexyl) peroxydicarbonate (0.005 ml) was injected into an addition tube under Argon flow, and the tube was quickly resealed. Carbon dioxide was charged into the cell slowly through the gas addition tube, carrying the initiator into the cell. The acetic acid was expanded by carbon dioxide. The pressure was approximately 8.5 MPa when the reactor was fully filled with the one-phase mixture. The reaction was held for 4 hours before being cooled down to room temperature. The final pressure was approximately 8.0 MPa. Carbon dioxide, excess TFE and other volatiles were slowly vented. The reactor was opened and a suspension of PTFE in acetic acid was collected. The polymer suspension was used directly in preparation of the samples for scanning electron microscope and dynamic light scattering measurements. Since not much polymer was collected, and a significant amount was used for analysis, the yield, apparently very low, was not measured. The polymer suspension was dried under vacuum for 12 hrs. The polymer was used for further characterization without purification.

Results and Discussion

This study was focused on the homopolymerization of TFE to make PTFE, a fluoropolymer with the largest production volume and the most widely established applications among all the fluoropolymers. The TFE feedstock used in this study was in an azeotropic mixture with CO_2 (50/50 by weight) which makes the handling of TFE much safer than the previous methods (16). **Caution is advised for any TFE-related reaction.** Polymerization of TFE in supercritical CO_2 was reported in the 1990s (5,17). In order to compare with the polymerization in CXLs, we carried out the homopolymerization of TFE in supercritical CO_2 at different pressures following the literature procedure (5). At 17 MPa, a relatively high yield of PTFE, 67.9%, was obtained using bis(perfluoro-2-propoxypropionyl) peroxide (3P), an initiator commonly used in fluoropolymer production (Entry 1, Table 1). In sharp contrast, the yield was much lower, approximately 6% when the polymerization pressure was reduced to 8.5 MPa, slightly above the critical point (Entry 2, Table 1). Since no chain transfer agent was present in the system, the low yield appears to be a direct result of the low reaction pressure. The polymer prepared at low pressure had inferior thermal stability compared to the polymer prepared at high pressure (Figure 1) suggesting it was of very low molar mass.

Table 1. Polymerization of TFE in scCO$_2$

Batch	P (MPa)	Yield (%)	$M_n{}^a$ (g/mol)	T^b (°C)	T_m (°C)
1	17	67.9	1.1×10^5	514.5	328.4
2	8.5	6.1	N/A	355.3	N/A

Conditions: TFE percent solid: 5 wt%; Initiator: bis(perfluoro-2-propoxypropionyl) peroxide; [I]: 0.01 mol%; Temperature:35 °C; Time: 4 hrs; ^{a}: Calculated based on the reported method (15); ^{b}: Temperature for 5% weight loss on thermogravimetric profile; T_m: melting temperature.

Previously reported Gas-expanded liquids (GXL) studies were generally used at conditions below the critical point of the mixture (9), which for many systems were in the range of 2-4 MPa (10), although near-critical pressure or higher pressures are sometimes used for certain applications (18). Usually, both gas and liquid phases with a clear boundary were present. In this study, we intended to use only the liquid phase of the expanded solvent in order to simplify the research. This was realized by expanding the solvent to the full volume of the reactor and then further filling the reactor with CO_2 to the desired pressure range which was typically from 5.9 to 17 MPa. The organic solvents

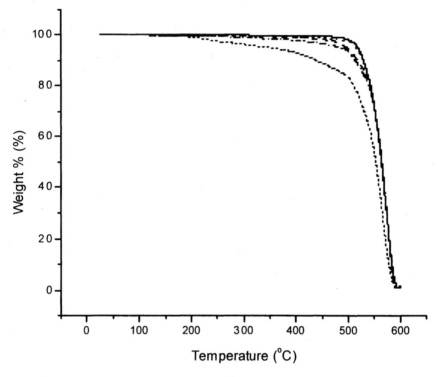

Figure 1. TGA of the PTFE samples prepared in scCO₂ at 17 MPa
(Entry 1 in Table 1, solid), in 4/96 AA/CO₂ at 8.5 MPa (Entry 5 in Table 3, dot),
in 40/60 AA/CO₂ at 17 MPa (Entry 4 in Table 3, dash), in 40/60 AA/CO₂ at 7.0
MPa (Entry 8 in Table 4, dash dot), in scCO₂ at 8.5 MPa (Entry 2 in Table 1,
short dash).

were all Class II liquids (9) and the molar ratio of CO_2 to the solvent was in the range of 30 – 98%. For convenience, the solvent ratios reported here are the volume ratio between the solvent and CO_2. It should be pointed out that the same volume ratio of two solvents reported at different pressures represents different molar ratios.

A wide variety of solvents have been reported to be compatible with or expandable by CO_2 (8,9). In this study, solvents that are more environmentally-beneficial, and with potentially smaller chain transfer constants in fluoropolymer synthesis were preferred. The chain transfer constants of various common organic solvents were determined previously in presence of perfluoroalkyl radicals (19). The transition state of hydrogen abstraction by highly electrophilic perfluorinated radicals favors hydrogen on carbons adjacent to an electron donating group (19,20). Therefore ether linkages generally facilitates this transition state while halogen atoms such as fluorine and chlorine usually disfavor it (19,20). Some of the better organic solvents screened in this study are listed in Table 2. These solvents all contain hydrogen and electron-withdrawing groups in their structures. Perfluorinated solvents were not considered in this study due to their persistence in the environment and their high global warming potential. Solvents containing other halogens such as chlorine and bromine are potentially ozone-depleting and were not considered. Extremely volatile compounds were also not studied, as their ability to lower the pressure of CO_2 required for effective polymerization was expected to be limited.

Table 2. Polymerization of TFE in various CO_2 expanded organic solvents

Batch	Solvent	Structure	Yield (%)	M_n^a (g/mol)	T^b (°C)	T_m (°C)
1	Acetone	$(CH_3)_2C=O$	17.4	N/A	387.5	321.1
2	Acetonitrile	CH_3CN	20.1	N/A	456.6	327.9
3	Acetic acid	CH_3COOH	63.7	5.2×10^4	514.0	328.0
4	Solkane® 365 mfc	$CF_3CH_2CF_2CH_3$	77.6	1.0×10^5	522.2	327.2
5	Vertrel® XF	$CF_3CHFCHFCF_2CF_3$	71.3	3.2×10^5	510.3	326.5
6	HFE-449	$CH_3OC_4F_9$	59.1	1.5×10^5	522.2	328.4
7	HFE-356	$CF_3CH_2OCH_2CF_3$	59.3	1.2×10^5	475.5	328.2

Conditions: TFE percent solids: 5 wt%; Solvent/CO_2 (V/V): 14/86; Initiator: bis(perfluoro-2-propoxypropionyl) peroxide; [I]: 0.01 mol%; Pressure: 16-17 MPa; Temperature: 35 °C; Time: 4 hrs; [a]: Calculated based on the reported method; [b]: Temperatures for 5% weight loss on thermalgravimetric profiles; T_m: melting temperature.

Among all the tested hydrocarbon solvents, only a few were able to produce an appreciable yield of PTFE. With acetone or acetonitrile, low yields were obtained (Entry 1 & 2, Table 2). Moreover, thermal degradation of the polymers produced in these two solvents began at much lower temperatures than reported for commercial PTFE samples, either due to high hydrocarbon content or existence of low molar mass polymer. Low yields obtained in these polymerizations were probably due to the inactive solvent radicals (generated by chain transfer reaction to solvent) or a high proportion of low-mass products (oligomers) which were lost in subsequent purification steps. In sharp contrast, high yields of PTFE with comparable physical properties to that prepared in pure scCO$_2$ were obtained in CO_2 expanded acetic acid (Entry 3, Table 2). This result was not originally expected, since a high percentage of such a hydrocarbon-based solvent was expected to produce either polymers with poor physical properties, or no polymer at all (21-23). The PTFE sample prepared in CO_2-expanded acetic acid had a slightly lower molecular weight than that prepared in pure CO_2 at 17 MPa, but the thermal stability was comparable.

Further experiments using hydrofluorocarbon solvents gave comparable results (Entry 4 – 7, Table 2). In this case, molecular weights of the polymers were slightly higher than that obtained with CO_2-expanded acetic acid (Entry 3, Table 2), presumably due to the lower chain transfer constants associated with fewer hydrogen atoms and the presence of multiple fluorine atoms to destabilize the transition state of the chain transfer reaction. Improvements in yield and molecular weight, compared with PTFE prepared in pure $scCO_2$ at the same pressure and temperature (Entry 5 of Table 2 vs Entry 1 of Table 1), was also observed. Lower yields were obtained using hydrofluoroethers compared with hydrofluoroalkanes, a result of the enhanced chain transfer to hydrofluoroethers (19,20). In the case of HFE-356, a noticeable compromise on thermal stability was also observed (Entry 7, Table 2). Hydrofluoroethers have been previously used as chain transfer agents in fluoropolymer synthesis (23).

The subsequent experiments were focused on acetic acid (AA), which was expected to be more environmentally friendly and economical than halogenated solvents. Table 3 details the polymerizations of TFE in CO_2-expanded acetic acid at various AA/CO_2 ratios and total pressures. Two sets of polymerizations were carried out at different pressures. In the first set, a constant pressure of approximately 17 MPa was maintained (Entry 1-4, Table 3). In the second set, we first visually expanded the solvent to the full volume of the reactor and then raised the pressure further to assure the absence of a gas phase (Entry 5-9, Table 3). This pressure varied from approximately 8.5 to 5.9 MPa depending on the ratio of acetic acid to CO_2. The yields were always higher than 50% when acetic acid was kept at 20 vol% level or below at 17 MPa (Entry 1-3, Table 3). Doubling the acetic acid concentration to 40 vol% at this pressure gave a lower yield (Entry 4, Table 3). When the polymerization pressure was further lowered to 7 and 8.5 MPa (Entry 6-8, Table 3), the yields were still comparable to those at a higher pressure (Entry 1-3, Table 3). This is unlike the reaction behavior in pure $scCO_2$, where the low pressure resulted in a substantially lower yield (Entry 2, Table 1). However, the yield was somewhat lowered when acetic acid concentration was reduced to only 4 vol% at 8.6 MPa (41.7% in Entry 5 vs 68.1% in Entry 1, Table 3). An even lower pressure of 5.9 MPa (approximately 870 psi) was still effective for polymerization in the co-solvent system with a 60/40 AA/CO_2 (Entry 9, Table 3), although the yield was further lowered. Polymerization is not expected in the pure CO_2 at this pressure since it would be a gas.

The thermogravimetric profiles of the PTFE prepared in above experiments showed comparable thermal stability to the PTFE prepared in neat $scCO_2$ at 17 MPa (see Figure 1). It is interesting to see that only 4% of acetic acid can dramatically improve the thermal stability of the resulting PTFE at 8.5 MPa. Thermal stabilities of the PTFE samples were somewhat compromised when increasing acetic acid ratio were used. However, there was no apparent degradation at around 200 °C even when the acetic acid ratio was increased to 40% at both 17 and 7.0 MPa. The CO_2-expanded acetic acid could be similarly

Table 3. Polymerization of TFE in CO_2-expanded acetic acid

Batch	AA/CO$_2$ (V/V)	P (MPa)	Yield (%)	M_n^a (g/mol)	T^b (°C)	T_m (°C)
1	4/96	17	68.1	8.5 x 10^4	506.8	328.0
2	8/92	17	60.2	7.6 x 10^4	502.4	328.1
3	20/80	17	53.7	5.7 x 10^4	492.2	328.3
4	40/60	17	22.0	4.6 x 10^4	489.8	327.9
5	4/96	8.5	41.7	6.8 x 10^4	507.7	327.2
6	8/92	8.5	71.1	5.3 x 10^4	508.2	328.0
7	20/80	7.0	76.3	7.6 x 10^4	503.5	328.6
8	40/60	7.0	52.9	4.0 x 10^4	472.8	327.2
9	60/40	5.9	38.1	3.0 x 10^4	483.4	327.4
10	20/80	8.5	N/A	N/A	309.6	N/A

Conditions: Entry 1-9: TFE percent solids: 5 wt%; Initiator: bis(perfluoro -2-propoxypropionyl) peroxide; [I]: 0.01 mol%; Temperature: 35 °C; Time: 4 hrs; Entry 10: TFE percent solid: 5 wt%; Initiator: di(2-ethylhexyl) peroxydicarbonate; [I]: 0.01 mol%; Temperature: 50 °C; Time: 20 hrs; 1 wt% of SDS; [a]: Calculated based on the reported method; [b]: Temperatures for 5% weight loss on thermogravimetric profiles; T_m: melting temperature.

applicable for the polymerization of many other fluoroolefins. Preliminary results show that the copolymerization of TFE and hexafluoropropene (HFP) was successful, and copolymer with comparable properties to commercial FEP was obtained.

Another benefit to using CO_2-expanded liquids, compared with scCO_2, is the adjustable polarity of the solvent mixture (10). Carbon dioxide is intrinsically non-polar and has limited solubility for polar solutes and compounds with relatively high molecular weights (24). As we have shown, the solvent quality of CO_2 can be improved through the use of co-solvents. The utility of carbon dioxide can also be expanded through the use of surfactants, however, for pure CO_2, this usually demands specially designed surfactants that contain a "CO_2-philic" domain such as fluorinated (4,26-29), siloxane (30,31), and some hydrocarbon compounds with special structures (32-34). Using the CXL system, such as the CO_2-expanded acetic acid used in this study, we expected a wider polarity range could be obtained (10) which in turn should substantially alter the interaction between CO_2 and conventional surfactants. As such, we set out to see if conventional hydrocarbon surfactants may become effective in CXLs which would increase the potential applicability of CO_2 as a reaction medium.

With this in mind, we studied the use of a conventional hydrocarbon surfactant, sodium dodecyl sulfate (SDS), to run the emulsion polymerization of TFE in the CO_2-expanded acetic acid (Entry 10, Table 3). After testing a few

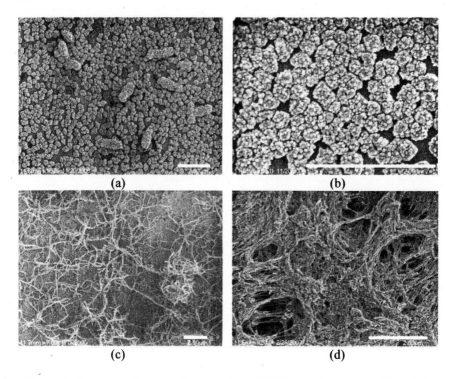

Figure 2. Scanning electron micrographs of PTFE samples prepared in scCO₂ and CO₂ expanded acetic acid. (a), (b) Colloidal PTFE prepared in CO₂ expanded acetic acid in presence of SDS (Entry 10, Table 3); (c) PTFE fibrils prepared in scCO₂; (d) PTFE fibrils prepared in CO₂ expanded acetic acid; The scale bar is 2 μm.

solvent ratios, we collected some colloidal PTFE particles as confirmed by scanning electron microscopy (SEM) (Figure 2a). As a comparison, both the polymers prepared in scCO₂ and in CO₂-expanded acetic acid in absence of any surfactant are fibrillar (Figure 2c & d). The particles shown in Figure 2a are comprised of much smaller primary particles (Figure 2b). Dynamic light scattering measurement of the particles shown in Figure 2a gave a mean diameter of approximately 220 nm by diluting the PTFE/acetic acid mixture with approximately 10 times of water. As a comparison, no particles were detected after the mixture was diluted with acetic acid. A similar study using PFPE carboxylate surfactants to assist the polymerization of TFE in scCO₂ was reported recently (35). Colloidal PTFE particles were observed with SEM in the presence of surfactants. However, the polymer particles appeared to be much larger than those obtained in this study (35). TGA measurement of the resulting polymer shows a two-step degradation (Figure 3): the second curve which occurred at approximately 500 °C appears to be the characteristic degradation of PTFE. The yields in presence of SDS, however, were much lower than those prepared in CO_2-expanded acetic acid without surfactant. We estimated the yields to be lower than 10%.

A few other more polar solvents, such as DMSO or DMF, were expected to offer a wider adjustable range of polarity and may be more effective for the above emulsion-like polymerization using conventional surfactants. These solvents were tested in this study without generating appreciable PTFE, probably due to having higher chain transfer constants in TFE polymerization. However, they may be suitable for other polymer systems, such as many hydrocarbon polymers. Combining CO_2-expanded solvents with conventional surfactants may represent an alternative to designing special surfactants for $scCO_2$.

In conclusion, we report for the first time the polymerization of a fluoroolefin, TFE, in CO_2-expanded liquids to synthesize high molar mass PTFE. Both hydrocarbon and hydrofluorocarbon solvents were successful in producing PTFE with satisfactory yields with good thermal stability. Use of a common hydrocarbon solvent, acetic acid, was able to dramatically lower the

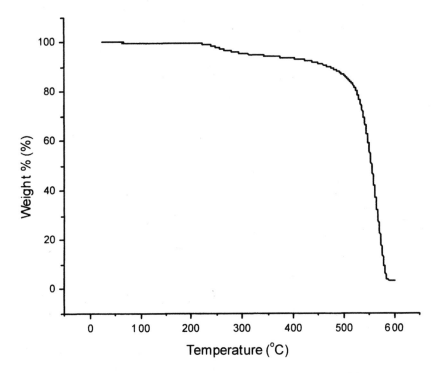

Figure 3. TGA graph of the PTFE from Entry 10 in Table 3.

pressure required for the polymerization of TFE in CO_2. Successes with a wide range of solvent ratios, from 4 to 60 vol% of acetic acid, and pressure, from 5.9 to 17 MPa, were demonstrated. The chain transfer reaction to the hydrogen-containing solvents, especially acetic acid, was not as severe as expected. Colloidal PTFE samples were obtained by using a conventional hydrocarbon surfactant, SDS, in CO_2-expanded acetic acid. This may represent an alternative way to run emulsion polymerizations in CO_2.

Acknowledgements

We thank DuPont and NSF Science and Technology Center (NSF-STC) for Environmentally Responsible Solvents and Processes for support of this study.

References

1. *Modern Fluoropolymers*, Scheirs, J., Eds. Wiley: New York, 1997.
2. Wood, C. D.; Yarbrough, J. C.; Roberts, G. W.; DeSimone, J. M. In *Supercritical Carbon Dioxide in Polymer Reaction Engineering*; Kemmere, M. F.; Meyer, T., Eds.; Wiley-VCH: Wrinheim, 2005; p 189-204.
3. EPA. 2006; http://www.epa.gov/oppt/pfoa/.
4. DeSimone, J. M.; Guan, Z.; Elsbernd, C. S. *Science* 1992, *257*, 945-947.
5. Romack, T. J.; DeSimone, J. M.; Treat, T. A. *Macromolecules* 1995, *28*, 8429-8431.
6. DeSimone, J. M. *Science* 2002, *297*, 799-803.
7. Qian, J.; Timko, M. T.; Allen, A. J.; Russell, C. J.; Winnik, B.; Buckley, B.; Steinfeld, J. I.; Tester, J. W. *J. Am. Chem. Soc.* 2004, *126*, 5465-5474..
8. Hallett, J. P.; Kitchens, C. L.; Hernandez, R.; Liotta, C. L.; Eckert, C. A. *Acc. Chem. Res.* 2006, *39*, 531-538.
9. Jessop, P. G.; Subramaniam, B. *Chem. Rev.* 2007, *107*, 2666-2694.
10. Ford, J. W.; Lu, J.; Liotta, C. L.; Eckert, C. A. *Ind. Eng. Chem. Res.* 2007, ASAP Article.
11. Xu, D.; Carbonell, R. G.; Roberts, G. W.; Kiserow, D. J. *J. Supercrit. Fluids* 2005, *34*, 1-9.
12. Xu, D.; Carbonell, R. G.; Roberts, G. W.; Kiserow, D. J. *Polym. Prep.* 2004, *45*, 502-503.
13. Xu, D.; Carbonell, R. G.; Kiserow, D. J.; Roberts, G. W. *Ind. Eng. Chem. Res.* 2005, *44*, 6164-6170.
14. Kemmere, M. F.; Kuijpers, M. W. A.; Keurentjes, J. T. F. *Macromol. Symp.* 2007, *248*, 182-188.
15. Suwa, T.; Takehisa, M.; Machi, S. *J. Appl. Polym. Sci.* 1973, *17*, 3253-3257.

16. Van Bramer, D. J.; Shiflett, M. B.; Yokozeki, A.: US Patent 5, 345, 013, 1994.
17. Romack, T. J.; Kipp, B. E.; DeSimone, J. M. *Macromolecules* **1995**, *28*, 8432-8434.
18. Nunes, R. M. D.; Arnaut, L. G.; Solntsev, K. M.; Tolbert, L. M.; Formosinho, S. J. *J. Am. Chem. Soc.* **2005**, *127*, 11890-11891.
19. Shtarev, A. B.; Tian, F.; Dolbier, W. B., Jr.; Smart, B. E. *J. Am. Chem. Soc.* **1999**, *121*, 7335-7441.
20. Dolbier, W. B., Jr. *Chem. Rev.* **1996**, *96*, 1557-1584.
21. Feiring, A. E.; Krespan, C. G.; Resnick, P. R.; Smart, B. E.; Treat, T. A.; Wheland, R. C., US Patent 5, 182, 342, 1993.
22. Feiring, A. E.; Hung, M.-H.; Rodriguez-Parada, J. M.; Zipfel, R. J., US Patent 5, 763, 552, 1998.
23. Farnham, W. B.; Feiring, A. E.; Lindner, P. E., US Patent 6, 399, 729 B1, 2002.
24. Kemmere, M. F. In *Supercritical Carbon Dioxide in Polymer Reaction Engineering*; WIley-VCH: Weinheim, 2005; p 1-14.
25. Consani, K. A.; Smith, R. D. *J. Supercrit. Fluids* **1990**, *3*, 51-56.
26. Kho, Y. W.; Conrad, D. C.; Knutson, B. L. *Langmuir* **2004**, *20*, 2590-2597.
27. Johnston, K. P.; DaRocha, S. R. P.; Holmes, J. D.; Jacobson, C. B.; Lee, C. T.; Yates, M. Z. In *Green Chemistry Using Liquid and Supercritical Carbon Dioxide*; DeSimone, J. M.; Tumas, W., Eds.; Oxford University Press US, 2003; p 134-148.
28. McClain, J. B.; Betts, D. E.; Canelas, D. A.; Samulski, E. T.; DeSimone, J. M.; Londono, J. D.; Cochran, H. D.; Wignall, G. D.; Chillura-Martino, D.; Triolo, R. *Science* **1996**, *274*, 2049-2052.
29. Muller, P. A.; Storti, G.; Morbidelli, M.; Costa, I.; Calia, A.; Scialdone, O.; Filardo, G. *Macromolecules* **2006**, *39*, 6483-6488.
30. Fink, R.; Beckman, E. J. *J. Supercrit. Fluids* **2000**, *18*, 101-110.
31. DeSimone, J. M.; Keiper, J. S. *Current Opinion in Solid State and Materials Science* **2001**, *5*, 333-341.
32. Eastoe, J.; Gold, S. *Phys. Chem. Chem. Phys.* **2005**, *7*, 1352-1362.
33. Eastoe, J.; Gold, S.; Steytler, D., C. *Langmuir* **2006**, *22*, 9832-9842.
34. Eastoe, J.; Paul, A.; Nave, S.; Steyler, D. C.; Robinson, B. H.; Rumsey, E.; Thorpe, M.; Heenan, R. K. *J. Am. Chem. Soc.* **2001**, *123*, 988-989.
35. Giaconia, A.; Scialdone, O.; Apostolo, M.; Filardo, G.; Galia, A. *J. Polym. Sci. Polym. Chem.* **2007**, *46*, 257-266.

Chapter 15

Green Methods for Processing and Utilizing Metal Complexes

**Joseph G. Nguyen[1], Chad A. Johnson[2,3], Sarika Sharma[1,3],
Bala Subramaniam[2,3,*], and A. S. Borovik[4,*]**

Departments of [1]Chemistry and [2]Chemical and Petroleum Engineering,
University of Kansas, Lawrence, KS 66045
[3]Center for Environmentally Beneficial Catalysis, University of Kansas,
1501 Wakarusa Drive, Lawrence, KS 66047
[4]Department of Chemistry, 1102 Natural Science II, University
of California, Irvine, CA 92697–2021

The use of supercritical carbon dioxide ($scCO_2$) and CO_2-expanded liquids (CXLs) in metal-mediated processes is described. Two examples are examined: 1) heterogeneous oxidation catalysis and 2) processing nanoparticles of metal complexes. Oxidation processes with dioxygen as the terminal oxidant are often limited because of the low O_2 solubility in common organic solvents. The complete miscibility of dioxygen and $scCO_2$ alleviates this problem and provides improved mass transfer properties. These factors are exploited in the oxidation of substituted phenols with porous solids containing immobilized metal complexes that serve as catalytic sites. The development of nanoparticles composed entirely of metal complexes has lagged behind atom-based systems. Precipitation with compressed antisolvent, a method that uses $scCO_2$ as the precipitant, was employed to process nanoparticles of metal complexes. A correlation between molecular shape and particle morphology was discovered, in which planar compounds gave nanoparticles with rodlike structures. In contrast, three-dimensional molecular species produced spherical nanoparticles. Investigations into the binding of dioxygen and nitric oxide to the rod-like nanoparticles composed of cobalt complexes showed pronounced differences compared to their unprocessed analogs. Dioxygen reversibly binds to the nanoparticles, whereas nitric oxide is disproportionated into other N_xO_y species. This type of nitric oxide reactivity has not been observed for other cobalt complexes at room temperature and illustrates the promise of nanoparticles of metal complexes.

Introduction

It is now readily accepted that newer, cleaner chemical processes are necessary in order to prolong human health. A variety of factors have led to this realization, many of which are intertwined. Economic and environmental concerns are the two most obvious examples of factors that are linked. For example, there is a continual rise in the number of hazardous chemical waste sites in the U.S.—the Environmental Protection Agency estimates that 217,000 new hazardous waste sites will be needed by 2033, costing an estimated $250 billion dollars (1). The environmental impacts of the present day sites are at threatening levels, with many leaking chemicals into the groundwater. Moreover, there are increasing data to show that humans are accumulating industrial chemicals at an increased rate. These chemicals are often toxic and can lead to serious human health issues.

Conventional methods for solving the above problems often involve reduction of chemical exposure and environmental pollutants. These approaches have modest effectiveness and are costly. Based on this lack of success, newer methods are being developed that will produce goods in a safer and healthier manner. Green chemistry is one such approach, in which environmental and health hazards are removed from the chemical process at the outset of their formulation, before they are introduced to the public (2). One challenge in this field is to develop technologies that are comparable to existing ones in cost and yield, but use reagents that are significantly less harmful. Finding the correct combination of solvents and reagents is often difficult and usually requires interdisciplinary teams of chemists, engineers, and polymer scientists.

This chapter describes some of our recent findings in coupling supercritical carbon dioxide ($scCO_2$) technology with transition metal chemistry to develop new heterogeneous oxidation catalysts. In addition, we illustrate how $scCO_2$ can be used to process a new class of nanoparticles composed of metal complexes whose function differs from their molecular precursors.

Green Oxidation Chemistry

The potential of green chemistry is exemplified with oxidation chemistry, especially those processes related to commercial chemicals, in which megatons of products are annually manufactured with a net worth in the tens of billions of dollars. Nearly 70% of the bulk chemical oxidations are performed with stoichiometric reagents, such as permanganate and chromate (*3*), which are environmentally and economically unfavorable. One solution to this problem is to design systems using dioxygen as the terminal oxidant: it is plentiful, inexpensive, and a "green oxidant" because water is the most common byproduct.

Harnessing the oxidizing power of O_2 has attracted the attention of chemists for years (*4*). While thermodynamically competent to react with a variety of substrates, dioxygen is relatively inert under ambient condition—this is why it can co-exist with combustible species such as carbohydrates and wood-products instead of reacting with them to produce water and CO_2. Its stability is attributed to a kinetic sluggishness, a property related to O_2 having a triplet ground state. This kinetic limitation can be overcome through interactions with species containing unpaired electrons. Substances containing transition metal ions are the most commonly employed, as illustrated in the many oxygenase enzymes that utilize metal ion cofactors.

There are additional factors that can reduce the oxidizing ability of O_2. A key issue is dioxygen's relatively low solubility in most commonly employed solvents. Therefore, the concentrations of dissolved O_2 are sufficiently low to hinder the outcome of even the most kinetically viable reactions. One method to circumvent this problem is to use *sc*CO_2 or CO_2–expanded liquids (CXLs) as solvents (*5*). The advantage of these media resides in the complete miscibility of O_2 and CO_2 to provide reaction conditions having O_2 concentrations that far exceed those found with more conventional solvent systems (*6*). Furthermore, *sc*CO_2 is a green solvent when derived from existing abundant sources, which when coupled with O_2, would produce environmentally beneficial methodologies.

Metal Bis(salicylaldehyde)ethylenediamine (Salen) Complexes

Typically, metal complexes have low solubilities in *sc*CO_2 that can be somewhat overcome by using CXL as the solvents. An excellent example of this approach is the work done with [Co^{II}(salen)] complexes as oxidation catalysts for substituted phenols (*7*). The highest yields of oxidation products were obtained in CXL media, which was attributed to the increased solubility of both dioxygen and the [Co(salen)] catalysts.

There are longstanding historical reasons for using [Co(salen)] complexes as oxidation catalysts. The reversible binding of dioxygen to [Co(salen)] has been known for over 100 years and is likely the first discrete coordination complex to exhibit this type of binding (8). The solution chemistry of [Co(salen)] is well defined and it is clear that the salen coordination to the Co(II) ion is not sufficient to obtain efficient dioxygen binding (Figure 1) (9, 10). An additional ligand is needed to bind to the Co(II) center, forming a five-coordinate, square pyramidal complex which is then active toward O_2 binding. Coordinating solvents, such as acetonitrile, can serve as the fifth ligand—this is why CXL with acetonitrile was used in the above [Co(salen)] studies.

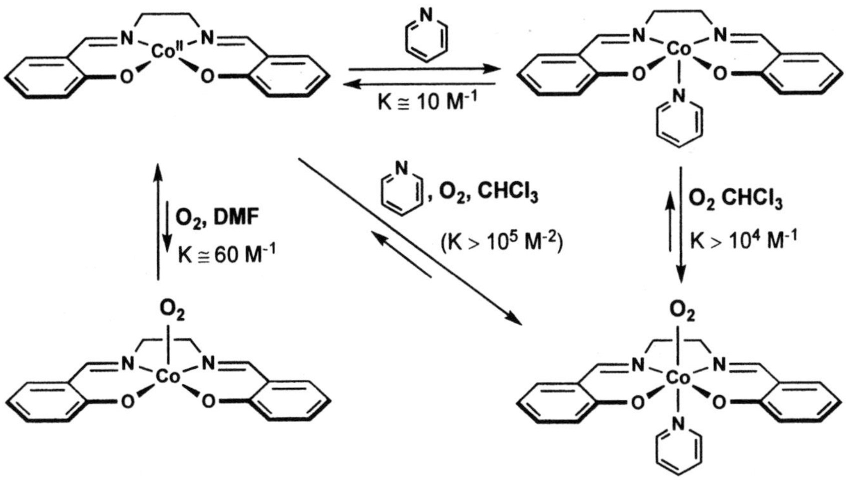

Figure 1. Reaction scheme for [Co(salen)] with O_2.

It is obvious that in order for dioxygen to be a useful oxidant, it needs to be reduced upon coordination to a metal complex. The cleavage of the O—O bond often leads to formation of metal oxides, whereby the oxo groups bridge between at least two metal centers. These species are usually unreactive and thus do not function as oxidation catalysts. One method to circumvent the problems is to immobilize the functional oxidation catalysts. Various methods have been developed, spanning the realm of homogeneous and heterogeneous systems. The goal of our work was to match green media with appropriate heterogeneous systems that reversibly bind O_2 to probe the factors for efficient oxidation of substrates.

278

Template Polymerizations with [Co(salen)] Complexes

We have found that immobilization of metal complexes within organic hosts can be accomplished using template copolymerization methods (*11, 12*). The materials have immobilized sites distributed randomly throughout porous organic hosts, which are highly crossed-linked network polymers that are usually made by free-radical polymerization methods (*13, 14, 15, 16, 17*). The immobilized sites are formed during polymerization through the use of template compounds that in our designs are stable transition metal complexes. These sites have similar architectures, corresponding to that of the template complex. A schematic of this method is shown in Figure 2. Our studies have shown that [Co(salen)] complexes can be immobilized within polymethacrylate hosts with

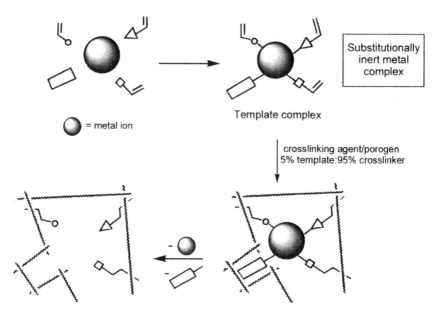

Figure 2. Schematic of the template copolymerization method.

good regio-control of the ligands surrounding the metal ions (*18, 19*). In addition, these materials have high site accessibility, allowing molecules such as dioxygen to bind to the Co(II) centers. Furthermore, the immobilized [Co(salen)] sites are sufficiently isolated to prevent undesirable intermolecular metal-metal interactions that can hinder dioxygen binding and subsequent oxidative processes. Details of the preparation of these materials have been reported (*18, 19, 20, 21, 22, 23*), and our polymer work has been recently reviewed (*11, 12*).

Figure 3 depicts three of these materials that are relevant to the oxidation catalysis. The salen ligands are modified to include styryloxy groups attached to the aromatic rings. These groups are polymerizable, providing covalent attachment of the [Co(salen)] complexes to the polymer backbone; this type of attachment limits leaching of the complexes during catalysis. We have shown that many of these [Co(salen)] immobilized materials bind O_2 and NO—in some cases the binding properties are superior to those found for their molecular analogs. (*18, 21, 22*) For instance, 90% of the immobilized Co(II) sites can reversibly bind O_2 in P-1.py[Co^{II}], a material with immobilized five-coordinate Co(II) sites. In contrast, the material with immobilized sites containing four-coordinate cobalt complexes, P-1[Co^{II}], has a relatively low affinity for dioxygen with only 10% of the sites forming O_2 adducts. These results illustrated how we can tune a functional property (O_2 binding) by adjusting the structure of the immobilized [Co(salen)] complexes. We were thus curious if these tunable properties would lead to heterogeneous catalysts with different activities.

In solution, molecular [Co(salen)] was shown to oxidize 2,6-di-*tert*-butylphenol (DTBP) to 2,6-di-*tert*-butyl-1,4-benzoquinone (DTBQ) and 3,5.3',5'-tetra-*tert*-butyl-4,4'-diphenoquinone (TTBDQ) (eq 1) (*7*). We recognized that the optimal experimental conditions for our materials could be different from those used for the homogeneous systems. In particular, *sc*CO$_2$ should be a more compatible solvent for heterogeneous oxidation systems because of its resistance to oxidation and complete O_2 miscibility. Supercritical CO$_2$ also has tunable transport properties that are important for porous heterogeneous catalysts in which mass transport factors can be rate limiting.

Heterogeneous Oxidation Catalysis in *sc*CO$_2$ Media

We have surveyed the catalytic oxidation of DTBP with the three polymers shown in Figure 3 (*24*). Three different solvent systems were employed: neat acetonitrile, CXL using acetonitrile, and neat *sc*CO$_2$. All the polymers had little catalytic activity in neat acetonitrile. For instance, only 11% of the oxidized products were observed with P-1[Co^{II}] at 35°C; this value only increased to 20% at 80°C (Table I). The largest oxidation conversions were found when the reactions were done in *sc*CO$_2$ with a 50% yield found at 80°C.

The findings with P-1[Co^{II}] illustrated the promise of doing heterogeneous oxidation catalysis with *sc*CO$_2$ media. However, we sought to improve on the conversions found with P-1[Co^{II}] by using catalysts with higher affinities for dioxygen. Our reasoning was that higher catalytic conversions would be obtained if the immobilized sites had greater affinities for dioxygen. Using the solution chemistry of [Co(salen)] as a guide, we employed materials whose immobilized sites contained five-coordinate Co(II) centers—these types of complexes are known to readily bind O_2 (Figure 1). With this as our goal, we probed the oxidation of DTBP using P-1·py[Co^{II}] and P-2[Co^{II}] (Figure 3). As

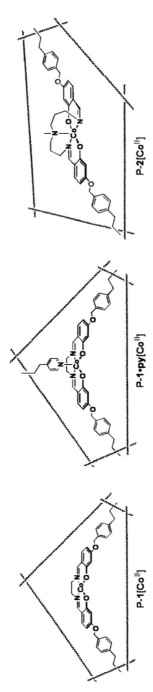

Figure 3. Active site structures of heterogeneous catalysts.

(1)

DTBP   DTBQ   TTBDQ

Table I. Substrate conversions for catalysts P-1[Co^{II}],
P-1·py[Co^{II}], and P-2[Co^{II}] at 35 and 80 °C in various
reaction media (error limits in parentheses).

Solvent	Catalyst	$x(CO_2)$	T(°C)	P (bar)	%X
CH$_3$CN	P-1[Co^{II}]	0	35	1	11(1)
CXL	P-1[Co^{II}]	0.695	35	55	11(1)
scCO$_2$	P-1[Co^{II}]	0.979	35	120	17(2)
CH$_3$CN	P-1[Co^{II}]	0	80	1	30(2)
CXL	P-1[Co^{II}]	0.695	80	94	43(2)
scCO$_2$	P-1[Co^{II}]	0.979	80	140	50(2)
CH$_3$CN	P-1·py[Co^{II}]	0	35	1	17(2)
scCO$_2$	P-1·py[Co^{II}]	0.979	35	120	23(3)
CH$_3$CN	P-1·py[Co^{II}]	0	80	1	28(1)
scCO$_2$	P-1·py[Co^{II}]	0.979	80	140	60(2)
CH$_3$CN	P-2[Co^{II}]	0	80	1	29(2)
CXL	P-2[Co^{II}]	0.695	80	96	39(2)
scCO$_2$	P-2[Co^{II}]	0.979	80	140	65(3)

shown in Table I, P-1·py[Co^{II}] gave substantially better conversion to products than P-1[Co^{II}]. For example, a 60% yield of products was observed at 80°C in scCO$_2$. Similar results were found for P-2[Co^{II}], with products conversion approaching 65%.

We also examined the reusability of the P-1·py[Co^{II}] and found essentially no change in activity between two catalytic cycles. However, there is a substantial loss of 30% in catalyst performance between the second and third cycles, which is attributed to the loss of cobalt from the immobilized sites. The causes for the decrease in cobalt content of the catalyst are not yet known.

These results illustrated several important points into the design of heterogeneous oxidation processes. Our findings demonstrate that porous solids with immobilized metal sites can be effective oxidation catalysts. The usual limitations caused by poor mass transfer through the pores have been somewhat mitigated through the use of CO_2-based solvents. This point is highlighted by the high conversion percentages found with the reactions in pure $scCO_2$, suggesting that it may be the best solvent for heterogeneous oxidations. A correlation was also obtained between the O_2 affinity of the catalysts and product conversion: higher O_2 affinity polymers gave higher yields of products. Finally, porous solids with immobilized metal sites have catalytic function over several repeated reactions, a property attributed to the protective effects of the polymer host that prevents deactivating metal-metal reactions.

Nanoparticles of Metal Salen Complexes

Control of Particle Morphologies

We have also been pursuing other opportunities to couple the chemistry of [M(salen)] complexes with $scCO_2$ media. One exciting opportunity was to use $scCO_2$ to process [M(salen)] nanoparticles. Processes for sub-micron particles of *molecules* have lagged behind atom-based systems because of the lack of convenient methods for particle preparation. Conventional techniques for particle size reduction, including milling, recrystallization from liquid cosolvents, freeze-drying and spray drying, may result in excessive use of organic solvent, thermal and chemical degradation of the solute, trace residues in the precipitate, and interbatch particle size variability (25). Precipitation with compressed antisolvents, or PCA (26, 27, 28, 29, 30), is an alternative approach to the production of nanoparticulates. PCA, also known as supercritical antisolvent (SAS) precipitation or SAS, is a semi-continuous process, which in theory can use a variety of solvent mixtures—in our processes $scCO_2$ is used as the precipitant. The nonpolar carbon dioxide dissolves into a solution of the polar compound causing substantial supersaturation and nucleation, affording uniform particles with length scales between nanometers to microns. Another similar process for dense gas assisted particle formation is gas antisolvent precipitation, or GAS, which is a batch process that initiates the precipitation by expanding the liquid phase with a dense gas (31). Both PCA and GAS technologies have several advantages including higher production output over conventional methods and a one-step route to pure, relatively dry particles that are formed under mild conditions in the absence of additional processing. However, PCA is preferred over GAS because the spraying of the solution as droplets in the PCA process provides more efficient mass transfer between the dense gas and solution, producing smaller particles (32). Further, because the extracted solvent is continuously removed by the $scCO_2$, separation and collection of the precipitated particles is much simpler in the PCA process.

PCA is most often used to process organic compounds, which are usually pharmaceuticals, such as taxol and insulin. Prior to our work this technology had not been used to make nanoparticles composed of metal complexes. However, we proposed that PCA could also be used to prepare nanoparticulate metal complexes, producing new materials that may have distinct structural and functional properties. To test this idea we prepared a series of nanoparticles of different metal salen complexes (*33*). A schematic of the instrumentation used to prepare the particles is shown in Figure 4. The procedure, in brief, involved spraying CH_2Cl_2 solutions of the appropriate complex and $scCO_2$ simultaneously through a coaxial nozzle (150 μm ID for the inner tube), causing instantaneous precipitation of the complexes as processed particles. Collection of the particles was achieved via a 0.2 μm filter.

The structures in Figure 5 illustrate some of the metal complexes we have examined as particles. All the complexes have salen ligands coordinated to the metal centers within the equatorial plane. They differ in the number of additional ligands bonded to the metal centers. Simple [M(salen)] complexes have square planar coordination geometries, resulting in nearly planar molecular units. Additional ligands bind perpendicular to the plane, which affords complexes having three-dimensionality.

The geometric differences between the complexes at the molecular level influence the morphology of the processed particles holding other experimental parameters relatively constant. Scanning electron microscopy (SEM) was used to examine the differences in the morphology of the processed particles. For instance, an SEM image of processed [Ru(salen)(NO)(Cl)] complex (Figure 6, right) shows aggregates of primary particles having spherical structures, with average diameters of 50 nm. In contrast, the SEM image of unprocessed [Ru(salen)(NO)(Cl)] depicted flat irregular shards with sizes ranging from microns to millimeters (Figure 6, left). Magnification of these shards did not reveal the presence of discrete primary particles; rather, only amorphous surfaces were observed.

The spherical morphology of the processed [Ru(salen)(NO)(Cl)] is typical of particles produced by PCA, in which nearly all the compounds investigated have nonplanar molecular structures. We found however that particles of planar [M(salen)] complexes have markedly different morphologies. The SEM images of [Ni(salen)] and [Co(salen)] (Figure 7) clearly show these nanoparticulates have rodlike structures, with average diameters and lengths of 85 and 700 nm, respectively. We have found a few reports describing formation of other rodlike nanoparticles using PCA methods, and in each case, planar compounds were employed (*34, 35*). We have thus proposed that rodlike morphologies are the structural motifs for particles arising from planar compounds.

The structural findings with the metal salen complexes have important consequences in nano-technology. Numerous parameters have been shown to

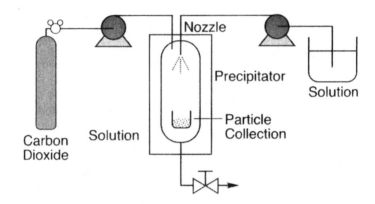

Figure 4. Schematic of the PCA instrumentation.
(See page 1 of color insert.)

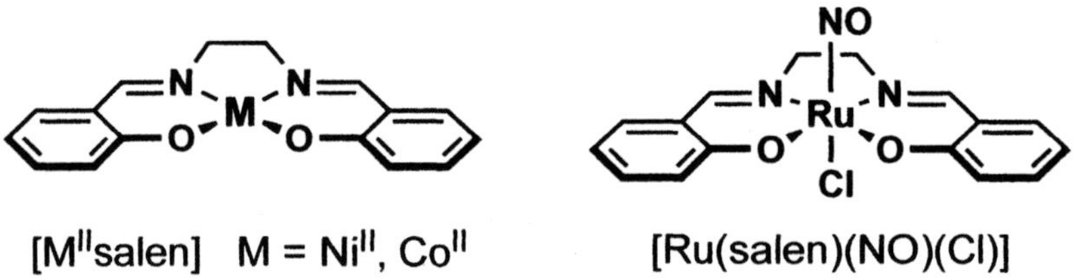

[M^{II}salen]   M = Ni^{II}, Co^{II} [Ru(salen)(NO)(Cl)]

Figure 5. Molecular structures of compounds used in the PCA studies.

Figure 6. SEM images of unprocessed [Ru(salen)(NO)Cl)] (left) and the spherical primary particles obtained from the PCA process (right).

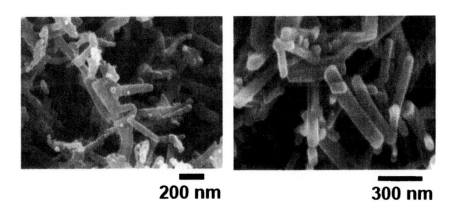

Figure 7. SEM images of [Co(salen)] (left) and [Ni(salen)] (right) nanoparticles.

affect the size of particles prepared via PCA, including temperature, pressure, and CO_2 flow rate. Until our work, there was no mention of a correlation between the molecular structures of the initial compounds being processed and the resulting morphologies of the nanoparticulates. We recognized that particle formation with PCA technology is a complex process with several variables influencing the final structures of the particles. Our findings suggest that a bottom-up approach based on molecular structure of the precursor compounds may provide enabling methods for control of structures of nanoparticles composed entirely of molecules. This possibility offers an exciting new approach to particle design, whereby structure-function relationships could be used to enhance existing function, or discover new ones that are unique to processed nanoparticles.

Conclusions and Vistas

We have described our efforts to incorporate $scCO_2$ within the framework of metal coordination chemistry. The use of $scCO_2$ as a solvent for heterogeneous oxidation catalysis enhanced the overall efficiency of the reactions because of improved O_2 solubility and greater mass transfer of the oxidant into the immobilized sites. Nanoparticles of metal complexes were prepared for the first time using PCA methods that utilize $scCO_2$ as the precipitant. Control of the particle morphology was linked to the molecular structure of the complexes.

The two projects discussed in this chapter are seemingly unrelated. The link between them is the coupling of $scCO_2$ media with metal salen complexes. In both cases, $scCO_2$ was used as an enabling medium to create new systems with different reactivities than the norm. Inorganic chemists have used metal salen complexes for over 100 years, but their chemical properties in $scCO_2$ are just emerging. The promising new findings obtained in our studies underscore the need to test even the oldest, most studied compounds in new media, especially those that are environmentally compatible.

The promise of these systems is illustrated in our current work on gas binding to the [M(salen)] nanoparticles (36,37). In particular, we have been investigating the binding of dioxygen and nitric oxide (NO) to the [Co(salen)] nanoparticles. Exposure of the nanoparticles to these gases imparts noticable color changes, as indicated in the photographs in Figure 8. The unprocessed [Co(salen)] does not show any color changes when treated with either gas. This simple test demonstrates that the [Co(salen)] nanoparticles have *functionally* different properties than their unprocessed analogs. Quantititative charac-terizations of these unique functional properties are provided in a recently completed dissertation (37).

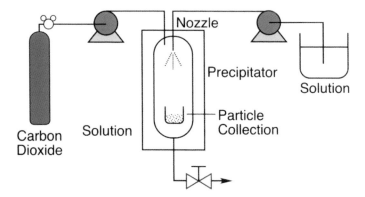

Figure 15.4. Schematic of the PCA instrumentation.

Figure 15.8. Photographic images of [Co(salen)] nanoparticles (A) and after exposure to NO (B) and O₂ (C).

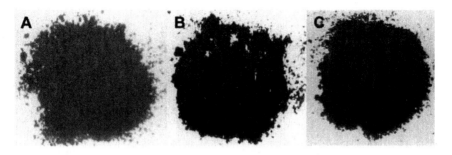

Figure 8. Photographic images of [Co(salen)] nanoparticles (A) and after exposure to NO (B) and O₂ (C). (See page 1 of color insert.)

More recently, we have been working on establishing the underlying structure-function relationships as to why the nanoparticles function so differently (*36*). We discovered that [Co(salen)] nanoparticles reversibly bind dioxygen at room temperature, a somewhat unusual result for Co(salen) systems which, in both in the condensed and the solid phases, often irreversibly bind O_2. Detailed spectroscopic studies on the processed and unprocessed [Co(salen)] revealed that this difference in O_2 binding may be rooted in the tetrahedral molecular structure of the [Co(salen)] complexes within the nanoparticles. [Co(salen)] normally adopt a square planar coordination geometry in both solution and the solid-state. We attribute this structural difference to the arrangement of the molecules within the nanoparticle, which is different than in an amphorous solid or a conventional crystal lattice (*37*).

In contrast, the [Co(salen)] nanoparticles do not reversibly bind NO. To our surprise, these nanoparticles appear to disproportionate NO into other N_xO_y species at room temperature (*37*). This is an exciting result because it is the first example of this type of reactivity by a solid-state cobalt system at room temperature. Normally, cobalt-based systems only disproportionate NO at temperatures above 400°C (*38, 39, 40, 41, 42, 43, 44, 45*). We are currently pursuing the structural reasons for this unusual reactivity, yet these results offer sufficient encouragement that other new functional properties await to be discovered with nanoparticles composed of metal complexes.

Acknowledgements

The financial support from the NSF-ERC program (NSF-EEC 0310689) and the NIH (GM58680 to A.S.B.) is gratefully acknowledged.

288

References

1. www.epa.gov/superfund/
2. See for example, Anastas, P. T.; Warner, J. C. *Green Chemistry: Theory and Practice*, Oxford University Press: New York, NY, 2000.
3. Davis, D. D.; Kemp, D. R. in *Kirk-Othmer Encyclopedia of Chemical Technology*. 4th ed., V1; Kroschwitz, J. I. Ed.; John Wiley & Sons, New York, NY, 1991; p 466.
4. Borovik, A. S.; Zart, M. K.; Zinn, P. J. in *Activation of Small Molecules: Organometallic and Bioinorganic Perspectives*; Tolman, W. B., Ed.; Wiley-VCH: Weinheim, 2006; p 187.
5. Musie, G. T.; Wei, M.; Subramaniam, B.; Busch, D. H. *Coord. Chem. Rev.,* **2001**, *789*, 219.
6. Arai, Y.; Sako, T.; Takebayashi, Y. *Supercritical Fluids. Molecular Interactions, Physical Properties, and New Applications,* Springer-Verlag: Berlin, 2002.
7. Wei, M.; Musie, G. T.; Busch, D, H.; Subramaniam, B. *J. Am. Chem. Soc.* **2002**, *124*, 2513.
8. Werner, A.; Mylius, A. Z. *Anorg. Allg. Chem.* **1898**, *16*, 245.
9. Cesarotti, E.; Gullotti, M.; Pasini, A.; Ugo, R. *J. Chem. Soc. Dalton Trans.* **1977**, 757.
10. Jones, R. D.; Summerville, D. A.; Basolo, F. *Chem. Rev.* **1979**, *79*, 139.
11. Borovik, A. S. *Proc. Indian Nat'n. Sci. Acad.* **2004**, *70*, 1.
12. Welbes, L. L.; Borovik, A. S. *Acc. Chem. Res.* **2005**, *38*, 765 and references therein.
13. Wulff, G. *Angew. Chem. Int. Ed. Engl.* **1995**, *34*, 1812.
14. *Molecular and Ionic Recognition with Imprinted Polymers*; ACS Symp. Ser. No. 703; Bartsch, R.; Maeda, M., Eds.; American Chemical Society: Washington, DC, 1998.
15. *Molecular Imprinted Polymers: Man-Made Mimics of Antibodies and Their Applications in Analytical Chemistry*; Sellergren, B., Ed.; Elsevier: Amsterdam, 2001.
16. *Molecular Imprinted Materials: Science and Technology*; Yan, M.; Ranström, O., Eds.; Marcel Dekker: New York, NY, 2005.
17. Conrad, P. G., III; Shea, K. J. in *Molecular Imprinted Materials: Science and Technology*; Yan, M.; Ranström, O., Eds.; Marcel Dekker: New York, NY, 2005; pp 123 and references therein.
18. Sharma, A. C.; Borovik, A. S. *J. Am. Chem. Soc.* **2000**, *122*, 8946.
19. Padden, K. M.; Krebs, J. F.; Trafford, K. T.; Yap, G. P. A.; Rheingold, A. H.; Borovik, A. S.; Scarrow, R. C. **2001**, *13*, 4305.
20. Krebs, J. F.; Borovik, A. S. *Chem. Commun.* **1998**, 553.
21. Padden, K. M.; Krebs, J. F.; MacBeth, C. E.; Scarrow, R. C.; Borovik, A. S. *J. Am. Chem. Soc.* **2001**, *123*, 1072.

22. Mitchell-Koch, J. T.; Reed, T. M.; Borovik, A. S. *Angew. Chem. Int. Ed.* **2004**, *43*, 2806.
23. Welbes, L. L.; Scarrow, R. C.; Borovik, A. S. *Chem. Commun.* **2004**, 2544.
24. Sharma, S; Kerler, B.; Subramaniam, B.; Borovik, A. S. *Green Chemistry* **2006**, *8*, 972.
25. Gonda, I. in *Pharmaceutical Inhalation Aerosol Technology*; Hickey, A. J., Ed.; Marcel Dekkar, New York, NY, 1992.
26. Dixon D. J.; Johnston K. P.; Bodmeier R. A. *AIChE J.* **1993**, *39*, 127.
27. Subramaniam, B.; Rajewski, R.; Snavely, K. *J. Pharm. Sci.* **1997**, *86*, 885.
28. Perrut, M. *STP Pharma Sciences* **2003**, *13*, 83.
29. Foster, N.; Mammucari, R; Dehghani, F.; Barrett, A.; Bezanehtak, K.; Coen, E.; Combes, G.; Meure, L.; Ng, A.; Regtop, H. L.; Tandya, A. *Ind. Eng. Chem. Res.* **2003**, *42*, 6476.
30. Rehman, M.; Shekunov, B. Y.; York, P.; Lechuga-Ballesteros, D.; Miller, D. P.; Tan, T.; Colthorpe, P. *Euro. J. Pharm. Sci.,* **2004**, *22*, 1.
31. Gallagher, P. M.; Coffey, M. P.; Krukonis, V. J.; Klasutis, N. in *Supercritical Science and Technology*; Johnston, K. P.; Penninger, J. M. L., Eds.; ACS Symposium Series 406; American Chemical Society: Washington, DC, 1989; pp 334-354.
32. Fusaro, F.; Hänchen, M.; Mazzotti, M.; Muhrer, G.; Subramaniam, B. *Ind. Eng. Chem. Res.* **2005**, *44*, 1502.
33. Johnson, C. A.; Sharma, S.; Subramaniam, B.; Borovik, A. S. *J. Am. Chem. Soc.* **2005**, *127*, 9698.
34. Reverchon, E. *J. Supercrit. Fluids* **1999**, *15*, 1.
35. Edwards, A. D.; Shekunov, B. Y.; Kordikowski, A.; Forbes, R. T.; York, P. *J. Pharm. Sci.* **2001**, *90*, 1115.
36. Johnson, C.; Long, B.; Nguyen, J.; Day, V.; Borovik, A.; Subramaniam, B.; Guzman, J. *J. Phys. Chem. C* (in press)
37. Johnson, C. A.; Ph.D. Dissertation, University of Kansas, 2008.
38. Chang, Y.; McCarty, J. G. *J. Catal.* **1998**, *178*, 408.
39. Bell, A. T. *Catal. Today* **1997**, *38*, 151.
40. Armor, J. N. *Catal. Today* **1995**, *26*, 147.
41. Iwamoto, M.; Yokoo, S.; Sakai, K.; Kagawa, S. *J. Chem. Soc., Faraday Trans. 1* **1981**, *77*, 1629.
42. Hilderbrand, S. A.; Lippard, S. J. *Inorg. Chem.* **2004**, *43*, 4674.
43. Potter, W. T.; Cho, J. Sublette, K. L. *Fuel Process. Technol.* **1994**, *40*, 355.
44. Rossi, M.; Sacco, A. *Chem. Commun.* **1971**, 694.
45. Miki, E.; Tanaka, M.; Sarro, K.; Maejima, T.; Mizumachi, K.; Ishimori, T. *Bull. Chem. Soc. Jpn.* **1985**, *58*, 1642.

Chapter 16

Application of Gas-Expanded Liquids for Nanoparticle Processing: Experiment and Theory

Christopher L. Kitchens[1], Christopher B. Roberts[2], Juncheng Liu[2], W. Robert Ashurst[2], Madhu Anand[2], Gregory Von White II[1], Kendall M. Hurst[2], and Steven R. Saunders[2]

[1]Department of Chemical and Biomolecular Engineering, Clemson University, Clemson, SC 29634
[2]Department of Chemical Engineering, Auburn University, Auburn, AL 36849

The numerous contributions to this ACS symposium series describe in detail the unique attributes of Gas eXpanded Liquids (GXLs) which provide for a novel class of solvents. Our contribution has applied GXLs as a processing medium for the fractionation and deposition of metallic nanoparticles (Pt, Pd, Ag, Au) and quantum dots (CdSe/ZnS). The tunable nature of the GXLs provides advantages over conventional solvents, and enables rapid, precise, and scalable size-dependant fractionation of nanoparticles into uniform populations (less than ± 0.5 nm in diameter). The wide range of accessible solvent strengths and facile removal of the CO_2, substantially reduces the amount of solvent needed for conventional anti-solvent size fractionation techniques and facilitates solvent recycling. We have also taken advantage of the reduced surface tension and interfacial forces exhibited by GXLs to produce wide-area, low defect nanoparticle arrays. Each of these nanoparticle processing applications is governed by the inter-particle interactions; dispersive and steric repulsion forces, resulting from the solvation forces between the nanoparticle stabilizing ligands and CO_2 expanded liquid medium. To better understand this system, we have developed an interaction energy modeling approach to determine the interparticle attractive and repulsive forces that control nanoparticle dispersibility through variations in the properties of the GXL medium.

Introduction

Advancements in nanotechnology have led to a wide-spread emergence of different techniques for nanoparticle synthesis and application. A vital and often overlooked component in the field is the development of efficient, reduced-cost, and large-scale methods for nanoparticle synthesis and processing. Solution based techniques have been demonstrated as effective and robust means for the controlled, "bottom-up" synthesis of metallic nanoparticles with varying size, shape, composition, and surface chemistry. The major advantage of solution-based nanoparticle synthesis is the single-pot system consisting of a metal salt precursor, reducing agent, stabilizing ligand, and solvent; each of which plays an integral role in the nanoparticle synthesis and processing. Considerable efforts have gone into developing solution-based methods to produce nanomaterials that possess unique, composition and morphology dependent properties; however, these methods are often limited when it comes to controlled deposition and size fractionation. We contend that gas expanded liquids (GXLs) will provide significant advancement in the processing of nanomaterials, providing sustainable means for nanomaterial applications. In the interest of this work, we shall limit our discussion to ligand-stabilized, spherical metallic nanoparticles less than 10 nm in diameter, which display size-dependent properties.

We have developed methods for the post-synthesis processing of spherical nanoparticles using GXLs.[1-3] The advantages of using GXLs[4,5] as a tunable processing medium for different nanoparticle applications include:

- Wide range of accessible solvent properties that can be achieved with the expansion of virtually any conventional organic solvent
- Ability to reversibly and controllably fine-tune the solvent properties by simply adjusting the solvent medium composition, via applied gas pressure
- Diminishing interfacial forces, low viscosity, and high diffusivity that lead to excellent surface wetting of wide areas and sub-micron features
- Green properties that provide the potential for sustainable, cost-effective, large scale methods that are vital for the future of nanotechnology applications

GXLs provide certain advantages over conventional fluids and supercritical fluids[6] for the deposition and size fractionation of ligand-stabilized metal nanoparticles. Controlled deposition of nanoparticles for the formation of uniform wide-area thin films, multi-layer assemblies and targeted coverage of sub-micron features requires a solvent medium that 1) is strong enough to effectively disperse the ligand-stabilized nanoparticles in solution, 2) has tunable solvent strength for controlled deposition of particles onto a surface, and 3) has favorable fluid properties for effective deposition and removal. Conventional liquid solvents are very effective in dispersing nanoparticles, but the deposition and solvent removal often involve evaporative phase changes and dewetting effects.[7-9] Property tunable fluids, such as near- and supercritical fluids, have excellent wetting characteristics and can be easily removed from the system, but their solvent strength is very limited and often requires the use of specialized nanoparticle ligands or the use of co-solvents.[6,10-13]

Significant progress has been made in the synthesis of innovative nanomaterials with unique properties that are very often size-dependent. In order to take advantage of the size-dependent properties, a size-dependent fractionation, extraction, or separation must be performed to isolate the desired nanoparticles. Many synthesis techniques have been popularized based on the uniformity of the nanomaterials produced;[14] however, many applications require further processing.

We have demonstrated the ability to take a polydisperse population of nanoparticles and fractionate them into monodisperse populations using GXLs.[2,3,15-18] The highly tunable solvent strength coupled with the facile pressure-induced control of GXL properties enables the precise and rapid fractionation of ligand-stabilized nanoparticles. Previously demonstrated methods of size fractionation rely on liquid anti-solvent techniques that are more costly, solvent intensive, laborious, subjective, and not conducive for large scale application.[19-24] The use of GXLs can ameliorate each of these drawbacks and when coupled with the deposition capabilities, provides for an attractive medium for the processing of nanomaterials.

We have also implemented an interaction energy model to predict the deposition and size-dependent fractionation of ligand-stabilized metallic nanoparticles as a function of the system thermo-physical properties. The model accounts for the attractive and repulsive forces that exist between particles dispersed within a GXL solvent medium, thus providing a fundamental understanding of the nanoparticle dispersibility as a function of the nanoparticle, stabilizing ligand, and solvent properties. This enables the ability to design an appropriate nanoparticle processing system that incorporates virtually any nanoparticle size, shape, and composition, as well as, stabilizing ligand structure and composition.

Nanoparticle Deposition

Following nanoparticle synthesis, the issue remains as to how to isolate the nanoparticles from solution and apply them to a surface for use as a catalyst, optical coatings, multi-layer assemblies, etc. Conventional solvent drying techniques often lead to dewetting instabilities resulting from capillary forces and surface tension at the liquid – vapor interface.[8,9,25] Surface dewetting effects lead to the formation of particle islands, voids, aggregates, and other non-uniformities; thus inhibiting defect-free, long-range ordering. Han and coworkers initially used gas anti-solvent (GAS) precipitation techniques to isolate ZnS nanoparticles from an isooctane – AOT reverse micelle solution, where the nanoparticles precipitated and the excess surfactant remained in solution.[26] GAS and other tunable fluid anti-solvent techniques have been used extensively to produce uniform particles of pharmaceuticals, polymers, and other materials which go through a controlled nucleation, growth, and precipitation process.[27,28] Similarly, Han used the tunable solvent strength of CO_2 expanded isooctane to effectively isolate the ZnS nanoparticles by adding 5.5 MPa of CO_2 at 308.15 K to the AOT reverse micelle system. Han also demonstrated these techniques for surfactant stabilized Ag and TiO_2 nanoparticles.[29,30]

Our approach has applied the tunable solvent properties of GXLs for the controlled deposition of ligand-stabilized Ag, Au, Pt, and Pd nanoparticles.[1,3,16] The experimental apparatus consists of a pressure vessel containing a carbon-coated TEM grid substrate immersed in a hexane dispersion of stabilized nanoparticles.[3] The pressure vessel is charged with CO_2 to expand the hexane and induce precipitation of the ligand-stabilized nanoparticles from solution. The vessel can then be pressurized and heated beyond the critical point of the hexane – CO_2 mixture, flushed with pure CO_2, and subsequently depressurized for supercritical drying. The GXL deposition process avoids dewetting effects by circumventing the surface tension and/or phase transition on solvent removal. The result is increased surface coverage in wide-area, uniform monolayers with a high degree of ordering.[3]

Figure 1 displays the contrast between CO_2 expanded hexane deposition and hexane evaporation deposition for dodecanethiol and lauric acid stabilized Ag nanoparticles. The nanoparticles deposited by hexane evaporation exhibit significant voids, islands, agglomerations and circular structures that are characteristic of dewetting effects during solvent evaporation. The nanoparticles deposited from CO_2 expanded hexane demonstrate significantly improved surface coverage and uniformity over the entire deposition area. All images in Figure 1 are characteristic of the entire sample, and for the case of the GXL depositions, some imperfections do appear but the occurrences are minimal and can be reduced further by adjusting the particle concentrations appropriately.

Figure 2 displays uniform monolayers of dodecanethiol stabilized Pt nanoparticles deposited on carbon from CO_2 expanded hexane. The Pt nanoparticles synthesized by an aqueous phase glucose mediated growth[1], are very monodispersed and when coupled with the GXL deposition, hexagonal packing into an ordered array occurs. Hexagonal ordering of nanoparticle arrays requires very monodispersed particle diameters as well as controlled deposition processes, each of which are attainable with GXLs. Comparable degrees of ordering have been achieved with liquid evaporation techniques, but this requires the use of significant excesses of stabilizing ligands which likely slow the rate of solvent removal allowing for the particles to order on the surface.[31]

We have also demonstrated the ability to form uniform, multi-layer assemblies using the controlled deposition afforded by GXLs.[1] Multi-layer nanoparticle assemblies and three-dimensional architectures are a critical component for the future of "bottom up" formation of nano-scale devices.[7,21,32] Current technologies are close to the limits of lithographic and other "top down" processing techniques, and in order to create devices on the nano-scale, directed assembly methodologies must be applied. GXLs have enormous potential as a

294

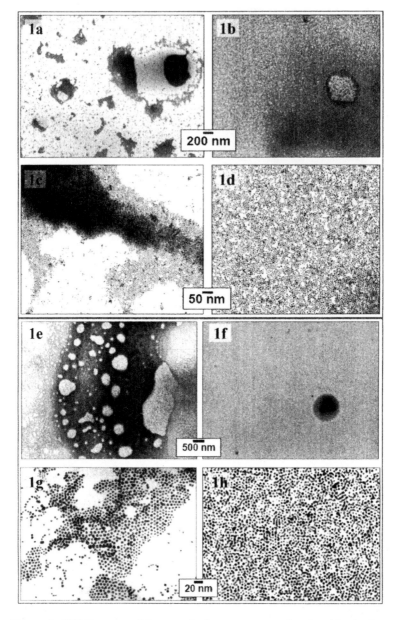

Figure 1. TEM images of deposited silver nanoparticles. Samples a-d are dodecanethiol stabilized Ag particles synthesized by the Brust method.[14] Images e-f are lauric acid stabilized Ag particles synthesized in an reverse micelle system.[3] All particles were dispersed in hexane; a,c,e,g were deposited by hexane evaporation; b,d,f,h were deposited by CO_2 expanded hexane.

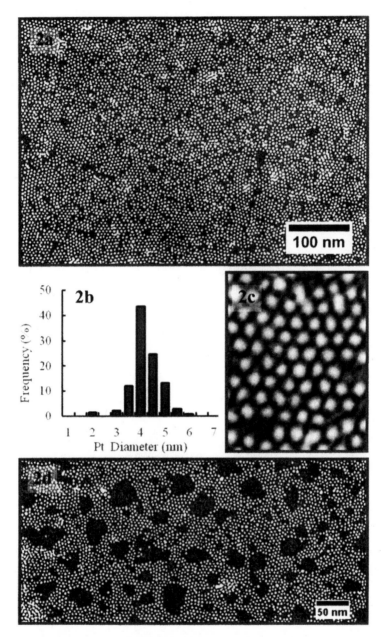

Figure 2. TEM images of dodecanethiol stabilized Pt nanoparticles synthesized by an aqueous glucose method.[1] Images 2a and 2c are particles deposited from CO_2 expanded hexane. 2b is the particle size distribution; 4.2 nm ave. dia. and 0.58 std. dev. 2d is an image of the Pt particles deposited by hexane evaporation.

296

processing medium for directed assembly of nano-devices and the formation of multi-layer assemblies is a first step in that direction.

One application we are currently investigating is the controlled deposition of Au nanoparticles onto MEMS devices and small feature silica surfaces using GXLs. The goal is to produce Micro-Electro-Mechanical Systems (MEMS) devices with advanced anti-stiction properties and increased wear resistance. For this application, the lack of solvent evaporation and the uniform conformal surface coverage are imperative for successful modification without detrimentally impacting the MEMS devices.[33]

GXL deposition of nanoparticles is a robust technique, applicable to a wide range of nanoparticle compositions, stabilizing ligands, and expanding fluids, where each contributes to the particle dispersibility. For example, cyclohexane at ambient temperature is too good a solvent for dodecanethiol stabilized silver nanoparticles and CO_2 addition does not induce precipitation below the vapor

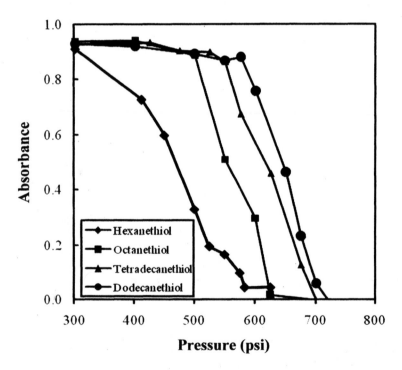

Figure 3. Concentration-corrected UV-vis absorbance measurements for alkanethiol-stabilized silver nanoparticles in hexane as a function of pressure for different stabilizing ligand lengths. This demonstrates the impact of the ligand length on particle dispersibility (Reproduced from reference 18. Copyright 2005 American Chemical Society.)

pressure. Similarly, ethane, an alternative gas anti-solvent, does not alter the solvent strength enough to induce precipitation of the same particles from ethane expanded hexane. Figure 3 displays the effect of ligand length on the particle dispersibility as evidenced by UV-vis absorbance spectroscopy. As the solvent is expanded with CO_2, nanoparticles precipitate from solution, causing a decrease in absorbance (corrected for volume expansion). The particle dispersibility increases with alkanethiol ligand length until a maximum around C12. The nanoparticle diameter and composition also influence the relative dispersibility and are applied in our size fractionation methods discussed below.

In summary, GXLs provide unparalleled ability to form uniform, wide-area nanoparticle arrays with additional benefits of process scalability, reduced processing time, and use of CO_2 as a green anti-solvent that can be easily removed, providing for solvent recycle. We are currently investigating sustainable methods to scale-up the synthesis and fractionation of ligand-stabilized metallic nanoparticles into uniform, monodisperse populations using GXLs.

Nanoparticle Fractionation in GXL's

Nanomaterials with dimensions less than 10 nm, often exhibit unique size-dependent properties. Solvent based synthesis methods are attractive due to their relatively low polydispersity; however many applications require post-synthesis size fractionation to obtain highly monodispersed populations. Liquid anti-solvent precipitation methods consisting of progressive anti-solvent addition and intermediate centrifugation with precipitate collection have been used to size fractionate nanoparticle populations.[22,23] Example solvent / anti-solvent systems include 1-butanol / methanol and chloroform / methanol. While size fractionation is achieved, this method is time consuming, solvent intensive, limited to small scale, and suffers from poor quantitative reproducibility.

GXLs are ideal for nanoparticle fractionation. The wide range of achievable solvent strengths that can be finely tuned with system pressure, coupled with the advantages pertaining to particle deposition, provides a precise and rapid method for size fractionations. We have demonstrated this application with a unique assembly that consists of a spiral tube within a pressure vessel, shown in Figure 4.[2] The process begins by placing ligand-stabilized nanoparticles, dispersed in an organic solvent, into the front of the spiral tube where it is isolated. The vessel is then pressurized with the gas anti-solvent (in the case of dodecanethiol stabilized silver nanoparticles in hexane, CO_2 is added to a pressure of 500 psi; 550 psi for gold particles). At this point, instability occurs and the largest diameter nanoparticles will precipitate onto the surface of the spiral tube.

The spiral tube is then rotated 180° and the remaining liquid is moved down the length of the tube. The first fraction containing the largest nanoparticles is left behind affixed to the tube surface, via van der Waals forces. The pressure is then increased by an incremental amount of 25 or 50 psi, over which a second instance of instability occurs, resulting in the precipitation of the next largest fraction of nanoparticles. This process is repeated to obtain multiple fractions with decreasing average particle diameters. Following depressurization, solvent can be placed in specific areas of the tube to redisperse the desired size fraction. Table 1 displays the average particle diameters from each pressure fraction for both gold and silver nanoparticles stabilized by dodecanethiol ligands and dispersed in hexane. Particle diameters were determined from TEM images as shown for the silver system.

Table 1. Average diameters for gold and silver nanoparticles obtained from the CO_2 expanded hexane size fractionation process; Ag diameters correspond to Figure 4.

Operating Pressure Range	Average Gold Diameter (nm)	Average Silver Diameter (nm)
Unprocessed	-	5.5
0 - 500 psi	-	6.7
500 - 550 psi	5.9	6.6
550 - 600 psi	5.8	5.8
600 - 625 psi	4.7	5.3
625 - 650 psi	4.2	4.8
after 650psi	3.2	4.1

Effective fractionation can be achieved by allowing as little as 20 minutes at each pressure for equilibration and precipitation. Thus, a separation with six size fractions can be achieved in ~2 hours, as opposed to several hours for the liquid anti-solvent technique previously mentioned. Additional benefits of the GXL system include minimal solvent use, facile solvent recovery for reuse (simply by depressurization), high degree of solvent tunability with pressure, quantitative repeatability, and applicability to a wide range of nanoparticle systems.

We have found that CO_2 expanded hexane works very well and have demonstrated the fractionation of Au, Ag, Pt, and Pd nanoparticles with different stabilizing ligands, including dodecanethiol, tetradecanethiol, octanethiol, hexanethiol, lauric acid, and laurylamine.[16] Hexane is a good solvent for nanoparticles with aliphatic stabilizing ligands and has a strong affinity for CO_2, which provides for a wide range of accessible solvent strengths.

CO_2 expanded hexane was also used to size fractionate CdSe / ZnS quantum dots, core-shell semiconductor nanocrystals stabilized by trioctylphosphine oxide (TOPO).[17] Quantum dots are well known for their unique, size-dependent optical properties, which are used for optoelectronic devices, sensors, and biomedical fluorescent labeling. To demonstrate the GXL size fractionation,

TOPO-stabilized CdSe/ZnS core-shell quantum dots with diameters of 5.2 nm (red), 3.2 nm (orange), 2.4 nm (yellow), and 1.9 nm (green) were obtained from Evident Technologies. The received CdSe/ZnS particles were originally dispersed in toluene and precipitation did not occur with the addition of CO_2 up to the GXL limit. Thus, CO_2 expanded toluene is "too good" of a solvent for the TOPO stabilized particles. The toluene was removed by evaporation and the nanoparticles were redispersed in hexane. The four nanoparticle dispersions were mixed together into a single hexane dispersion and then fractionated into four populations using the spiral tube apparatus. Each of the four size fractions exhibited the same optical properties as the original nanoparticle samples, which were characterized using UV-vis absorbance spectroscopy and TEM imaging.[17]

Recently, a new apparatus has been constructed to take advantage of this GXL anti-solvent separation technique to separate nanoparticles in much larger quantities. This new process has been shown to successfully fractionate 20 mL of a concentrated nanoparticle dispersion into monodisperse fractions.[34]

Interaction Energy Model

The ability to process nanoparticles into uniform wide-area arrays or fractionate them into monodispersed populations using GXLs is strongly dependent on the particle dispersibility. In order to gain a fundamental understanding of these processes, we have applied an interaction energy model that accounts for the interparticle interactions that exist between two ligand-stabilized nanoparticles dispersed in a fluid. This modeling approach has been used previously to predict the maximum particle size synthesized or stabilized in liquid alkane solvents,[35,36] supercritical ethane,[37] compressed propane,[35] and supercritical carbon dioxide.[38] The interaction energy model takes a soft-sphere approach, where the interparticle interactions (attractive and repulsive forces) of nanoparticles, dispersed in a solvent medium, are calculated to predict the nanoparticle dispersibility.[39] The total interaction energy, Φ_{total}, involves a summation of the van der Waals attractive forces and steric repulsive forces, which are calculated as a function of the nanoparticle composition and diameter, interparticle separation distances (h), ligand length (l), ligand surface coverage, and solvent properties. This model allows the prediction of the maximum particle diameter that can be dispersed in a given solvent system. For example, Figure 5 presents Φ_{total} as a function of surface-to-surface interparticle separation distance, h, for 6 to 12 nm diameter Ag nanoparticles stabilized by AOT surfactant and dispersed in hexane at 25°C. As the nanoparticle diameter increases, the potential minimum decreases and when the minimum crosses the -$3/2 \, k_B T$ stabilization threshold, instability occurs, leading to precipitation. Thus the maximum size particle that can be dispersed by AOT in hexane is 7.5 nm, which corresponds well with the experimental reverse micelle synthesis.[36] The potential minimum is also a function of the solvent properties, ligand properties, and type of nanoparticle, thus each contributes to the maximum particle size that can be dispersed or synthesized in a specific system.

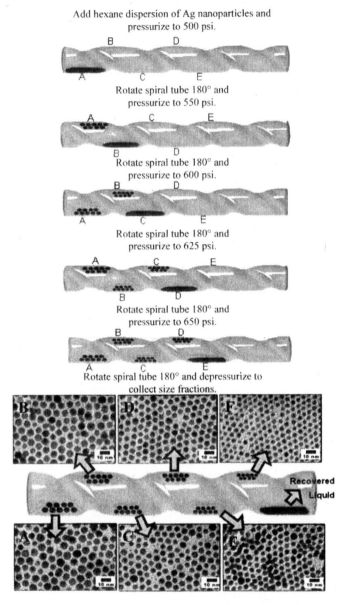

Add hexane dispersion of Ag nanoparticles and
pressurize to 500 psi.

Rotate spiral tube 180° and
pressurize to 550 psi.

Rotate spiral tube 180° and
pressurize to 600 psi.

Rotate spiral tube 180° and
pressurize to 625 psi.

Rotate spiral tube 180° and
pressurize to 650 psi.

Rotate spiral tube 180° and depressurize to
collect size fractions.

*Figure 4. Schematic of the GXL size fractionation method using the spiral
tube assembly, which occurs within a cylindrical pressure vessel. TEM images
are from the corresponding fractions for dodecanethiol stabilized silver
nanoparticles deposited from hexane.[2] (Reproduced from references 18 and 2.
Copyright 2005 American Chemical Society.)*

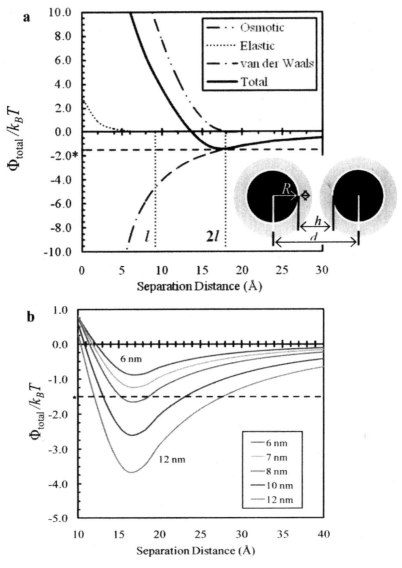

Figure 5. The total interaction energy potential plotted as a function of the interparticle separation distance, h. 6a shows the relative contributions of the van der Waals attractive forces (neg.) and the osmotic and elastic repulsive forces (pos). 6b is Φ_{total} for increasing diameters of silver nanoparticles stabilized by AOT surfactant in hexane at 25°C. * represents the 3/2 k_BT stabilization threshold. (Reproduced from reference 36. Copyright 2003 American Chemical Society.).

The total interaction energy potential, Φ_{total}, consists of the attractive van der Waals potential, Φ_{vdW}, and the repulsive osmotic, Φ_{osm}, and elastic, Φ_{elas}, contributions.

$$\Phi_{total} = \Phi_{vdW} + \Phi_{osm} + \Phi_{elas} \tag{1}$$

The van der Waals attractive force, eq. 2, increases in magnitude with increasing particle radius, R, and decreasing center-to-center particle separation distance between the particles, d. [40]

$$\Phi_{vdW} = -\frac{A_{131}}{6}\left[\frac{2R^2}{d^2-4R^2}+\frac{2R^2}{d^2}+\ln\left(\frac{d^2-4R^2}{d^2}\right)\right] \tag{2}$$

A_{131} is the Hamaker constant and is a proportionality factor that accounts for the interaction between two nanoparticles of the same material (component 1) through a solvent medium (component 3). In the case of a GXL solvent medium, the solvent contribution to the Hamaker constant can be weighted by the respective volume fractions[15], given by eq. 3

$$A_{131} \approx \left[\sqrt{A_{11}} - \left(\tilde{\phi}_{3'}\sqrt{A_{(33)'}} + \tilde{\phi}_{3''}\sqrt{A_{(33)''}}\right)\right]^2 \tag{3}$$

For the repulsive contribution to the soft-sphere model, Vincent et. al.[41] proposed an osmotic term to account for the free energy of the solvent-ligand interactions and an elastic term that accounts for the entropic loss due to ligand compression and deformation. Previous applications of the interaction energy model considered ligand-stabilized nanoparticles dispersed in a single component solvent system; however, for GXLs, the multi-component solvent medium must be taken into account. If we consider dodecanethiol-stabilized silver nanoparticles dispersed in CO_2 expanded hexane, the hexane will have a favorable interaction with the dodecanethiol ligands and the CO_2 anti-solvent will have an unfavorable interaction. The osmotic term of the interaction energy model is dominated by the solvent – ligand and interparticle ligand–ligand interactions, thus we have modified the original osmotic term to account for the binary solvent system.[15]

In addition to the multi-component solvent media, three phenomenological aspects are unaccounted for in the model. 1) Preferential solvation of the nanoparticle ligands, where the local solvent composition differs from the bulk composition; 2) geometric configuration of the particle ligands, where the ligands are either fully extended into the solvent or collapsed on the particle surface; and 3) the degree of solvent penetration into the stabilizing ligand layer. Figure 6 provides a cartoon of these scenarios. We are currently investigating experimental methods to measure the local solvent composition and ligand behavior as a function of the bulk solvent properties with increasing CO_2 pressure; however, we have investigated this behavior theoretically.[15]

We have performed a sensitivity analysis on the contributing adjustable parameters within the model and have determined that the most influential parameters are the degree of ligand surface coverage, Hamaker constant, solvent composition (pressure), and effective ligand length. The degree of ligand surface coverage was measured using thermogravimetric analysis (TGA) and found to be consistent with our 75% surface coverage estimation. The Hamaker constant is dominated by the metal nanoparticle contribution and less by the solvent composition, resulting in the differences observed for Au nanoparticles versus Ag nanoparticle, etc. Hamaker constants have been measured for some nanoparticles.[42] The influence of solvent composition is evident in the pressure effects on the nanoparticle dispersion; however, the model does not account for local preferential solvation.

The behavior of the stabilizing ligands as a function of the solvent properties significantly influences the maximum dispersible particle size predicted by the model. Changes in ligand configurations in response to local solvent effects have been observed in latex and colloidal dispersions.[39,43] Potential ligand configurations include a fully extended solvation model (ELSM), a condensed or collapsed phase model (CPM), and a limited ligand solvation model (LLSM).

The ELSM configuration (Figure 6a,d) would be expected in a favorable solvent and the model over predicts the maximum particle dispersible in CO_2 expanded hexane, assuming a 15 Å ligand length and 75% ligand coverage. The CPM configuration (Figure 6c) would be expected in a poor solvent; solvent conditions conducive for particle precipitation. The CPM configuration only slightly over predicts the maximum particle size. The calculated effective ligand length is dependent on the particle diameter and 75% ligand surface coverage. If the CPM configuration assumes an adjustable ligand surface coverage, the experimental results can be fit as shown in Figure 7. This approach suggests that the ligand surface coverage would decrease with increasing CO_2 pressure from 70% to 58%; which are acceptable values, see Table 2.

An alternate approach is to fit the experimental results assuming a LLSM where the ligands are fully extended, as with ELSM, and the degree to which the solvent molecules penetrate into nanoparticle ligand region is limited, Figure 6e. This configuration assumes a constant ligand length from the nanoparticle surface, and an effective ligand length that is used to fit the model to the experimental results. Table 2 displays the LLSM results for dodecanethiol stabilized Ag nanoparticles with 75% surface coverage where the solvated effective ligand length decreases with the addition of CO_2 anti-solvent from 8.2 Å at 500 psi to 5.9 Å at 700 psi.

The LLSM model can effectively predict the diameter of Ag nanoparticles precipitated from the CO_2 expanded hexane, as well as the larger range of diameters observed as a function of pressure, shown in Figure 7 and Table 2. The LLSM configuration suggests that the effective ligand length changes in response to solvent strength fluctuations are largely responsible for nanoparticle precipitation. This behavior is yet to be observed, however we hope to probe the local solvation effects of the nanoparticle ligands in GXLs using small angle neutron scattering.

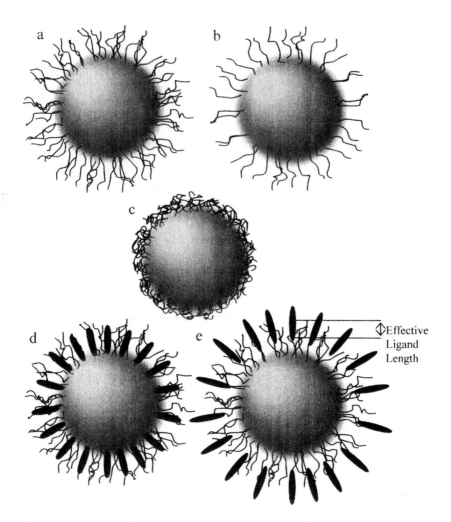

Figure 6. Conceptual representation of the stabilizing ligand configuration on the nanoparticle surface. 6a is the fully extended ELSM with a high degree of surface coverage. 6b is the fully extended ELSM with low surface coverage. 6c is the collapsed ligand structure, CPM, representative of poor solvation. 6d is the fully extended, completely solvated configuration where the entire ligand length is the effective ligand length as in the ELSM, representative of excellent solvation. 6e is the LLSM where the outer regions of the ligand shell are solvated, providing an effective ligand length less than the total ligand length.

Table 2. Comparison of the experimental and modeling results for the maximum diameter of dodecanethiol stabilized silver nanoparticle in CO_2 expanded hexane as a function of system pressure. *Experimental particle diameters are averages and correspond to Table 1.[15]

Pressure (psi)	500	550	600	625	650	700
*Experimental (nm)	6.7	6.6	5.8	5.3	4.8	4.1
ELSM Dia. (nm)	12.6	12.5	12.4	12.3	12.2	12
CPM with 75 % surface coverage						
CPM Dia. (nm)	7.2	7.1	6.9	6.7	6.5	6.1
Effective Length (Å)	8.8	8.7	8.5	8.4	8.2	7.9
CPM with variable surface coverage						
CPM Dia. (nm)	6.7	6.6	5.8	5.3	4.8	4.1
Surface Coverage	69.3%	69.2%	63.2%	60.3%	60.2%	58.3%
Effective Length (Å)	8.2	8.2	7.4	7	6.9	6.5
LLSM w/ 75% surface coverage						
CPM Dia. (nm)	6.7	6.6	5.8	5.3	4.8	4.1
Effective Length (Å)	8.2	8.1	7.3	6.8	6.3	5.9

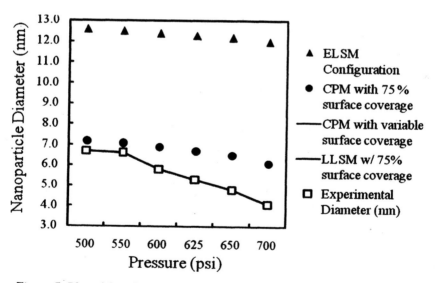

Figure 7. Plot of the silver nanoparticle diameters as a function of pressure in the CO_2 expanded hexane system for the experimental observations and model predictions. Values correspond to those in Table 2.[15]

Conclusions

This work presents the opportunity that GXLs hold for nanoparticle processing and application. The controlled deposition techniques provide significant enhancements over alternative solution based methods for the formation of uniform, wide-area thin films or multilayer assemblies. Specific applications include targeted deposition into small MEMS features, effective and efficient size fractionation into monodispersed populations, and potential for cost-effective large-scale production of monodispersed, ligand-stabilized nanoparticles. The wide range of accessible solvent strengths, pressure tunable solvent properties, facile solvent recovery, attractive interfacial wetting properties, and potential for large scale make GXLs a very attractive processing fluid to ensure a sustainable future of nanotechnology.

References

1. Liu, J. C.; Anand, M.; Roberts, C. B. *Langmuir* **2006**, *22*, 3964-3971.
2. McLeod, M. C.; Anand, M.; Kitchens, C. L.; Roberts, C. B. *Nano Lett.* **2005**, *5*, 461-465.
3. McLeod, M. C.; Kitchens, C. L.; Roberts, C. B. *Langmuir* **2005**, *21,* 2414-2418.
4. Jessop, P. G.; Subramaniam, B. *Chem. Rev.* **2007**, *107*, 2666-2694.
5. Hallett, J. P.; Kitchens, C. L.; Hernandez, R.; Liotta, C. L.; Eckert, C. A. *Accounts Chem. Res.* **2006**, *39*, 531-538.
6. Johnston, K. P.; Shah, P. S. *Science* **2004**, *303*, 482-483.
7. Lin, X. M.; Jaeger, H. M.; Sorensen, C. M.; Klabunde, K. J. *J. Phys. Chem. B* **2001**, *105*, 3353-3357.
8. Korgel, B. A.; Fitzmaurice, D. *Phys. Rev. Lett.* **1998**, *80*, 3531-3534.
9. Ohara, P. C.; Gelbart, W. M. *Langmuir* **1998**, *14*, 3418-3424.
10. Anand, M.; Bell, P. W.; Fan, X.; Enick, R. M.; Roberts, C. B. *J. Phys. Chem. B* **2006**, *110*, 14693-14701.
11. Fan, X.; McLeod, M. C.; Enick, R. M.; Roberts, C. B. *Ind. Eng. Chem. Res.* **2006**, *45*, 3343-3347.
12. Saunders, A. E.; Shah, P. S.; Park, E. J.; Lim, K. T.; Johnston, K. P.; Korgel, B. A. *J. Phys. Chem. B* **2004**, *108*, 15969-15975.
13. Shah, P. S.; Holmes, J. D.; Doty, R. C.; Johnston, K. P.; Korgel, B. A. *J. Am. Chem. Soc.* **2000**, *122*, 4245-4246.
14. Brust, M.; Walker, M.; Bethell, D.; Schiffrin, D. J.; Whyman, R. *J. Chem. Soc.-Chem. Commun.* **1994**, 801-802.
15. Anand, M.; You, S.-S.; Hurst, K. M.; Saunders, S. R.; Kitchens, C. L.; Ashurst, W. R.; Roberts, C. B. *Ind. Eng. Chem. Res.* **2008**, *47*, 553-559.
16. Roberts, C. B.; McLeod, M. C.; Anand, M.; (Auburn University, USA). Application: WO, 2007, p 100.

17. Anand, M.; Odom, L. A.; Roberts, C. B. *Langmuir* **2007**, *23*, 7338-7343.
18. Anand, M.; McLeod, M. C.; Bell, P. W.; Roberts, C. B. *J. Phys. Chem. B* **2005**, *109*, 22852-22859.
19. Sigman, M. B.; Saunders, A. E.; Korgel, B. A. *Langmuir* **2004**, *20*, 978-983.
20. Rogach, A. L.; Kornowski, A.; Gao, M. Y.; Eychmuller, A.; Weller, H. *J. Phys. Chem. B* **1999**, *103*, 3065-3069.
21. Korgel, B. A.; Fullam, S.; Connolly, S.; Fitzmaurice, D. *J. Phys. Chem. B* **1998**, *102*, 8379-8388.
22. Vossmeyer, T.; Katsikas, L.; Giersig, M.; Popovic, I. G.; Diesner, K.; Chemseddine, A.; Eychmuller, A.; Weller, H. *J. Phys. Chem.* **1994**, *98*, 7665-7673.
23. Murray, C. B.; Norris, D. J.; Bawendi, M. G. *J. Am. Chem. Soc.* **1993**, *115*, 8706-8715.
24. Fischer, C. H.; Weller, H.; Katsikas, L.; Henglein, A. *Langmuir* **1989**, *5*, 429-432.
25. Shah, P. S.; Novick, B. J.; Hwang, H. S.; Lim, K. T.; Carbonell, R. G.; Johnston, K. P.; Korgel, B. A. *Nano Lett.* **2003**, *3*, 1671-1675.
26. Zhang, J.; Han, B.; Liu, J.; Zhang, X.; Liu, Z.; He, J. *Chem. Commun. (UK)* **2001**, 2724-2725.
27. Cooper, A. I. *J. Mater. Chem.* **2000**, *10*, 207-234.
28. Hutchenson, K. W.; Foster, N. R. *Innovations in Supercritical Fluids*; Amer Chemical Soc: Washington, 1995; Vol. 608, p 1-31.
29. Liu, D.; Zhang, J.; Han, B.; Chen, J.; Li, Z.; Shen, D.; Yang, G. *Colloids Surf., A* **2003**, *227*, 45-48.
30. Zhang, J. L.; Han, B. X.; Liu, J. C.; Zhang, X. G.; He, J.; Liu, Z. M.; Jiang, T.; Yang, G. Y. *Chem.-Eur. J.* **2002**, *8*, 3879-3883.
31. Liu, J.; Sutton, J.; Roberts, C. B. *J. Phys. Chem. C* **2007**, *111*, 11566-11576.
32. Motte, L.; Billoudet, F.; Lacaze, E.; Douin, J.; Pileni, M. P. *J. Phys. Chem. B* **1997**, *101*, 138-144.
33. Hurst, K. M.; Roberts, C. B.; Ashurst, W. R. *Proc. IEEE Solid-State Sensors and Actuators Workshop* Hilton head, SC, 2008.
34. Saunders, S. R.; Anand, M.; Roberts, C. B. *AICHE Annual International Conference* Salt Lake City, Utah, 2007.
35. Kitchens, C. L.; Roberts, C. B. *Ind. Eng. Chem. Res.* **2004**, *43*, 6070-6081.
36. Kitchens, C. L.; McLeod, M. C.; Roberts, C. B. *J. Phys. Chem. B* **2003**, *107*, 11331-11338.
37. Shah, P. S.; Holmes, J. D.; Johnston, K. P.; Korgel, B. A. *J. Phys. Chem. B* **2002**, *106*, 2545-2551.
38. Shah, P. S.; Husain, S.; Johnston, K. P.; Korgel, B. A. *J. Phys. Chem. B* **2001**, *105*, 9433-9440.
39. Israelachvili, J. N. *Intermolecular and surface forces: with applications to colloidal and biological systems;* Academic Press: London; Orlando, [Fla]. 1985.

40. Hamaker, H. C. *Physica IV* **1937**, 1058-1072.
41. Vincent, B.; Edwards, J.; Emmett, S.; Jones, A. *Colloids Surf.* **1986**, *18*, 261-281.
42. Eichenlaub, S.; Chan, C.; Beaudoin, S. P. *J. Colloid Interface Sci.* **2002**, *248*, 389-397.
43. Yates, M. Z.; Shah, P. S.; Johnston, K. P.; Lim, K. T.; Webber, S. *J. Colloid Interface Sci.* **2000**, *227*, 176-184.

Chapter 17

Development of a Novel Precipitation Technique for the Production of Highly Respirable Powders: The Atomized Rapid Injection for Solvent Extraction Process

Neil R. Foster and Roderick Sih

School of Chemical Sciences and Engineering, The University of New South Wales, Sydney, New South Wales 2052, Australia

A novel rapid injection technique was integrated into a processing platform for the generation of dry powders with high aerodynamic efficiencies. Cascade impactor classification of pure reprocessed bovine insulin indicated respirable fractions as high as 78%. The Atomized Rapid Injection for Solvent Extraction (ARISE) process is a 'bottom up' approach to drug micronization where organic solutions containing dissolved pharmaceutical ingredients are delivered as a single bolus injection energized under a pressure differential into a vessel of static dense or supercritical carbon dioxide to effect solute drying. The rapid injection technique eliminates the conventional use of low flow-rates and capillary nozzles or micro-orifices to achieve solution atomization. The inertia of the released solution also resulted in precipitation being distributed throughout the entire contained volume, facilitating the formation of fluffy powders with bulk densities as low as 0.01 g/ml.

Introduction

Despite their demonstrated abilities to generate novel and superior biomedical products, supercritical fluid (SCF) precipitation processes have yet to receive significant commercial attention. Possible factors inhibiting the commercialization of existing SCF precipitation processes may include their fragile nature and their scale-up complexity. The economics of operating a commercial scale SCF precipitation facility does not appear to be an impediment for industrialization of SCF precipitation technology, as evidenced by the continued operation of SCF extraction plants that process low value ingredients (1).

With the objective of developing a novel precipitation technique employing SCF anti-solvent mechanisms, ideal process embodiments of excess precipitation volume, instantaneous solution delivery and homogeneous phase precipitation were amalgamated to form the basis of the Atomized Rapid Injection for Solvent Extraction (ARISE) process. The ARISE process was designed to facilitate the production of dry powders suitable for inhalation delivery, while at the same time, dissociating from inherent disadvantages of conventional SCF precipitation processes. The ARISE process was successfully implemented to generate dry powders of bovine insulin with characteristics that were highly tunable as a function of anti-solvent pressures. Under certain experimental conditions, ARISE processed bovine insulin obtained directly from processing displayed extremely high aerodynamic efficiencies.

Improvement to Existing SCF Precipitation Technology

Central to the operation of all existing SCF anti-solvent platforms is the concept of gradual addition of either solute laden working solution into a contact chamber containing carbon dioxide (CO_2) as the anti-solvent phase, or vice-versa. The most fervently researched platform is the Aerosol Solvent Extraction System (ASES) process (2) due to its applicability over a wide spectrum of solutes and its relatively high product recovery rates. The ASES process is also known as the Precipitation with Compressed Anti-solvent (PCA) process (3) and the Supercritical Anti-solvent System (SAS) (4). The effect of operating conditions and system arrangements have been extensively described and reviewed (5-12). Specialty injection devices have also been developed for use within ASES processing to provide higher control over solvent dispersion and inter-phase mixing (13-19).

Such processes that operate with gradual elution suffer from several inherent impediments that may be further improved upon. Gradual elution brings about manifestations of concentration gradients, induces mixing controlled precipitation kinetics and lengthens processing times. Above all, conducting continuous precipitation over a prolonged period exposes precipitation kinetics to inevitable fluctuations in processing conditions, thus contributing to higher product variability.

Phase Homogeneity during Precipitation

Particle precipitations within the referenced processes are effected by the supersaturation of organic solution as a result of a decrease in the localized solvent : anti-solvent ratio. As a result of the consistent introduction of organic solution from a single aperture in the ASES process, concentration gradients manifest within the precipitation vessels during processing (*12, 19-21*). Concentration gradients imply that the degree of supersaturation within the precipitation stage of processing is ill-maintained and is a function of residence time as the two phases contact. Lack of phase homogeneity before particle nucleation and growth may account for the variation seen in products of GAS and ASES processing (*22-25*).

Use of Capillary Nozzles and Micro-orifices

The principal driving mechanism for the precipitation of micro-particles within the ASES process is the solvent extraction from a plume of finely aerosolized liquid organic solutions. Solvent extraction from a mist of well dispersed fine droplets creates an environment in which supersaturation rates are in excess of particle growth rates, thus facilitating the precipitation of particles with smaller physical sizes than the original droplets (*8*). Atomization within the ASES process is typically achieved by eluting working solution through capillary nozzles or micro-orifices at very low flow-rates.

The use of capillary nozzles and similar aperture devices presents several processing disadvantages. Capillary nozzles have the tendency to block and clog during the processing of viscous or concentrated solutions, either as a result of poor liquid flowability, or the premature precipitation of APIs as a result of pressure enhancement or solvent volatility (*26*). Precipitate adhering to the nozzle tip would also alter spray geometries and symmetry during operation. High pressure drops over the nozzle are also experienced, especially when working with viscous solutions. Excessive delivery pressures of organic solutions are sometimes required to force viscous liquid through nozzle capillaries (*26*). The low flow-rates at which organic solution is introduced through the capillary nozzles also serve to extend process times appreciably, thereby decreasing process unit efficiencies. Solute recovery with the continuous delivery of working solution from a single point also presents an additional disadvantage. Upon introduction of working solution into $ScCO_2$, solute precipitation occurs after a period of jet development. As such, solute precipitation occurs over only a very small locality relative to the entire volume of the precipitation vessel (20). The spatial volume in which solute precipitation occurs may be increased by operating ASES at higher flow-rates and hence, at increased jet development envelopes (27-29), but pressure drops over the capillary devices would be excessive.

Mixing Controlled Precipitation Kinetics of SCF Processes

Characteristics of ASES products processed above critical mixture points typically display insignificant variations as a function of different operating conditions of pressure, temperature and organic solution concentration (5, 10, 25, 30, 31). The effects of nozzle geometry and configuration however, exert substantial impact on product characteristics and morphology (5, 8). Performing ASES with specially designed injector devices and at increased dispersion efficiencies instead generated products that displayed an increased variance upon changed operating conditions at constant nozzle configuration, as compared with operation with a standard capillary nozzle at conservative flow-rates (3, 23, 28). Operating pressures, temperatures and solution concentrations affect inter-phase mass transfer at the thermodynamic level while droplet sizes and jet developments of atomized plumes affect inter-phase mass transfer at the mixing level. The varying effect of operating conditions on product characteristics as a function of atomization and degree of mixing indicates that the precipitation mechanism within the conventional ASES process is phase-mixing limited and therefore lacks thermodynamically controlled elements. When ASES was operated with higher inter-phase mixing using injection devices, the dominance of thermodynamic control was increased to the level where thermodynamic properties on product characteristics was influential. Mass transfer within a precipitation process with perfect inter-phase mixing would therefore be completely dependent on system thermodynamics, and hence display a greater degree of product tunability as a function of operating conditions.

Ideal Process Embodiments

In developing a progressive operating platform enabling the application of SCF anti-solvent precipitation, while not inheriting the existing impediments associated with current SCF precipitation processes, several heuristics may be applied. If solution introduction was conducted at very high flow-rates, inter-phase homogeneity and processing times would be greatly improved. Critical precipitation times would also be minimized for more effective control over precipitation kinetics (*17*). Conducting precipitation within a homogeneous phase would also encourage more uniform rates of precipitation and therefore, enhance product homogeneity.

Solution delivery at high flow-rates also eliminates the need for capillary nozzles or micro-orifices to effect solution atomization, thereby avoiding the associated complexities of process operation and translation. Further to that, the inertia of introduced solution may be harnessed to intensify inter-phase mixing over larger excess volumes. If precipitation was made to occur over a larger volume, 'nucleation densities' are lowered and this may allow the formation of product with lower bulk densities. Products of low bulk densities are particularly useful for applications in inhalation delivery technologies (*32-35*).

Layout of the ARISE Process

The schematic of the ARISE process is presented in Figure 1. Wetted parts were all of Grade 316 stainless steel construction. A 300 ml bolt closure vessel (Autoclave Engineers, Erie, Pennsylvania) was used as the precipitation vessel. An oversized precipitation vessel allowed precipitate formation over a larger spatial volume, thereby lowering 'nucleation density' for the formation of product with lower bulk densities. The wide access neck of the bolt closure vessel also allowed the visualization of precipitate deposition and hence, product distribution. Wider access also facilitated precipitate recovery. The bolt closure vessel came equipped with a 3.2 mm port in the top centre of the cap, two 3.2 mm ports at the side of the cap and a central 6.4 mm port at the base.

The injection chamber was fashioned out of a length of 12.7 mm tubing. The internal volume of the injection chamber was 10 ml. The internal surface of the 12.7 mm tubing was polished to a high mirror finish to minimise liquid adherence to the sides of the injection chamber. The back-pressure vessel (150 ml Swagelok sample cylinder 316L-50DF4-150) was connected directly to the injection chamber with 6.4 mm tubing to provide back-pressure during solution injection. A 150 ml back-pressure vessel was incorporated in the injection circuit to permit use of a lower injection pressure. A higher injection circuit volume would correspond to a lower injection pressure for a given volume of working solution injected. Increasing the back-pressure volume would also ensure that the entire charge of working solution is energetically released and not just during the initial stage of injection.

The injection chamber and precipitation vessel were connected by a straight 3.2 mm tubing 100 mm in length with internal diameter of about 1 mm. The 3.2 mm tubing extended past the cap into the precipitation vessel by about 30 mm, functioning as a conduit to direct solution during injection. The ends of the 3.2 mm tubing were squared, burr-free and polished to ensure symmetrical injection patterns. The contents of the injection chamber and the precipitation vessel

314

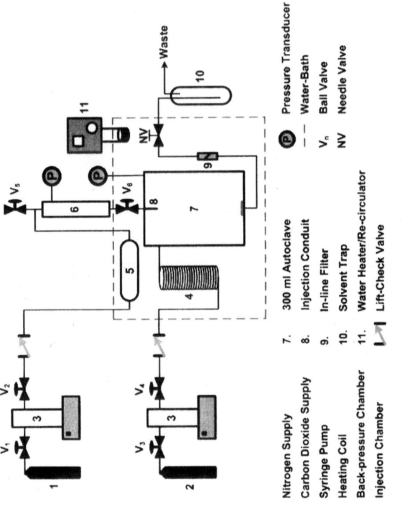

Figure 1. Schematic layout of the ARISE process.

1. Nitrogen Supply
2. Carbon Dioxide Supply
3. Syringe Pump
4. Heating Coil
5. Back-pressure Chamber
6. Injection Chamber

7. 300 ml Autoclave
8. Injection Conduit
9. In-line Filter
10. Solvent Trap
11. Water Heater/Re-circulator

Lift-Check Valve

Pressure Transducer
Water-Bath
V_n Ball Valve
NV Needle Valve

were separated by a ball valve V_6 (Swagelok Series 41, SS-41S2) connected in-line with the 3.2 mm tubing. When V_6 was open the injection chamber was directly connected to the precipitation vessel through the 3.2 mm tubing and when the valve was closed the injection chamber was isolated from the precipitation vessel. Internal pressures of the injection chamber and precipitation vessel were separately monitored by two factory calibrated pressure transducers (Druck DPI 280, UK). Critical components of the process were immersed in a temperature controlled water bath (Thermoline Unistat 130, Australia).

Materials and Methods

Materials

Chemicals and reagents employed in this study are listed in Table I. All compounds were used as received without any further treatment or purification.

Table I. Chemical Entities Consumed during ARISE Experimentation

Entity	Supplier	Lot No.	Purity / Activity
Carbon Dioxide	Linde Aust.	-	$\geq$ 99.5%
Nitrogen	Linde Aust.	-	$\geq$ 99.999%
Dimethyl Sulfoxide (DMSO)	Ajax Finechem Aust.	AH412153	$\geq$ 99.9%
Insulin from Bovine Pancreas	Sigma-Aldrich GmbH	054K1375	28USP units/mg

ARISE Process Method

Results reported in this study were obtained by first completely dissolving bovine insulin in DMSO to form a working solution. The precipitation vessel was next purged of atmospheric air with CO_2 before supercritical carbon dioxide ($ScCO_2$) was introduced into the precipitation vessel through a spiral heating coil with a syringe pump (ISCO 500D) to the desired working pressure. The precipitation vessel was then sealed. $ScCO_2$ within the precipitation vessel was next allowed about 30 minutes to achieve thermal equilibrium.

Working solution was next introduced into the injection chamber with a syringe through V_5 (Swagelok Series 41, SS-41S2). The injection chamber and back-pressure vessel were then charged with Nitrogen (N_2) with a syringe pump (ISCO 500D) to a pressure in excess of the precipitation vessel pressure and sealed. N_2 was selected as the medium to achieve the pressure differential because of its low solubility in the working solution (*36*) so as to prevent undesirable expansion and hence precipitation prior to injection. In addition, N_2 does not undergo any phase transition and remains gaseous under experimental conditions (*37*). The injection circuit consisting of the injection chamber and back-pressure vessel was next allowed about 5 minutes to achieve equilibrium.

With the system at rest, V_6 was quickly flicked open for a period of 5 seconds to allow the working solution to be energetically injected into the precipitation vessel through the 3.2 mm tubing. After injection, the contents of the precipitation vessel were allowed to rest for 10 minutes to achieve a stable pressure. Fresh $ScCO_2$ was next passed through the precipitation vessel under isobaric and isothermal conditions to flush the system of organic solution. Precipitated bovine insulin was retained in the precipitation vessel with a 0.5 μm frit mounted at the base. The precipitation vessel was next disassembled to facilitate recovery of product in the form of dry powders.

Scanning Electron Microscopy

Scanning Electron Microscope (SEM) imaging was used to examine and record the morphology and appearance of ARISE processed bovine insulin. Powders were mounted on carbon stubs and coated with chromium under vacuum at 150 mA. Coated stubs were then imaged with a Hitachi S900 FESEM at various levels of magnification.

Laser Diffraction Analysis

The relative geometrical particle size of processed bovine insulin was evaluated with integrated laser diffraction. The Malvern Mastersizer 2000 (Malvern Instruments, UK) was equipped with the Hydro 2000μP 18 ml wet sample dispersion unit and ethanol was used as the wet dispersant. The circulating pump of the Hydro 2000μP was run at 2000 rpm and internal dispersion ultrasound was turned off. Prior to analysis, 10 mg of processed bovine insulin was dispersed in 2 ml of ethanol for 1 minute in an external ultrasound bath (Unisonics Model FXP8M, Australia). The dispersed sample was next added into the measurement cell to attain an obscuration of between 4 to 6%. Reported data was the calculated average of 5 successive determinations taken 10 seconds apart.

Cascade Impactor Analysis

The aerodynamic properties of processed bovine insulin were assessed with an unmodified 8-stage Non-Viable Cascade Impactor (Andersen Instruments,

Atlanta, Georgia) equipped with a preseparator stage and a glass fiber filter stage. Samples of ARISE processed bovine insulin were assessed as obtained from processing, without any additional dilution or formulation. The cascade impactor was operated at a set flow-rate of 28.3 L/min and processed bovine insulin was drawn through the cascade impactor for a time duration corresponding to approximately 4 L of air to simulate *in-vivo* inspiration. Variable doses of processed bovine insulin were manually loaded into gelatin capsules (Size 3, Tyco Healthcare, Australia) and delivered with a HandiHaler device (Boehringer Ingelheim, Australia) into the induction port of the cascade impactor through a custom fitted adapter piece. The mass of bovine insulin loaded during each test was determined by weighing the gelatin capsule before and after it was loaded. Loaded doses generally ranged between 2 mg to 4 mg. The impaction plates of the cascade impactor were not coated prior to use and the low loaded doses of processed bovine insulin were not observed to contribute to particle bounce and re-entrainment.

Two independent cascade impactor analyses were conducted on each sample. After approximately 4 L of air was drawn through the cascade impactor, it was disassembled and the impactor plates of each stage were rinsed separately with a known volume of dilute hydrochloric acid. The inhaler device, the expired capsule, the preseparator and the glass fiber filter were also rinsed separately. The nozzle stages of the cascade impactor were not rinsed; inter-stage deposition of bovine insulin was considered to consist of uncharacterized material.

The quantities of bovine insulin dissolved into aliquots obtained from rinsing the respective cascade impactor components were determined with UV-Visible Spectrophotometric techniques (Agilent Technologies 8453 UV-Visible Spectrophotometer). Absorption peaks were observed and recorded at 276 nm wavelengths. Mass distributions within the cascade impactor were next presented as a nominal function of stages and auxillary components in order of component location.

Bulk Density Measurement

A quantity of bovine insulin equivalent to about 3 ml was loaded into a narrow stemmed glass funnel positioned on top of a pre-weighed dry 10 ml graduated glass measuring cylinder. By gently tapping the side of the glass funnel, bovine insulin was passed through the stem of the glass funnel evenly into the measuring cylinder. Ensuring that the bovine insulin in the measuring cylinder was level, the unsettled apparent volume was recorded and the weight of the loaded measuring cylinder was next measured. Recorded data was then normalized to present values of bulk density in g/ml.

Results and Discussion

The working principle and operability of the ARISE process was examined using bovine insulin as the model compound. Operating variables of pressure, differential pressure and injection conduit aperture size were systematically varied. When processing was successful at initial system volumes, injected volumes, solution concentration and rinse procedures, these parameters were held constant. Values of parameters held constant during experimentation are listed in Table II, and the sequence of experimental conditions evaluated is listed in Table III.

Table II. Values of Constant Processing Parameters

Parameter	Value
Injection Circuit Volume (ml)	182
Precipitation Circuit Volume (ml)	355
Volume of Injected Solution (ml)	10
Insulin Concentration in DMSO (mg/ml)	20
ScCO$_2$ Rinse Flowrate (ml/min)	15
ScCO$_2$ Rinse Volume (L)	3

Table III. Experimental Conditions for ARISE Precipitation of Bovine Insulin from DMSO

Expt. No.	Press. (bar)	Temp. (°C)	Pressure Differential (bar)	Conduit Aperture (mm)	Product Recovery (%)
BI01 [a]	90	40	50	1.0	90
BI02 [a]	90	40	50	1.0	93
BI03 [b]	120	40	50	1.0	96
BI04 [b]	120	40	50	1.0	99
BI05 [c]	150	40	50	1.0	96
BI06 [c]	150	40	50	1.0	100
BI07	120	25	50	0.762	90
BI08	120	40	50	0.2	Unsuccessful
BI09	90	40	50	0.762	95
BI10	120	40	50	0.762	99
BI11	150	40	50	0.762	100
BI12	120	40	80	1.0	100

[a, b, c] Reproducibility runs; conditions identical between each pair

Process Attributes

The injection chamber was disassembled and inspected after each experiment. Solution retention within the injection chamber was negligible. Amounts corresponding to about 0.05 ml were retrieved in very rare cases. The volume injected during experimentation was more than 10 ml in every circumstance, ascertained after working solution injection by determining the volume of N_2 required to pressurize the injection circuit back to pre-injection conditions.

A K-type thermocouple of 0.1°C resolution was inserted in the volume of the precipitation vessel to monitor the temperature of its interior. The thermocouple did not register any temperature fluctuations during or after working solution injection, although it was capable of detecting temperature increases while the precipitation vessel was being pressurized with $ScCO_2$ to experimental conditions. The constant temperature observed during and after working solution injection allows the assumption of quasi-isothermal injection conditions.

A stated objective of this process development was to effect precipitation homogeneously over a larger volume, preferably throughout the entire precipitation vessel. As evidenced by photographs in Figure 2, precipitation does indeed occur throughout the entire precipitation vessel, with particles precipitating and depositing uniformly over the entire internal surface, even above the point of injection. A confident assumption made at this stage is that product deposition on the internal surfaces of the precipitation vessel had not been adversely altered when fresh $ScCO_2$ was passed through to rinse the vessel of residual DMSO.

Scanning Electron Microscopy

Electron microscope images of bovine insulin obtained from ARISE processing under BI01, BI03 and BI05 are presented in Figure 3. Images are presented at three magnification levels of 1 K, 10 K and 100 K.

Bovine insulin recovered from ARISE processing under all three sets of experimental conditions consisted of highly aggregated sub-micron individual particles. At a pre-injection pressure of 90 bar $ScCO_2$ (BI01), bovine insulin precipitated as microspheres ranging between 50 to 100 nm. These individual particles appear to have fused together to form aggregates about 10 μm. The 10 μm aggregates are in turn loosely connected to each other.

As the pre-injection pressure of $ScCO_2$ employed was increased from 90 bar (BI01) to 120 bar (BI03), the degree of aggregation appeared to decrease with more interstitial voids sighted within the aggregates. Primary particles were sighted between 20 and 50 nm. The primary particles were not as severely fused to each other and this resulted in the primary particles taking on a more spherical

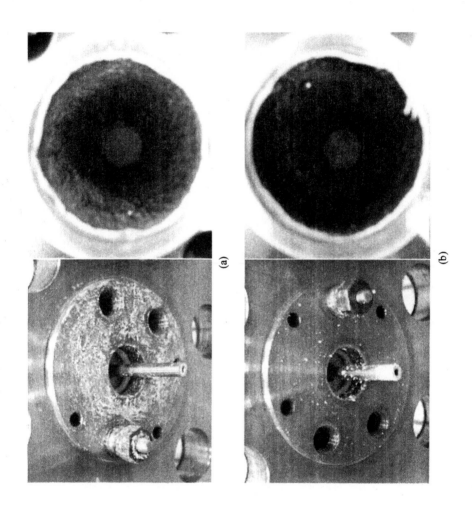

(a)

(b)

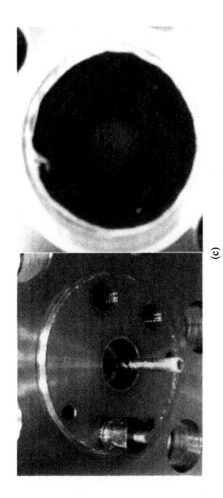

(c)

Figure 2. Deposition of bovine insulin within precipitation vessel imaged immediately after ARISE processing under (a) BI01, (b) BI03 and (c) BI05.

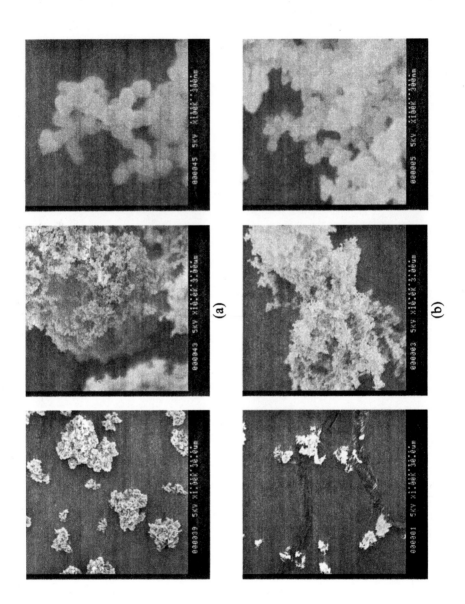

(a)

(b)

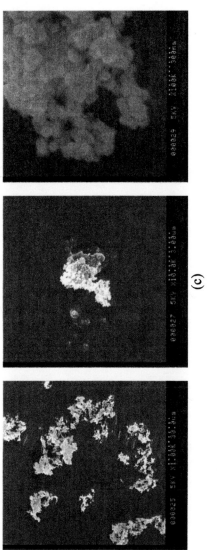

(c)

Figure 3. SEM images of ARISE processed bovine insulin obtained from (a) BI01, (b) BI03 and (c) BI05.

regularity. The primary particles were seen to have formed plate-like aggregates about 4 μm in size and these platelet aggregates were in turn, delicately attached together in three dimensions to form aggregates of about 10 μm.

When the pre-injection pressure of $ScCO_2$ was increased to 150 bar (BI05), the primary particles of bovine insulin appeared to lose their shape regularity. Although primary particle size had decreased compared to BI03, the degree of aggregation appeared to increase slightly. Electron microscope images indicate that bovine insulin may be precipitated as particles of differing degrees of agglomeration and agglomerate size, depending on anti-solvent pressures.

Particle Size Measurement with Laser Diffraction

The particle size distribution of ARISE processed bovine insulin are presented as frequency size curves extracted from Mastersizer 2000 Version 5.22 software (Malvern Instruments) in Figure 4.

Bovine insulin obtained from injecting the working solution into $ScCO_2$ at 90 bar (BI01) had a mean geometrical diameter of about 6 μm. Laser diffraction particle sizing revealed that a product with a tight particle size distribution had been achieved. The calculated span, which is a dimensionless measure of the width of particle size distribution, was determined to be 0.991.

At a pre-injection pressure of 120 bar $ScCO_2$ (BI03), a bimodal product was achieved. Primary particles smaller than 40 nm were detected with laser diffraction. The existence of these particles was supported by SEM images, as shown in Figure 3(b). Noteworthy is that almost 100% of the particles obtained from ARISE processing under the experimental conditions outlined in BI03 possessed geometrical diameters less than 5 μm, the prescribed upper size limit for pulmonary delivery of dry powder formulations.

When the pre-injection pressure of $ScCO_2$ was increased to 150 bar (BI05), laser diffraction revealed similar results to that obtained at 120 bar. The population of the finer particles remained similar, but particles of about 600 nm had remained aggregated in the dispersant to form a very small fraction of much larger particles up to 100 μm. Laser diffraction results indicate that bovine insulin may be precipitated as particles of differing size and distribution, depending on anti-solvent pressures.

Geometric particle size statistics from laser diffraction analysis of bovine insulin obtained from ARISE processing at pressures of 90 bar, 120 bar and 150 bar (BI01 to BI06) are also summarized in Table IV. Geometric particle sizes quoted are at 10%, 50% and 90% of cumulative distributions acquired. The variations in size distributions of product obtained from two independent experiments under identical operating conditions are expressed as the percentage difference in observed particle sizes divided by the average of these two values. Inter-batch variations were observed to be between zero and 10%, and process repeatability appeared to be improved at higher pre-injection pressures of

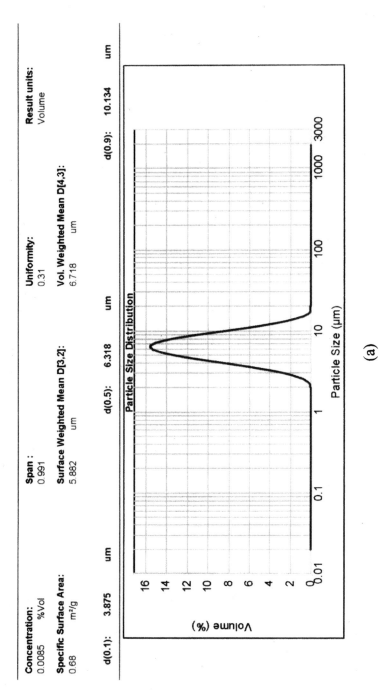

Figure 4. Particle size distribution of ARISE processed bovine insulin obtained from (a) BI01, (b) BI03 and (c) BI05. Continued on next page.

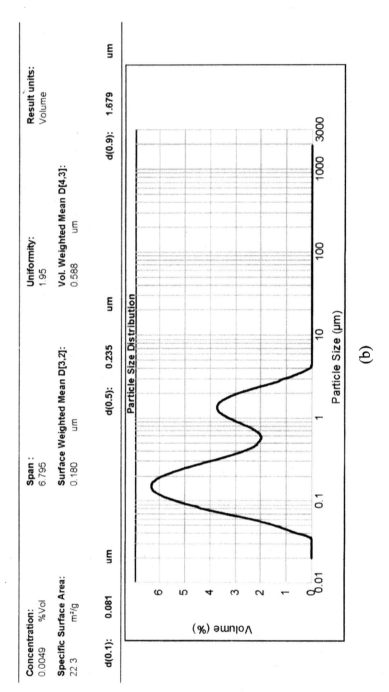

(b)

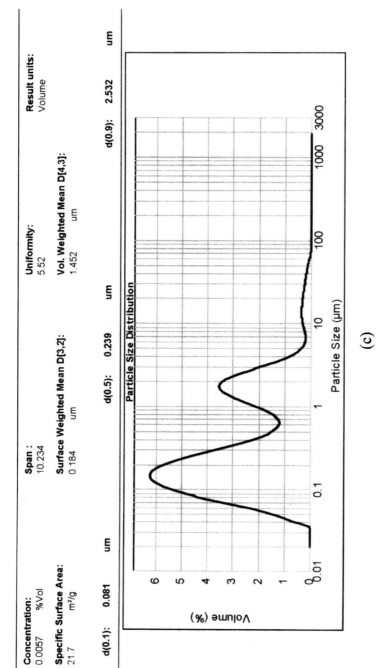

Figure 4. Continued.

Table IV. Geometrical Particle Size Statistics of ARISE Processed Bovine Insulin

Expt. No.	Laser Diffraction Results					
	D(0.1)[a]		D(0.5)[b]		D(0.9)[c]	
	Particle Size (μm)	Percent Variation[d] (%)	Particle Size (μm)	Percent Variation[d] (%)	Particle Size (μm)	Percent Variation[d] (%)
01	3.875	9.0	6.318	6.5	10.134	3.7
02	4.241		6.745		10.516	
03	0.083	1.2	0.250	2.0	2.103	8.0
04	0.082		0.245		1.942	
05	0.081	0.0	0.239	0.0	2.532	3.5
06	0.081		0.239		2.621	

[a] Geometrical Particle Diameter at 10th Percentile of Cumulative Distribution
[b] Geometrical Particle Diameter at 50th Percentile of Cumulative Distribution
[c] Geometrical Particle Diameter at 90th Percentile of Cumulative Distribution
[d] Percent Variation = (Particle size difference / Particle size average) x 100%

ScCO$_2$. Reproducibility of product obtained from independent experiments at identical operating conditions with the ASES process may also be deduced from previous work (*30*). When insulin was recovered from DMSO with the ASES process at 86.2 bar and 35°C, variation in the mean particle size of powders obtained from three separate batches was calculated to be between 13.6% and 29.8%. It appears that compared to the ASES process, ARISE processing is capable of generating product with lower inter-batch variation in particle size.

Cascade Impactor Analysis

The mass depositions of ARISE processed bovine insulin recovered from cascade impactor classification in the order of component location are collectively presented in Figure 5. Values of the loaded doses, emitted doses, respirable fractions and mass median aerodynamic diameters (MMAD) of individual determinations are presented in Table V.

Table V. Aerodynamic Statistics of ARISE Processed Bovine Insulin

Expt. No.	BI01		BI03		BI05	
Loaded Dose (mg)	*4.0*	*4.1*	*1.8*	*3.0*	*2.0*	*2.3*
Emitted Dose (%)	94.4	93.5	94.1	94.1	96.3	90.7
Respirable Fraction (%)	24.7	32.7	75.2	78.0	73.2	70.1
MMAD (μm)	6.9	6.3	0.4	0.4	0.4	0.4
Recovered Dose (%)	104.2	102.6	97.4	91.2	107.6	102.1

Upon being introduced into the cascade impactor, the bulk of ARISE processed bovine insulin from BI01 was retained in the preseparator. The preseparator of the cascade impactor was designed to retain particles with aerodynamic diameters greater than 9 μm, typically comprising of the larger carrier particles used to improve drug dispersibility in dry powder formulations. Mass distribution results within the cascade impactor indicate however, that ARISE processed bovine insulin from BI01 was still effectively released from the capsule and inhaler device and entrained into the airstream.

ARISE processed bovine insulin from BI03 displayed much higher aerodynamic efficiencies when compared to product from BI01. More than 40% of the loaded dose remained entrained in the airstream and was captured only in the glass fiber filter after the last impaction plate stage of the cascade impactor. Very similar aerosol statistics were also observed for different loaded doses of 1.8 mg and 3.0 mg. Samples from BI05 exhibited similar deposition profiles to that of BI03.

Analytical results on the aerodynamic properties of dry powders obtained under all three experimental conditions demonstrate superior dry dispersibility properties. The high emitted doses indicate the absence of static charge within ARISE processed bovine insulin. During the formulation of ARISE processed bovine insulin for pulmonary end-use, a high emitted dose would also imply lower 'overfill' requirements, leading to lower consumption of high value insulin. The relatively high respirable fractions of product obtained from BI03 and BI05 also attest to the suitability of ARISE processed bovine insulin as a candidate for further development into a pulmonary delivery formulation.

Bulk Densities of ARISE Processed Bovine Insulin

The aerodynamic performance of dry powder formulations are influenced not only by their geometrical diameters. Aerodynamic diameters of particles are also a function of their bulk densities, shape and morphology (38-44). The bulk densities of ARISE processed bovine insulin are listed in Table VI. Reported values have not been obtained with sophisticated techniques and are only presented to serve as a qualitative indication of processed powder characteristics.

Table VI. Bulk Density Values of ARISE Processed Bovine Insulin

Expt. No.	Anti-solvent Pressure (bar)	Bulk Density (g/ml)
BI01	90	0.053
BI03	120	0.010
BI05	150	0.013

The low bulk densities of ARISE processed bovine insulin obtained from BI03 and BI05 are also observed in Figures 6(a) through 6(d) displaying images of 200 mg of bovine insulin in 50 ml sample bottles. Figure 6(a) captures 200 mg of lyophilized bovine insulin while Figure 6(b) indicates 200 mg of product collected from ARISE processing under experimental conditions of BI01. Figure 6(c) shows 200 mg of bovine insulin recovered from BI03 while the sample bottle in Figure 6(d) contains 200 mg of bovine insulin obtained from BI05.

ARISE processing under experimental conditions prescribed in BI03 to BI06 have yielded dry powders of extremely low bulk density. Bulk density reduction is believed to be the result of effecting precipitation simultaneously over a larger volume. Powders of low bulk densities are known to display aerodynamic diameters substantially lower than their geometrical dimensions. Fractal powders deviating from the shape of a perfect sphere are also known to

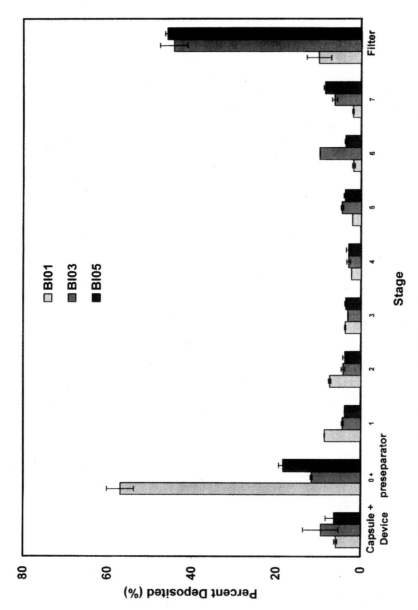

Figure 5. Cascade impactor mass distribution of ARISE processed bovine insulin obtained from BI01, BI03 and BI05.

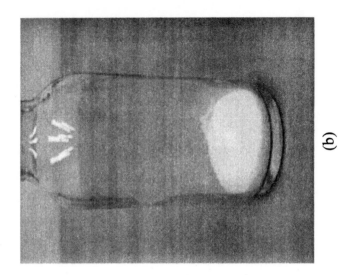

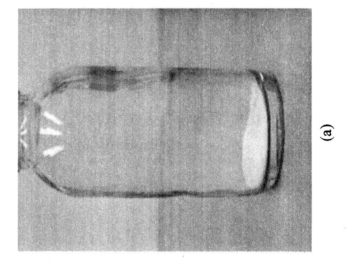

(a)

(b)

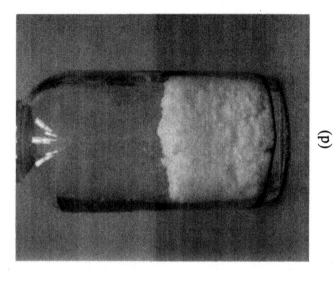

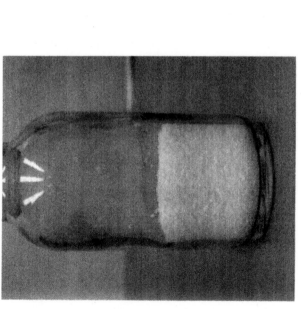

(c)

(d)

Figure 6. Appearance of 200mg bovine insulin in 50ml sample bottles (a) lyophilized, processed with ARISE under (b) BI01, (c) BI03 and (d) BI05.

have decreased aerodynamic diameters due to an increase in drag forces experienced. SEM images in Figures 3(b) and 3(c) indicate the fluffy nature of bovine insulin obtained from ARISE processing with BI03 and BI05.

Powders of lower bulk density also experience lower interparticulate van der Waals attractive forces due to the increased interstitial spaces and lower contact area between adjacent particles (42). Lower interparticulate attractive forces translate to increased powder dispersibility and flowability as well as decreased powder cohesiveness. Reasons for the less dramatic bulk density reduction experienced in BI01 will be offered in the following section.

Effect of Anti-solvent Pressure

Physical properties of $ScCO_2$ under ARISE experimental conditions BI01, BI03 and BI05 are presented in Table VII. The mild pressurization conditions of BI01 offered a $ScCO_2$ anti-solvent system with a low viscosity and density. Consequently, when $ScCO_2$ at higher pressures were used, anti-solvent viscosity and density increased. Although the pressure setpoints between the three separate cases had been increased by equal amounts, the increase in viscosity and density were more dramatic going from BI01 to BI03 and the degree of change was reduced when comparing BI03 and BI05.

Table VII. Physical Properties of $ScCO_2$ under Various ARISE Experimental Conditions

Expt. No.	Temp. (°C)	Pressure (bar)	$ScCO_2$ Viscosity (cP)	$ScCO_2$ Density (g/cm^3)	Moles $ScCO_2$
BI01	40	90	0.035	0.486	3.92
BI03	40	120	0.059	0.718	5.79
BI05	40	150	0.068	0.780	6.30

NOTE: Values obtained from the NIST REFPROP Version 7.0 Database

Jet atomization and mixing efficiencies are highly influenced by system viscosity and density. Atomization and mixing efficiencies are lower with systems of higher viscosities and densities. Simultaneously, supersaturation and precipitation rates within $ScCO_2$ anti-solvent systems are impacted by the density of $ScCO_2$ used. It is well known that higher supersaturation and precipitation rates in $ScCO_2$ anti-solvent processes are easily achieved by using $ScCO_2$ at higher pressures and hence higher densities. The pre-injection stage of the ARISE process also maintains the $ScCO_2$ anti-solvent in the precipitation vessel under stagnant conditions. In an isochoric system, $ScCO_2$ charged at higher pressure implies an increase in the amount of $ScCO_2$ added. Postinjection of a fixed volume and mass of DMSO into the precipitation vessel under higher pressure of ScCO$_2$ would indicate a higher anti-solvent to solvent system ratio resulting. These phenomena contribute to the manifestation of competing effects of anti-solvent viscosities and densities during ARISE precipitation.

ARISE processing under low pressure anti-solvent conditions of BI01 portrays conditions that favor jet atomization and mixing while decreasing the rate at which solvated bovine insulin precipitates. As Figure 2(a) suggests, a homogeneous mixture was achieved within the precipitation vessel before bovine insulin had begun to precipitate, which facilitates the explanation of the high deposition of bovine insulin at the walls of the precipitation vessel – the vessel walls had been wet with a layer of DMSO containing solvated bovine insulin before the DMSO was extracted. Precipitation conditions suggested where homogeneity is achieved before precipitation would imply a product with a very narrow particle size distribution – this is supported with the laser diffraction results in Figure 4(a).

By using an increased pressure of ScCO$_2$, the viscosity and density of the anti-solvent in BI03 were substantially increased. Such a system implies a lower atomization and mixing efficiency but increased precipitation rates. Although deposition of particles implies that precipitation had taken place throughout the entire vessel, the much lighter wall deposition seen in Figure 2(b) indicates that precipitation had occurred primarily within the core of the precipitation vessel and DMSO had been extracted before it could effectively wet the vessel walls. This scenario may help explain the bi-modal nature of the laser diffraction results in Figure 4(b) – precipitation commenced before a homogeneous mixture was attained within the precipitation vessel, giving rise to non-uniformed precipitation rates.

Despite the care taken not to disturb vessel contents during disassembly, the powder cake of bovine insulin depicted in Figure 2(c) had inadvertently been inverted. Nevertheless, when the pressure of ScCO$_2$ was further increased as in BI05, there was much stronger indication that ARISE processing with ScCO$_2$ under higher pressure conditions induces bovine insulin to precipitate within the core of the precipitation vessel. Under this condition, particle deposition on the vessel wall was almost non-existent. The precipitation of bovine insulin was invariably effected well before the contents of the precipitation vessel were able to attain homogeneity. The exacerbation of non-uniformed precipitation rates gave rise to the even wider particle size distribution of Figure 4(c). It appears that ARISE processing under conditions prescribed in BI03 offered the optimum compromise between mixing efficiency and the rate of precipitation to generate bovine insulin in the form of dry powders of low bulk density and acceptable particle size distribution suitable for pulmonary administration. Experimental observations also indicate that bovine insulin may be precipitated as particles of differing degrees of agglomeration, size and aerosol efficiency simply by adjusting anti-solvent pressures to influence precipitation kinetics. Enhanced tunability of product implies that the precipitation kinetics of ARISE were not hindered by imperfect inter-phase mixing to a significant extent.

Effect of Anti-solvent Temperature

ARISE processing at the anti-solvent pressure of 120 bar was attempted at the lower temperature of 25°C (BI07) to assess the potential of processing at lower temperatures. SEM images and laser diffraction results of product from BI07 are presented in Figure 7. Product of inferior qualities was obtained when processing at 25°C as compared to processing at 40°C (BI03). Primary particles were highly agglomerated and aggregate size was as large as 50 μm. Processing at 25°C indicated that fluid phase in the precipitation vessel resulted in a liquid and vapor separation subsequent to solution injection. During the rinse stage, the flow-rate of $ScCO_2$ underwent a dramatic step increase after about 200 ml of $ScCO_2$ was passed. The liquid phase would have been forced through the sintered frit at the initial stage and while remaining at constant pressure, resistance to flow would have been dramatically reduced upon complete purge of liquid.

Effect of Injection Conduit Aperture

Laser diffraction results of the experimental subset of batches operated under similar conditions but with different injection conduit apertures are listed in Table VIII. ARISE processing was experimented with apertures of 0.2 mm, 0.762 mm and 1.0 mm.

Table VIII. Results Obtained from ARISE Processing with Different Injection Conduit Apertures

Expt. No.	Pressure (bar)	Temp . (°C)	Conduit Aperture (mm)	Laser Diffraction Results (μm)		
				D (10%)	D (50%)	D (90%)
BI01	90	40	1.0	3.875	6.318	10.134
BI09	90	40	0.762	3.879	6.324	10.145
BI03	120	40	1.0	0.081	0.235	1.679
BI10	120	40	0.762	0.082	0.245	1.942
BI08	120	40	0.2		Unsuccessful	
BI05	150	40	1.0	0.081	0.239	2.532
BI11	150	40	0.762	0.081	0.239	2.621

ARISE injection during BI08 was attempted through a 0.2 mm diameter capillary nozzle. Solution delivery was however unsuccessful with the nozzle blocking immediately upon injection, despite repeated successive attempts at restoring flow through the nozzle. Operation with decreased conduit apertures under all other experimental conditions did not appear to affect product particle size dramatically. Even when conduit aperture was decreased by almost 25%, product particle size did not experience any appreciable improvement. A significant decrease in conduit apertures during ARISE processing therefore, did not impact on atomization efficiencies sufficiently to influence product characteristics. The efficiencies of conventional SCF precipitation techniques however, were found to be extremely dependent on nozzle geometries.

Effect of Backpressure Intensity

A higher backpressure intensity of 80 bar was attempted with BI12 at 120 bar anti-solvent pressure, with SEM and laser diffraction results appended in Figure 8. Cascade impactor deposition profiles of sample BI12 are presented in Figure 9.

Poorer product characteristics appeared to have been obtained from injecting at higher backpressure intensity. Primary particles appeared to have formed a higher population of larger agglomerates than product obtained at 50 bar backpressure (BI03), and there was a corresponding decrease in aerosol efficiency of product. Mechanistically, injecting at a higher intensity would result in improved atomization, thereby leading to smaller solution droplets and hence smaller sized precipitate. The reduction in conduit aperture size to improve atomization efficiencies however, was not shown to affect product characteristics by any appreciable amount. Instead, an increase in N_2 concentration within the precipitation vessel subsequent to solution injection in this case almost certainly lowered particle supersaturation and nucleation rates during solvent extraction and hence, encouraged particle growth and bridging.

Conclusion

The ARISE process was designed to ameliorate the disadvantages of existing SCF precipitation techniques by introducing solution via a single bolus injection without the use of capillary nozzles or orifices. In doing so, the inertia of the injected stream of solution was also harnessed to intensify mixing throughout the entire volume of the precipitation vessel. The ARISE process incorporated a novel method of energetic solution delivery via a single bolus injection directly through a 1.0 mm bore injection conduit to achieve homogeneous distribution over the entire vessel containing static ScCO$_2$. While the ASES process operates by controlling mixing over a localized precipitation envelope over a continuous period, the ARISE process demonstrated its ability to conduct precipitation within the entire contained volume for enhanced

338

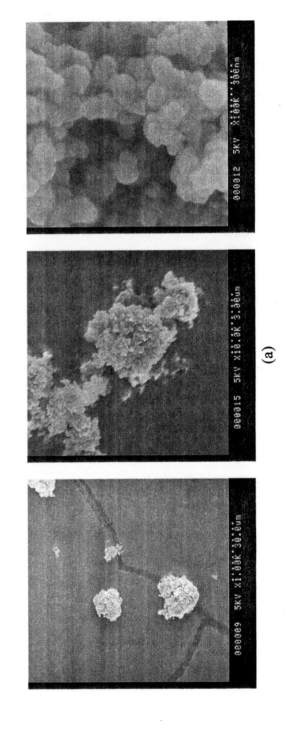

(a)

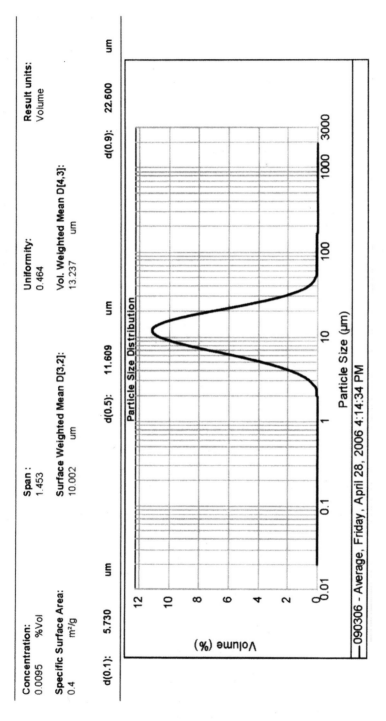

Figure 7. Analytical results (a) SEM and (b) Laser Diffraction on ARISE insulin processed at 120 bar and 25°C (BI07)

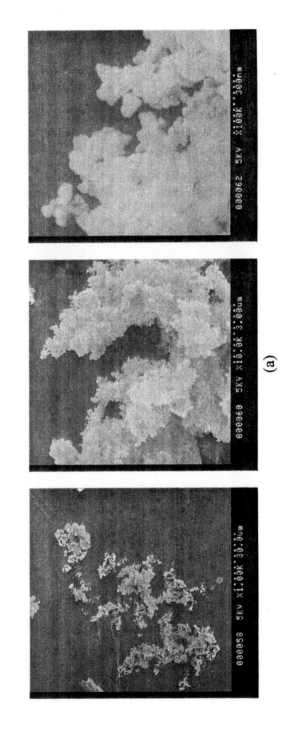

(a)

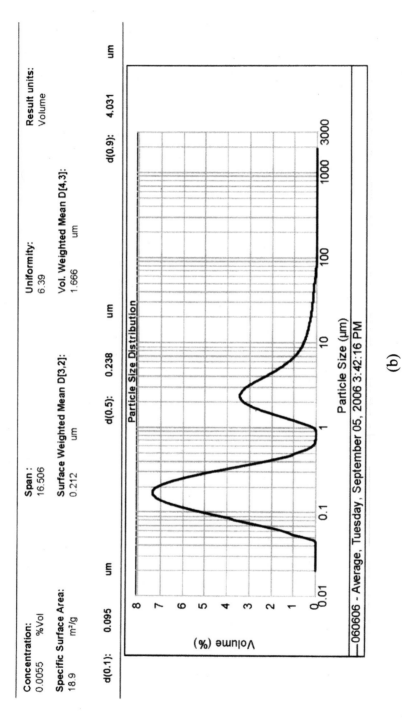

Figure 8. Analytical results (a) SEM and (b) Laser Diffraction on product obtained from BI12.

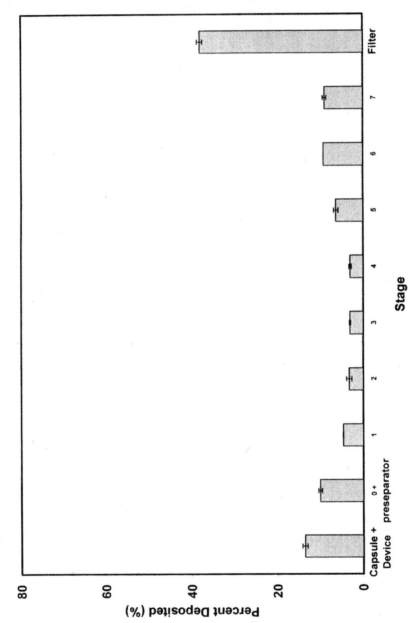

Figure 9. Cascade impactor mass distribution of ARISE processed bovine insulin obtained from BI12.

thermodynamic control over product characteristics. The ARISE process utilizes the concept of energetic discharge to atomize solutions with a rapid injection technique to achieve bulk phase homogeneity before solute precipitation is effected by solvent extraction.

The ARISE process was successfully implemented for the production of low bulk density bovine insulin displaying high suitability for pulmonary delivery applications. Product characteristics were highly tunable as a function of anti-solvent pressures, and were relatively unaffected by atomization efficiencies. Atomization efficiencies improved through the use of smaller conduit aperture size and higher injection back-pressure did not appear to influence product characteristics dramatically, indicative that adequate solution atomization under experimental conditions was achieved using a 1 mm bore conduit and a 50 bar back-pressure.

Insulin obtained directly from ARISE processing displayed extremely high aerodynamic efficiencies, without traditional need for formulation and bulking with excipients. Apart from its ability to generate insulin of extremely low bulk densities by precipitating over a much larger volume, additional processing benefits of the ARISE process include decreased processing times resulting from quasi-instantaneous working solution introduction, a high degree of thermodynamically-based product tunability, high inter-batch product reproducibility, and high modular process scalability. The lack of capillary nozzles within processing also makes the ARISE process suitable for applications involving slurries and suspensions.

References

1. McHugh, M. A.; Krukonis, V. J., *Supercritical Fluid Extraction. Principles and Practice.* 2nd ed.; Butterworth-Heinemann: Newton, MA, 1994.
2. Bustami, R. T.; Chan, H.-K.; Dehghani, F.; Foster, N. R., Generation of Microparticles of Proteins for Aerosol Delivery Using High Pressure Modified Carbon Dioxide. *Pharmaceutical Research* **2000**, *17*, (11), 1360-1366.
3. Perez de Diego, Y.; Pellikaan, H. C.; Wubbolts, F. E.; Witkamp, G. J.; Jansens, P. J., Operating Regimes and Mechanism of Particle Formation During the Precipitation of Polymers Using the PCA Process. *Journal of Supercritical Fluids* **2005**, *35*, (2), 147-156.
4. Dukhin, S. S.; Shen, Y.; Dave, R.; Pfeffer, R., Droplet Mass Transfer, Intradroplet Nucleation and Submicron Particle Production in Two-Phase Flow of Solvent-Supercritical Antisolvent Emulsion. *Colloids and Surfaces, A: Physicochemical and Engineering Aspects* **2005**, *261*, (1-3), 163-176.

5. Reverchon, E., Supercritical Antisolvent Precipitation of Micro- and Nano-Particles. *Journal of Supercritical Fluids* **1999**, *15*, (1), 1-21.

6. Reverchon, E.; Caputo, G.; De Marco, I., Role of Phase Behavior and Atomization in the Supercritical Antisolvent Precipitation. *Industrial & Engineering Chemistry Research* **2003**, *42*, (25), 6406-6414.

7. Yeo, S.-D.; Kiran, E., Formation of Polymer Particles with Supercritical Fluids: A Review. *Journal of Supercritical Fluids* **2005**, *34*, (3), 287-308.

8. Perrut, M.; Clavier, J.-Y., Supercritical Fluid Formulation: Process Choice and Scale-up. *Industrial & Engineering Chemistry Research* **2003**, *42*, (25), 6375-6383.

9. Harjo, B.; Ng, K. M.; Wibowo, C., Synthesis of Supercritical Crystallization Processes. *Industrial & Engineering Chemistry Research* **2005**, *44*, (22), 8248-8259.

10. Thiering, R.; Dehghani, F.; Dillow, A.; Foster, N. R., Solvent Effects on the Controlled Dense Gas Precipitation of Model Proteins. *Journal of Chemical Technology & Biotechnology* **2000**, *75*, (1), 42-53.

11. Thiering, R.; Dehghani, F.; Dillow, A.; Foster, N. R., The Influence of Operating Conditions on the Dense Gas Precipitation of Model Proteins. *Journal of Chemical Technology & Biotechnology* **2000**, *75*, (1), 29-41.

12. Mawson, S.; Kanakia, S.; Johnston, K. P., Coaxial Nozzle for Control of Particle Morphology in Precipitation with a Compressed Fluid Antisolvent. *Journal of Applied Polymer Science* **1997**, *64*, (11), 2105-2118.

13. Chattopadhyay, P.; Gupta, R. B., Supercritical CO2 Based Production of Magnetically Responsive Micro- and Nanoparticles for Drug Targeting. *Industrial & Engineering Chemistry Research* **2002**, *41*, (24), 6049-6058.

14. Thote, A. J.; Gupta, R. B., Formation of nanoparticles of a hydrophilic drug using supercritical carbon dioxide and microencapsulation for sustained release. *Nanomedicine* **2005**, *1*, (1), 85-90.

15. Snavely, W. K.; Subramaniam, B.; Rajewski, R. A.; Defelippis, M. R., Micronization of Insulin from Halogenated Alcohol Solution Using Supercritical Carbon Dioxide as an Antisolvent. *Journal of Pharmaceutical Sciences* **2002**, *91*, (9), 2026-2039.

16. Bustami, R. T.; Chan, H.-K.; Sweeney, T.; Dehghani, F.; Foster, N. R., Generation of Fine Powders of Recombinant Human Deoxyribonuclease Using the Aerosol Solvent Extraction System. *Pharmaceutical Research* **2003**, *20*, (12), 2028-2035.

17. Jarmer, D. J.; Lengsfeld, C. S.; Randolph, T. W., Scale-up Criteria for an Injector with a Confined Mixing Chamber During Precipitation with a Compressed-Fluid Antisolvent. *Journal of Supercritical Fluids* **2006**, *37*, (2), 242-253.

18. Shekunov, B. Y.; York, P., Crystallization Processes in Pharmaceutical Technology and Drug Delivery Design. *Journal of Crystal Growth* **2000**, *211*, (1-4), 122-136.

19. Bristow, S.; Shekunov, T.; Shekunov, B. Y.; York, P., Analysis of the Supersaturation and Precipitation Process with Supercritical CO2. *Journal of Supercritical Fluids* **2001**, *21*, (3), 257-271.

20. Elvassore, N.; Bertucco, A.; Caliceti, P., Production of Protein-Loaded Polymeric Microcapsules by Compressed CO2 in a Mixed Solvent. *Industrial & Engineering Chemistry Research* **2001**, *40*, (3), 795-800.

21. Chavez, F.; Debenedetti, P. G.; Luo, J. J.; Dave, R. N.; Pfeffer, R., Estimation of the Characteristic Time Scales in the Supercritical Antisolvent Process. *Industrial & Engineering Chemistry Research* **2003**, *42*, (13), 3156-3162.

22. Lobo, J. M.; Schiavone, H.; Palakodaty, S.; York, P.; Clark, A.; Tzannis, S. T., SCF-engineered Powders for Delivery of Budesonide from Passive DPI Devices. *Journal of Pharmaceutical Sciences* **2005**, *94*, (10), 2276-2288.

23. Perez, Y.; Wubbolts, F. E.; Witkamp, G. J.; Jansens, P. J.; de Loos, T. W., Improved PCA Process for the Production of Nano- and Microparticles of Polymers. *AIChE Journal* **2004**, *50*, (10), 2408-2417.

24. Jovanovic, N.; Bouchard, A.; Hofland, G. W.; Witkamp, G.-J.; Crommelin, D. J. A.; Jiskoot, W., Stabilization of Proteins in Dry Powder Formulations Using Supercritical Fluid Technology. *Pharmaceutical Research* **2004**, *21*, (11), 1955-1969.

25. Thiering, R.; Dehghani, F.; Foster, N. R., Current Issues Relating to Anti-Solvent Micronisation Techniques and their Extension to Industrial Scales. *Journal of Supercritical Fluids* **2001**, *21*, (2), 159-177.

26. Todo, H.; Iida, K.; Okamoto, H.; Danjo, K., Improvement of Insulin Absorption from Intratracheally Administrated Dry Powder Prepared by Supercritical Carbon Dioxide Process. *Journal of Pharmaceutical Sciences* **2003**, *92*, (12), 2475-2486.

27. Reverchon, E.; Della Porta, G.; Di Trolio, A.; Pace, S., Supercritical Antisolvent Precipitation of Nanoparticles of Superconductor Precursors. *Industrial & Engineering Chemistry Research* **1998**, *37*, (3), 952-958.

28. Reverchon, E.; De Marco, I.; Torino, E., Nanoparticles production by supercritical antisolvent precipitation: A general interpretation. *Journal of Supercritical Fluids* **2007**, *43*, (1), 126-138.

29. Adami, R.; Reverchon, E.; Jaervenpaeae, E.; Huopalahti, R., Supercritical AntiSolvent micronization of nalmefene HCl on laboratory and pilot scale. *Powder Technology* **2008**, *182*, (1), 105-112.

30. Yeo, S. D.; Lim, G. B.; Debenedetti, P. G.; Bernstein, H., Formation of Microparticulate Protein Powders Using a Supercritical Fluid Antisolvent. *Biotechnology and Bioengineering* **1993**, *41*, (3), 341-346.

31. Mammucari, R.; Dehghani, F.; Foster, N. R., Dense Gas Processing of Micron-Sized Drug Formulations Incorporating Hydroxypropylated and Methylated Beta-Cyclodextrin. *Pharmaceutical Research* **2006**, *23*, (2), 429-437.

32. Ben-Jebria, A.; Eskew, M. L.; Edwards, D. A., Inhalation System for Pulmonary Aerosol Drug Delivery in Rodents Using Large Porous Particles. *Aerosol Science and Technology* **2000**, *32*, (5), 421-433.

33. Dunbar, C.; Scheuch, G.; Sommerer, K.; DeLong, M.; Verma, A.; Batycky, R., In vitro and in vivo Dose Delivery Characteristics of Large Porous Particles for Inhalation. *International Journal of Pharmaceutics* **2002**, *245*, (1-2), 179-189.

34. Edwards, D. A.; Ben-Jebria, A.; Langer, R., Recent Advances in Pulmonary Drug Delivery Using Large, Porous Inhaled Particles. *Journal of Applied Physiology* **1998**, *85*, (2), 379-385.

35. Edwards, D. A.; Hanes, J.; Caponetti, G.; Hrkach, J.; Ben-Jebria, A.; Eskew, M. L.; Mintzes, J.; Deaver, D.; Lotan, N.; Langer, R., Large Porous Particles for Pulmonary Drug Delivery. *Science (Washington, D. C.)* **1997**, *276*, (5320), 1868-1871.

36. Weng, W.-L.; Chen, J.-T.; Yang, J.-W.; Chang, J.-S., Isothermal Vapor-Liquid Equilibria of Binary Mixtures of Nitrogen with Dimethyl Sulfoxide, N-Methyl-2-pyrrolidone, and Diethylene Glycol Monobutyl Ether at Elevated Pressures. *Journal of Chemical and Engineering Data* **2007**, *52*, (2), 511-516.

37. Vos, W. L.; Schouten, J. A., Improved Phase Diagram of Nitrogen up to 85 kbar. *Journal of Chemical Physics* **1989**, *91*, (10), 6302-6305.

38. Chan, H.-K., Dry Powder Aerosol Delivery Systems: Current and Future Research Directions. *Journal of Aerosol Medicine* **2006**, *19*, (1), 21-27.

39. Tang, P.; Chan, H. K.; Raper, J. A., Prediction of Aerodynamic Diameter of Particles with Rough Surfaces. *Powder Technology* **2004**, *147*, (1-3), 64-78.

40. Brain, J. D., Inhalation, Deposition, and Fate of Insulin and Other Therapeutic Proteins. *Diabetes Technology & Therapeutics* **2007**, *9*, (S1), 4-15.

41. Crowder, T. M.; Rosati, J. A.; Schroeter, J. D.; Hickey, A. J.; Martonen, T. B., Fundamental Effects of Particle Morphology on Lung Delivery: Predictions of Stokes' Law and the Particular Relevance to Dry Powder Inhaler Formulation and Development. *Pharmaceutical Research* **2002**, *19*, (3), 239-245.

42. Chew, N. Y. K.; Chan, H.-K., Use of Solid Corrugated Particles to Enhance Powder Aerosol Performance. *Pharmaceutical Research* **2001**, *18*, (11), 1570-1577.
43. Mohamed, F.; van der Walle, C. F., PLGA Microcapsules with Novel Dimpled Surfaces for Pulmonary Delivery of DNA. *International Journal of Pharmaceutics* **2006**, *311*, (1-2), 97-107.
44. Ganser, G. H., A Rational Approach to Drag Prediction of Spherical and Nonspherical Particles. *Powder Technology* **1993**, *77*, (2), 143-152.

Indexes

Author Index

Ahosseini, Azita, 218
Anand, Madhu, 290
Arzhantzev, Sergei, 95
Ashurst, W. Robert, 290
Bálint, A.-M., 235
Banet-Osuna, Anna, 191
Bodor, A., 235
Bogel-Łukasik, Ewa, 191
Bogel-Łukasik, Rafal, 191
Borovik, A. S., 274
Busch, Daryle H., 145
Cai, Hu, 202
Chialvo, Ariel A., 66
Chialvo, Sebastian, 66
Cooper, Andrew I., 243
Cuiper, A. D., 235
da Ponte, Manuel Nunes, 191
DeSimone, Joseph M., 259
Du, Libin, 259
Eckert, Charles A., 81, 131
Fábos, V., 235
Foster, Neil R., 309
Gohres, John L., 81
Hallett, Jason P., 131
Hernandez, Rigoberto, 81
Horváth, I. T., 235
Houndonougbo, Yao A., 41
Hurst, Kendall M., 290
Hutchenson, Keith, 3
Jin, Hong, 202
Johnson, Chad A., 274
Kitchens, Christopher L., 290

Kuczera, Krzysztof, 41
Laird, Brian B., 41
Lantos, D., 235
Lee, Jun-Young, 243
Li, Hongping, 95
Liotta, Charles L., 81, 131
Liu, Juncheng, 290
Maroncelli, Mark, 95
Mika, L. T., 235
Najdanovic-Visak, Vesna, 191
Nguyen, Joseph G., 274
Ren, Wei, 112, 218
Roberts, Christopher B., 290
Roberts, George W., 259
Saunders, Steven R., 290
Scurto, Aaron M., 3, 112, 218
Serbanovic, Ana, 191
Sharma, Sarika, 274
Shiflett, Mark B., 112
Sielcken, O. E., 235
Sih, Roderick, 309
Simonson, J. Michael, 66
Subramaniam, Bala, 3, 145, 202, 274
Swalina, Chet, 95
Tan, Bien, 243
Tunge, Jon A., 202
Wang, Ruihu, 202
Wood, Colin D., 243
Von White, Gregory, II, 290
Xie, ZhuanZhuan, 202
Yokozeki, Akimichi, 112

Subject Index

A

Acetal formation, alkylcarbonic acids
as catalysts, 137–139
Acetic acid (AA)
colloidal poly(tetrafluoroethylene)
(PTFE) particles, 269–270
hydrogen bonding in CO_2-
expanded, 58–59, 60f
parameters for non-bonded
interactions, 46t
polymerization of TFE in CO_2-
expanded AA, 263–264, 268,
269t
pressure-composition diagram for
CO_2-expanded, 48, 51f
pressure-density curves for CO_2-
expanded, 51, 53f
vapor-liquid coexistence curves for
single component system, 47f
volume expansion for CO_2-
expanded, 48, 49f
See also Fluoropolymers
Acetone
Coumarin 153 (C153) absorption
maxima vs. CO_2 concentration
for gas-expanded, 88, 90f
Coumarin 153 emission maxima
vs. CO_2 concentration for gas-
expanded, 88, 90f
diazodiphenylmethane (DDM)
measuring relative acidities, 134
Kamlet–Taft parameters, 13t
local compositions by simulation
and experiment for C153 in
varying concentration, 91t
mixture critical points of, with
CO_2, 8t

parameters for non-bonded
interactions, 46t
polymerization of
tetrafluoroethylene (TFE) in
CO_2-expanded, 266, 267t
TrAPPE (transferable potentials for
phase equilibria force field), 87
vapor-liquid coexistence curves for
single component system, 47f
volume expansion for CO_2-
expanded, 48, 49f
Acetonitrile
Kamlet–Taft parameters, 13t
mixture critical points of, with
CO_2, 8t
orientational correlations in CO_2-
expanded, 56–58
oxidation of di-t-butylphenol
(DTBP), 279, 281t
parameters for non-bonded
interactions, 46t
polymerization of
tetrafluoroethylene (TFE) in
CO_2-expanded, 266, 267t
pressure-composition diagram for
CO_2-expanded, 48, 50f
pressure-density curves for CO_2-
expanded, 50, 52f
simulation and experimental for
rotational correlation times, 53,
54f
simulation and experimental for
translational diffusion constants,
52, 54f
simulation vs. experimental shear
viscosity, 53, 55f
vapor-liquid coexistence curves for
single component system, 47f

353

volume expansion for CO_2-expanded, 48, 49*f*
volumetric properties of CO_2-expanded, 97*f*
See also Solvatochromism and solvation dynamics in CO_2-expanded liquids
Acid catalysis, gas-expanded liquids (GXLs), 22–23
Acidity, polarity parameter, 13, 15*f*
Acids, catalysts in industry, 132
Aerosol solvent extraction system (ASES)
mixing controlled precipitation kinetics of SCF processes, 312
particle formation, 24–25
phase homogeneity during precipitation, 311
supercritical fluid anti-solvent platform, 310
use of capillary nozzles and micro-orifices, 311–312
See also Atomized rapid injection for solvent extraction (ARISE)
Aligned porous materials, application and interest, 245
Alkylcarbonic acids
acetal formation reactions, 137–139
analysis of dielectric constant of gas-expanded liquids (GXLs), 137, 138*f*
applications, 137–142
β-pinene conversion and product distribution at different methanol/water ratios, 142, 143*f*
characterization, 133–137
diazodiphenylmethane (DDM) results using acetone as diluent, 134
diazotization reaction, 139, 141–142
effect of CO_2 concentration on DDM reactions, 134, 136
formation of carbonic acid, alkyl carbonic acid and

peroxycarbonic acid in GXLs, 132
generation in situ, 132–133
Hammett-type analysis of DDM reactions, 136–137
measurement of relative acidities using DDM, 133–137
pseudo first rate constant for reaction of DDM with methylcarbonic acid vs. CO_2 pressure, 135*f*
rate constants in Hammett correlation of p-substituted benzyl alcohols, 137*t*
rate of DDM decomposition in CO_2-expanded solvents, 134, 135*f*
solvolysis reactions, 142
synthesis and mechanism for formation of methyl yellow from aniline, 139, 140
Anions, common, in ionic liquids, 221*f*
Applications of gas-expanded liquids (GXLs)
acid catalysis, 22–23
ASES (aerosol solvent extraction system), 24–25
catalytic reactions, 17–23
DELOS (depressurization of expanded liquid organic solution), 23
gas antisolvent solvent (GAS) technique, 23, 24*f*
H_2O_2-based oxidation, 20–21
homogeneous catalysis, 17–21
hydroformylations, 18–19, 22
hydrogenations, 17–18, 21–22
multiphase catalysis, 21–23
non-catalytic reactions, 17
O_2 oxidations, 19–20
particle formation, 23–26
particle fractionation and coating, 25–26

particle precipitation from reverse
emulsions, 25
PCA (precipitation with
compressed antisolvents), 24–25
PGSS (particles from gas-saturated
solution), 24f, 25
polymerizations, 21
polymer processing, 26
selective oxidations, 22
separations, 26–27
See also Gas-expanded liquids
(GXLs)
ARISE. See Atomized rapid
injection for solvent extraction
(ARISE)
ASES (aerosol solvent extraction
system), particle formation, 24–25
Atomized rapid injection for solvent
extraction (ARISE)
aerodynamic statistics of, processed
bovine insulin, 331t
anti-solvent pressure, 334–336
anti-solvent temperature, 336, 338f,
339f
appearance of bovine insulin
samples, 332f, 333f
backpressure intensity, 337, 340f,
341f, 342f
bulk densities of, processed bovine
insulin, 330, 334
bulk density measurement, 317–
318
cascade impactor analysis, 329–330
cascade impactor analysis method,
317
cascade impactor mass distribution
of, processed bovine insulin,
331f, 342f
chemical entities consumed during,
experimentation, 315t
deposition of bovine insulin within
precipitation vessel, 319, 320f,
321f
experimental conditions for, of
bovine insulin, 318t

geometrical particle size statistics
of, processed bovine insulin,
324, 328t
injection conduit aperture, 336–
337
jet atomization and mixing
efficiencies, 334–335
laser diffraction analysis method,
316
layout of process, 313, 314f, 315
low pressure anti-solvent
conditions, 335
materials and methods, 315–318
particle size distribution of,
processed bovine insulin, 325f,
326f, 327f
particle size measurement with
laser diffraction, 324, 329
process attributes, 319, 324, 329–
330, 334–337
process method, 315–316
scanning electron microscopy
(SEM) method, 316
SEM of bovine insulin from, 319,
322f, 323f, 324
values of constant processing
parameters, 318t
working principle and operability,
318
Axial distribution functions (ADFs)
equation, 85–86
ground-state around Coumarin 153
(C153), 89f

B

Basicity, polarity parameter, 13, 15f
Beckmann rearrangement, commercial
processes, 236
Benzaldehyde, toluene oxidation using
Co/Zr/ketone catalyst, 165–166,
167t
Benzene, diffusivity of, in methanol
and CO_2, 11, 12f

Benzoic acid, toluene oxidation using Co/Zr/ketone catalyst, 165–166, 167t

Benzyl alcohol
rate constants in Hammett correlation of p-substituted, 136, 137t
rate of diazodiphenylmethane (DDM) decomposition, 134, 135f

Binary systems, dilute, as special cases of dilute ternary systems, 73

Biodegradable, CO_2-philes, 248–249

Biphasic ionic liquid/carbon dioxide systems
analysis, 225–226
background, 220–223
effect of pressure on reaction rates, 227, 228f
effects of phase equilibrium on catalytic reaction rate, 230, 232
experimental, 224–226
idealized multiphase catalysis, 219f
literature survey of hydrogenation in, 223
mass transfer effects, 226–227
materials, 226
mixture critical points of 1-octene, octane and CO_2 with H_2, 231t
multiphase homogeneous catalysis, 219
phase behavior, 222f
phase behavior of reactant/product, 228, 230, 231f
phase equilibria, 224–225, 228–230
reaction catalysis, 224
safety, 224
synthesis of 1-hexyl-3-methylimidazolium bis(trifluoromethylsulfonyl)imide ([HMIm][Tf_2N]), 225
volume expansion of IL, 228, 229f, 229t

Bovine insulin
appearance of various samples, 332f, 333f
bulk densities, 330, 334
bulk density measurement, 317–318
cascade impactor analysis, 329–330, 331f, 342f
deposition within precipitation vessel, 320f, 321f
experimental conditions for precipitation, 318t
geometrical particle size statistics, 324, 328t, 329
laser diffraction, 339f, 341f
particle size distribution, 324, 325f, 326f, 327f, 329
scanning electron microscopy (SEM) images, 319, 322f, 323f, 324, 338f, 340f
See also Atomized rapid injection for solvent extraction (ARISE)

Bridged macrocycle, manganese complex, 150, 151

t-Butanol, rate of diazodiphenylmethane (DDM) decomposition, 134, 135f

C

Calculated adiabatic reaction temperature (CART), method estimating vapor phase hazardous behavior, 28–29

Capillary nozzles, precipitation of micro–particles, 311–312

ε-Caprolactam
protonation to caprolactamium hydrogen sulfate, 236, 237
temperature-dependent vapor pressures of mixtures, 238f

Caprolactamium hydrogen sulfate rearrangement mixture, 236, 237

strong interactions with sulfur
trioxide, 236, 237, 239
Carbon dioxide (CO_2)
high throughput solubility
measurements in, 247–248
Kamlet–Taft parameters by density,
13t
MidCentury (MC)-like process in
dense, 178
parameters for non-bonded
interactions, 46t
polymer solubility in, 246–254
simulation and experimental for
rotational correlation times, 53,
54f
simulation and experimental for
translational diffusion constants,
52, 54f
simulation vs. experimental shear
viscosity, 53, 55f
vapor-liquid coexistence curves for
single component system, 47f
See also Biphasic ionic
liquid/carbon dioxide systems;
Oxidations of gas-expanded
liquids
Carbonic acid
CO_2 interaction with water, 132–
133
formation in gas-expanded liquids,
132
See also Alkylcarbonic acids
4-Carboxybenzaldehyde (4-CBA)
contaminant in terephthalic acid
production, 162, 163f
occlusion into precipitating TPA,
173
Catalysis, heterogeneous oxidation, in
$scCO_2$ media, 279, 281–282
Catalyst design
definability, 150–151
durability, 149–150
efficacy, 149
principles, 148–151

selectivity, 149, 150f
sustainability, 148–149
Catalysts, acids, 132
Catalytic oxidations
background in catalytic studies in
$scCO_2$, 151–152, 156
See also Oxidations of gas-
expanded liquids
Catalytic reactions of gas-expanded
liquids (GXLs)
acid catalysis, 22–23
H_2O_2-based oxidation, 20–21
homogeneous catalysis, 17–21
hydroformylations, 18–19, 22
hydrogenations, 17–18, 21–22
multiphase catalysis, 21–23
O_2 oxidations, 19–20
polymerizations, 21
selective oxidations, 22
Cations, common classes in ionic
liquids, 221f
Center for Environmentally Beneficial
Catalysis (CEBC), CO_2-expanded
liquids (CXL), 42
Chlorohydrin, comparing
methyltrioxorhenium (MTO)
process, 182, 183t
Citral
hydrogenation in carbon dioxide,
193
vapor-liquid equilibrium behavior
of binary mixtures of, with CO_2,
200
See also Terpene hydrogenations
Classification, gas-expanded liquids
(GXLs), 4–5, 6t
CO_2-expanded liquids (CXLs)
advantages, 260–261
beneficial properties, 4
biphasic ionic liquids/CO_2 systems,
220, 222–223
catalytic chemical processes, 147–
148
force field models, 44–45

hydrogen bonding in CXL methanol and CXL acetic acid, 58–59
mixtures, 48, 50–51
optimization for industrial reactor use, 43
orientational correlations in CXL acetonitrile, 56–58
Peng–Robinson equation of state (PR–EOS), 44, 48, 50–51, 61–63
pressure-composition diagrams, 50f, 51f
pressure vs. density curves, 51, 52f, 53f
promising alternative catalytic media, 42–43
simulation results for phase equilibrium, 45, 47–51
single component systems, 47
structural properties of, 55–59
translational diffusion constants in CO_2/acetonitrile, 52, 54f
transport properties, 51–53, 55
vapor-liquid coexistence curves for single component systems, 47f
volume expansion for various CXLs, 49f
volumetric properties, 97f
See also Fluoropolymers; Simulations; Solvatochromism and solvation dynamics in CO_2-expanded liquids
Coating, particle fractionation and, 25–26
Cobalt. See Metal complexes
Cobalt(salen*) studies
axial ligand (AL) basicity affecting catalyst activity, 158–159
catalytic studies in scCO_2, 151–152, 156
CO_2-expanded liquids (CXLs) as media for oxidations, 156, 158–159

enhancing solubility of O_2 in CXLs, 156, 157f
mechanism of oxidation of phenol, 154, 155
reactivities with different AL, 159f
reactivity of catalyst system Co(salen)(MeIM), 160t
relative efficacies of scCO_2 and CXL systems, 157f
saturation of O_2 conversion dependence in scCO_2, 153f
structure of Co(salen) complex, 158f
temperature dependence of conversion and selectivity in scCO_2, 152, 153f
See also Metal complexes; Oxidations in gas-expanded ligands
Cobalt acetate, beneficial factors, 165
Cobalt acetate with N-hydroxysuccinimide (NHSI) catalysis, 166, 169, 172–173
isothermal titration calorimetry, 172–173, 175f
Colloidal particles, poly(tetrafluoroethylene) (PTFE), 269–270
Commercialization, gas-expanded liquids (GXLs), 27–28
Compressible solvents. See Modeling dilute solutes in compressible solvents
Computer simulations. See Cybotactic structure of gas-expanded liquids (GXLs)
Condensed phase model (CPM), ligand configurations, 303, 304f
Coumarin 153 (C153)
charge distribution, 83f
molecular structure, 83f
probing cybotactic region, 82
See also Cybotactic structure of gas-expanded liquids (GXLs)

Crystal structure, diaquabis(N-hydroxysuccinimidato)cobalt(II), 173*f*

Cybotactic structure of gas-expanded liquids (GXLs)

absorption and emission spectroscopy probing, of chromophore Coumarin 153 (C153), 82

ADF (axial distribution function), 85–86

ADFs of ground-state C153 in acetone GXL, 89*f*

C153 absorption maxima at varying CO_2 concentrations, 90*f*

C153 emission maxima at varying CO_2 concentrations, 90*f*

calculated and experimental solvatochromic shifts, 85

calculated solvatochromic shifts from Onsager's reaction field theory (ORFT), 86*f*

charge distribution of ground- and lowest-excited states of C153, 83*f*

combining computer simulations and experiments, 82

defining cybotactic region of solvent, 82, 84, 96

experimental solvatochromic shifts for C153 absorption in CO_2-expanded methanol, 86*f*

local compositions by spectroscopic experiments and MD simulations at bulk-fluid acetone concentration, 91*t*

local compositions by spectroscopic experiments and MD simulations at bulk-fluid methanol concentration, 91*t*

Lorentz–Lorenz equation estimating refractive indices, 84–85

molecular dynamics (MD) simulations, 82, 85, 87–88

molecular structure of C153, 83*f*

ORFT as relation between solvatochromic shift and local polarity, 84

schematic of energy diagram, 92*f*

spectroscopic experiments, 88, 91–92

synergistic combination of simulations and experiments, 93

Cyclic olefins, ring cleavages, 169, 172

Cycloalkanes, oxidation using cobalt acetate/*N*-hydroxysuccinimide, 174

Cyclohexane, volumetric properties of CO_2-expanded, 97*f*

D

Definability, catalyst design, 150–151

DELOS (depressurization of expanded liquid organic solvent), particle formation, 23

Deposition

controlled, of Au nanoparticles, 296

uniform, multi-layer assemblies using controlled, 293

See also Nanoparticles

Depressurization of expanded liquid organic solvent (DELOS), particle formation, 23

Diazodiphenylmethane (DDM)

Hammett-type analysis, 136–137

measurement of relative acidities, 133–137

rate of DDM decomposition in various CO_2-expanded solvents, 134, 135*f*

See also Alkylcarbonic acids

Diazotization reactions, alkylcarbonic acids as catalysts, 139, 141–142

Dielectric constants

gas-expanded liquids, 137, 138*f*

methanol and CO_2 solutions, 12*f*

Diffusivity
  benzene in methanol and CO_2
    solutions, 11, 12f
  Stokes–Einstein relationship, 11
  Dilute solutes. *See* Modeling dilute
    solutes in compressible solvents
Dipolarity/polarizability, polarity
  parameters, 13, 15f
Directional freezing, porous material
  preparation, 245, 246f
Durability
  catalyst design, 149–150
  methyltrioxorhenium (MTO)
    catalyst, 184, 185–186

E

Economics, gas-expanded liquids
  (GXLs), 27–28
Efficacy, catalyst design, 149
Emulsions
  CO_2-in-water (C/W) high internal
    phase, (HIPE), 249, 251f
  particle precipitation from reverse,
    25
  templating concentrated C/W, 252–
    253
Emulsion-templated porous materials.
  See Porous materials
Energy diagram, cybotactic region,
  92
Epoxidation
  first monophasic reaction using
    CO_2-expanded liquid media, 180
  higher olefins, 178–179
  See also Propylene oxide (PO)
Equations of state (EoS), modified
  Redlich–Kwong type of cubic,
  119–120
Equilibrium, multi-phase, 14, 16
Ethanol
  diazodiphenylmethane (DDM)
    measuring relative acidities, 133

  parameters for non-bonded
    interactions, 46t
  pressure-composition diagram for
    CO_2-expanded, 48, 50f
  pressure-density curves for CO_2-
    expanded, 50, 52f
  rate of DDM decomposition, 134,
    135f
  vapor-liquid coexistence curves for
    single component system, 47f
  volume expansion for CO_2-
    expanded, 48, 49f
1-Ethyl-3-methyl-imidazolium
  bis(trifluoromethylsulfonyl)imide
  [EMIm][Tf$_2$N]. *See* Ionic liquids
  (ILs) and refrigerants
Ethylene glycol, rate of
  diazodiphenylmethane (DDM)
  decomposition, 134, 135f
Expansion coefficients, molecular-
  based interpretation, 75–76
Extended solvation model (ELSM),
  ligand configurations, 303, 304f

F

Flammability, Gibbs free energy
  minimization method for
  predicting, 28–29
Fluoropolymers
  adjustable polarity of solvent
    mixture, 269
  applications, 260
  CO_2 solubility, 244, 247
  differential scanning calorimetry
    (DSC), 262
  dynamic light scattering (DLS),
    262
  experimental, 261–264
  experiments focusing on acetic acid
    (AA), 268, 269t
  gas-expanded liquids (GXLs), 260–
    261

homopolymerization of tetrafluoroethylene (TFE) to PTFE, 265
hydrocarbon solvents, 267–268
materials, 261
measurements, 262
polymerization in $scCO_2$, 260
polymerization of TFE in CO_2-expanded AA, 263–264
polymerization of TFE in CO_2-expanded AA using sodium dodecyl sulfate (SDS), 264
polymerization of TFE in sc CO_2, 263
polymerization of TFE in $scCO_2$, 265*t*
polymerization of TFE in various CO_2 expanded solvents, 267*t*
previous GXL studies, 265
scanning electron microscopy (SEM), 262
SEM of PTFE samples prepared in $scCO_2$ and CO-expanded AA, 269–270
solvent choices for CO_2 expansion, 266–267
synthesis procedure, 262–264
TGA of PTFE samples prepared in $scCO_2$, 266*f*
thermal stability, 266*f*, 268–269
thermogravimetric analysis (TGA) method, 262
two-step degradation of polymers, 270–271
using liquid phase of expanded solvent, 265–266
Force field models, CO_2-expanded solvents, 44–45
Fractionation
nanoparticle, in gas-expanded liquids, 297–299
particle, and coating, 25–26
spiral tube within pressure vessel, 297–298, 300*f*
See also Nanoparticles

Free volume, CO_2-expanded liquids (CXLs), 96
Fugacity coefficients, calculations, 120–121

G

Gas antisolvent solvent (GAS)
nanoparticle formation, 282
particle formation, 23, 24*f*
Gases
binding to metal(salen) nanoparticles, 286–287
multi-phase equilibrium, 16
Gas-expanded liquids (GXLs)
applications, 17–27
behavior of ionic liquids, water and polymers with CO_2, 7, 10
biphasic ionic liquids/CO_2 systems, 220, 222–223
classification, 4–5, 260–261
comparison of different solvent classes and expansion behavior, 6*t*
dielectric constant of methanol and CO_2 solutions, 12*f*
diffusivity of benzene in methanol and CO_2, 12*f*
economics, 27–28
effect of inerts on flammability envelope of methane, 29*f*
expansion of liquids vs. wt% CO_2 in liquid phase, 6*f*
future outlook, 30
gases in, 16
1-hexyl-3-methyl-imidazolium bis(trifluoromethylsulfonyl)imide ([HMIm][Tf_2N]), 7, 10
Kamlet–Taft (KT) parameters of polarity, 13, 15*f*
liquid density and molar volume with pressure for methanol and CO_2, 7, 9*f*
liquids in, 14, 16

362

lower-critical end point (LCEP), 14, 16

mass density increase, 96

mixture critical points of common organic liquids with CO_2, 8*t*

multi-component and multi-phase equilibrium, 14, 16

phase equilibrium and volume expansion, 5, 7

polarity, 11, 13–14

process engineering issues, 27–30

safety, 28–30

solid components in, 14

transport properties, 10–11

tunability of properties, 147–148

unique properties, 147

vapor-liquid equilibrium (VLE) of methanol and CO_2, 5, 8*f*

vapor-liquid-liquid equilibrium (VLLE), 14, 16

viscosity of methanol and CO_2 with pressure, 11*f*

VLE data of [HMIm][Tf_2N] and CO_2, 10*f*

volume expansion with pressure for methanol and CO_2, 7, 9*f*

See also Applications of gas-expanded liquids (GXLs); CO_2-expanded liquids (CXLs); Cybotactic structure of gas-expanded liquids (GXLs)

Gas-to-solution shifts

modeling emission frequency for acetonitrile + CO_2 mixtures, 98*f*

observed and simulated, of absorption and emission frequencies, 101*f*

See also Solvatochromism and solvation dynamics in CO_2-expanded liquids

Gibbs–Duhem relation

comparing consistent expressions and literature, 73–74, 75

dilute ternary systems, 69–70

Gibbs Ensemble Monte Carlo (GEMC)

simulation, 45

two-phase simulations, 48

Gibbs free energy minimization, predicting flammability, 28–29

Global phase behavior

data of ionic liquid and refrigerant system, 125*t*

liquids and gases, 116–117

See also Ionic liquids (ILs) and refrigerants

Gold nanoparticles

controlled deposition onto micro-electro-mechanical systems (MEMS), 296

See also Nanoparticles

Green chemistry

health hazards, 275

oxidation, 276

See also Metal complexes

H

Hammett correlation, diazodiphenylmethane (DDM) reactions, 136–137

Hazardous waste, human health, 275

Health hazards, chemical waste, 275

n-Hexane, Kamlet–Taft parameters, 13*t*

1-Hexyl-3-methyl-imidazolium bis(trifluoromethylsulfonyl)imide ([HMIm][Tf_2N])

behavior of, water and polymers with CO_2, 7, 10

synthesis, 225

volume expansion of, with CO_2 and CO_2 with H_2, 228, 229*t*

See also Biphasic ionic liquid/carbon dioxide systems

High internal phase emulsion (HIPE), CO_2-in-water (C/W) HIPE, 249, 251*f*

Homogeneous catalysis
gas-expanded liquids (GXLs), 17–21
GXL economics, 28

Hydrocarbon solvents, polymerization of tetrafluoroethylene (TFE) in CO_2-expanded, 267–268

Hydroformylations
bidentate phosphite ligands supporting polymer, 211
catalyst recycle, 210–211
CO_2-expansion effect of 1-octene, at constant volume and pressure, 208
comparing percent expansion in, of 1-octene with $Rh(CO)_2(acac)$, 207*f*
development of rhodium-catalyzed, 203
effect of CO_2 pressure on, of 1-octene using polymer supported catalyst, 215*t*
effect of solvent volume and temperature on, of 1-octene, 212*t*
formation of byproducts, 205
gas-expanded liquids (GXLs), 18–19, 22
general mechanism, 203, 204
higher olefins, 203–204, 208, 219
mass balance of internal octene isomers, 204, 207
mole fraction vs. fugacity of hydrogen and CO in acetone and CO_2-expanded acetone, 206*f*
1-octene in non-expanded solvent vs. CXL, 204
1-octene with polymer-supported monophosphite polymer, 210*t*
radical copolymerization of styrene with biaryl styrene cross-linker, 211, 212

recyclability of soluble polymer-supported Rh, 213, 214*t*
solubility in CXL-expanded toluene, 215
turnover frequencies (TOFs), 28
with soluble-polymer supported ligands, 208–215

Hydrogels, poly(vinyl alcohol) (PVA), 253

Hydrogenations
expanded liquid vs. supercritical phase, 199
gas-expanded liquids (GXLs), 17–18, 21–22
literature survey of, in biphasic ionic liquids/CO_2 systems, 223
mass transfer effects, 226–227
reaction rate for, of 1-octene, 227*f*, 228*f*
terpenes, 192–193
See also Biphasic ionic liquid/carbon dioxide systems; Terpene hydrogenations

Hydrogen bonding, CO_2-expanded methanol and acetic acid, 58–59

Hydrogen peroxide propylene oxidation (HPPO)
propylene to propylene oxide, 179
See also Propylene oxide (PO)

I

Imperfect gases, describing behavior, 77

Insulin, bovine. *See* Atomized rapid injection for solvent extraction (ARISE)

Interaction energy
model, 299, 302–303
total, potential vs. interparticle separation distance, 301*f*

Ionic liquids (ILs)
   advantages of combining ILs and refrigerants, 114, 116
   background, 114, 115*f*
   behavior of ILs, water and polymers with CO_2, 7, 10
   common cation classes and anions, 221*f*
   1-ethyl-3-methyl-imidazolium bis(trifluoromethylsulfonyl)imide ([EMIm][Tf$_2$N]) as model IL, 113–114
   molecular design, 235–236
   structure of [EMIm][Tf$_2$N], 115*f*
   synthesis, 118–119
   See also Biphasic ionic liquid/carbon dioxide systems; Gas-expanded liquids (GXLs); Ionic liquids (ILs) and refrigerants
Ionic liquids (ILs) and refrigerants
   advantages of combining, 114, 116
   classifying [EMIm][Tf$_2$N] and R-134a, 125
   equation of state parameters and binary interaction parameters, 120*t*
   experimental, 117–119
   experimental data and model predictions for VLLE region by temperature, 121, 122*f*
   experimental global phase behavior for, with model predictions, 123*f*
   fugacity coefficient, 120–121
   global phase behavior, 116–117
   global phase behavior and phase equilibria of [EMIm][Tf$_2$N] and R-134a, 121–123
   global phase behavior data for [EMIm][Tf$_2$N] and R-134a, 125*t*
   materials, 119
   modeling, 119–121
   model predicting vapor-liquid-liquid equilibria (VLLE), 122–123

   modified Redlich–Kwong type cubic equations of state (EoS), 119–120
   phase behavior and equilibrium, 117–118
   pressure-composition diagram for VLE, VLLE, and mixture critical point, 126*f*
   safety, 117
   single- and multi-phase equilibria, 121–122
   synthesis of IL, 118–119
   three VLLE, 121, 122–123
   VLE data for [EMIm][Tf$_2$N] and R-134a, 124*t*
Isopropanol, rate of diazodiphenylmethane (DDM) decomposition, 134, 135*f*
Isothermal titration calorimetry, cobalt/*N*-hydroxysuccinimide complexes, 175*f*

J

Jacobson's catalyst, Co(salen*), 151–152

K

Kamlet–Taft (KT) parameters, polarity, 13–14, 15*f*
Kirkwood–Buff integral, solvation behavior of species in solution, 75–76

L

Ligand configuration, stabilizing, on nanoparticle, 302, 304*f*
Limited ligand solvation model (LLSM), ligand configurations, 303, 304*f*

Limonene
  chemical structure, 192*f*
  hydrogenation, 193, 198
  hydrogenation reaction rates in one
    and two phase mixtures, 199
  liquid-vapor equilibrium behavior
    in mixtures of CO_2 + hydrogen,
    198–199
  molar ratio of hydrogen to, in
    liquid phase, 195, 196*t*
  See also Terpene hydrogenations
Liquids
  expansion vs. wt% CO_2 in liquid
    phase, 6*f*
  multi-phase equilibrium, 14, 16
  See also Gas-expanded liquids
    (GXLs)
Lorentz–Lorenz equation, estimating
  refractive indices, 84–85
Lower-critical end point (LCEP)
  global phase behavior, 116–117
  liquids, 14, 16

M

Macrocyclic ligands, manganese
  complex, 150, 151
Mass density
  gas-expanded liquids (GXLs), 96
  region of maximum, vs.
    composition, 105
Mass transfer effects, hydrogenation
  reaction, 226–227
Maxwell relation, dilute ternary
  systems, 70, 74, 75
Metal complexes
  active site structures of
    heterogeneous catalysts, 280*f*
  binding of dioxygen and nitric
    oxide (NO) to, 286–287
  bis(salicylaldehyde)ethylenediamin
    e (salen), 276–277
  [Co(salen)] oxidizing 2,6-di-*t*-
    butylphenol (DTBP), 279, 281

controlling particle morphologies,
    282–283, 286
  environmental pollutants, 275
  gas antisolvent precipitation
    (GAS), 282
  geometric differences influencing
    morphology of particles, 283,
    285*f*
  green oxidation chemistry, 276
  heterogeneous oxidation catalysis
    in $scCO_2$ media, 279, 281–282
  nanoparticles of metal salen
    complexes, 282–283, 286
  PCA (precipitation with
    compressed antisolvents), 282–
    283
  reaction scheme for [Co(salen)]
    with O_2, 277*f*
  reusability of catalysts, 281
  rodlike structure of [Ni(salen)] and
    [Co(salen)] nanoparticles, 283,
    285*f*
  schematic of PCA instrumentation,
    284*f*
  schematic of template
    copolymerization method, 278*f*
  spherical morphology of
    [Ru(salen)(NO)(Cl)] from PCA
    process, 283, 285*f*
  structures of compounds in PCA
    studies, 284*f*
  substrate conversion for catalysts in
    various reaction media, 281*t*
  supercritical antisolvent (SAS)
    precipitation, 282
  template polymerizations with
    [Co(salen)], 278–279
Metal nanoparticles. *See*
  Nanoparticles
Methane, effect of inerts on
  flammability envelope, 29*f*
Methanol
  analysis of dielectric constant of
    gas-expanded liquid, 137,
    138*f*

C153 (Coumarin 153) absorption
maxima vs. CO_2 concentration
for gas-expanded, 88, 90f
C153 emission maxima vs. CO_2
concentration for gas-expanded,
88, 90f
diazodiphenylmethane (DDM)
measuring relative acidities, 133
dielectric constant as function of
CO_2 mole fraction for gas-
expanded, 137, 138f
dielectric constant of, and CO_2
solutions, 12f
diffusivity of benzene in, and CO_2,
11, 12f
hydrogen bonding in CO_2-
expanded, 58–59
Kamlet–Taft parameters, 13t
liquid density and molar volume
with pressure of, and CO_2, 7,
9f
local compositions by simulation
and experiment for C153 in
varying concentration, 91t
mixture critical points of, with
CO_2, 8t
parameters for non-bonded
interactions, 46t
pressure-composition diagram for
CO_2-expanded, 48, 51f
pressure-density curves for CO_2-
expanded, 51, 53f
pseudo first-order rate constant for
reaction of DDM with CO_2-
expanded, 134, 135f, 136
pseudo first-order rate constants of
cyclohexanone acetal formation
in gas-expanded, 138, 139f
rate of DDM decomposition, 134,
135f
TrAPPE (transferable potentials for
phase equilibria force field),
87
vapor-liquid coexistence curves for
single component system, 47f

vapor-liquid equilibrium (VLE) of,
and CO_2, 5, 7, 8f
viscosity of, and CO_2, 11f
volume expansion for CO_2-
expanded, 48, 49f
volume expansion with pressure of,
and CO_2, 7, 9f
volumetric properties of CO_2-
expanded, 97f
Methyltrioxorhenium (MTO)
comparing pressure intensified,
process with major industrial
processes, 183t
durability, 184, 185–186
organometallic catalyst, 179
pressure intensified MTO process,
180f
theoretical in-service lifetime of,
183–186
See also Propylene oxide (PO)
Micro-orifices, precipitation of micro-
particles, 311–312
Micro-particles, use of capillary
nozzles and micro-orifices, 311–
312
MidCentury (MC) process
benchmark data for, 173, 176
MC-like processes in dense CO_2,
178
terephthalic acid (TPA), 160
Mixture critical point
gas-expanded liquids (GXLs), 7, 8t
1-octene, octane and CO_2 with H_2,
230, 231t
Mixtures, simulations of CO_2-
expanded solvents, 48, 50–51
Modeling dilute solutes in
compressible solvents
coefficients of second species, 71
comparison between consistent
expressions and literature, 73–
75
dilute binary systems as special
cases of dilute ternary systems,
73

Gibbs–Duhem relation, 69–70, 73–74, 75
green chemistry, 67
homogeneous system of linear equations, 70–71
imperfect gases, 77
Kirkwood–Buff integral, 75
Maxwell relation, 70, 74, 75
molecular-based interpretation of expansion coefficients, 75–76
solubility enhancement of non-volatile solutes in near critical solvents, 67
solvation phenomenon, 67
thermodynamic consistency of truncated expansions, 76–78
truncated composition expansions, 67–68
truncated composition-expansions for fugacity coefficients of dilute ternary systems, 69–73
Models
interaction energy, 299, 302–303
ionic liquid and refrigerant systems, 119–121
predicting global phase behavior, 122–123
See also Solvatochromism and solvation dynamics in CO_2-expanded liquids
Molecular dynamics (MD)
simulations for cybotactic region, 85, 87–88
simulations in gas-expanded liquids (GXLs), 97–98
simulations of CO_2-expanded solvents, 43–44
solvation in GXLs, 82
TrAPPE (transferable potentials for phase equilibria force field), 87
See also Cybotactic structure of gas-expanded liquids (GXLs); Solvatochromism and solvation dynamics in CO_2-expanded liquids

Monte Carlo
Gibbs Ensemble, simulation (GEMC), 45
simulations, 43
Morphology, nanoparticles of metal salen complexes, 283, 285*f*
Multiphase catalysis
gas-expanded liquids (GXLs), 21–23
idealized system, 219
See also Biphasic ionic liquid/carbon dioxide systems

N

Nanoparticles
Ag nanoparticle diameters vs. CO_2 expanded hexane, 305*f*
alkanethiol-stabilized Ag, in hexane vs. pressure, 296*f*
average diameters for Au and Ag, from CO_2-expanded hexane size fractionation process, 298*t*
behavior of stabilizing ligands, 302, 303, 304*f*
controlled deposition of ligand-stabilized Ag, Au, Pt and Pd, using GXLs, 293
controlling deposition of Au, onto micro-electro-mechanical systems (MEMS), 296
deposition, 292–297
dodecanethiol and lauric acid stabilized Ag, 293, 294*f*
dodecanethiol stabilized Pt, on carbon, 293, 295*f*
effect of ligand length for particle dispersibility, 296*f*, 297
experimental and models for maximum dodecanethiol stabilized Ag, in CO_2 expanded hexane vs. system pressure, 305*t*
fractionation in GXLs, 297–299

interaction energy model, 299, 302–303
metal
  bis(salicylaldehyde)ethylenediamine (salen) complexes, 282–283, 286
  quantum dots, 298–299
  schematic of GXL size fractionation method using spiral tube assembly, 300f
  TEM (transmission electron microscopy) images of deposited Ag, 294f
  TEM images of dodecanethiol stabilized Pt, 295f
  total interaction energy potential vs. interparticle separation, 301f
  uniform, multi-layer assemblies using controlled deposition, 293
Nanotechnology, advances, 291
N-hydroxysuccinimide (NHSI), catalysis of cobalt acetate with, 166, 169, 172–173
Nitric oxide (NO), binding to metal(salen) nanoparticles, 286–287
Non-bonded interactions, parameters, 46t
Non-catalytic reactions, gas-expanded liquids (GXLs), 17

O

1-Octene
  challenge of rhodium-catalyzed hydroformylation, 203, 205
  effect of CO_2-expansion on hydroformylation at constant volume and pressure, 208
  effect of solvent volume and temperature on hydroformylation, 212t
  hydroformylation in non-expanded solvent vs. CXL, 204

parameters for non-bonded interactions, 46t
phase behavior of, and octane and CO_2 as percent of initial amount of octane, 230, 231f
pressure-composition diagram for CO_2-expanded, 48, 51f
pressure-density curves for CO_2-expanded, 51, 53f
reaction rate for hydrogenation of, 227f, 228f
selectivity of hydroformylation, 204, 208
vapor-liquid coexistence curves for single component system, 47f
volume expansion for CO_2-expanded, 48, 49f
See also Hydroformylations
Olefins
  hydroformylation of higher, 203–204, 208, 219
  See also Hydroformylations
Onsager's relation field theory (ORFT)
  calculating absorption/emission spectrum, 85
  solvatochromic shift and solvent polarity, 84
  solvatochromic shifts for Coumarin 153, 86f
  See also Cybotactic structure of gas-expanded liquids (GXLs)
Organic-aqueous tunable solvents (OATS), separations, 27
Organic liquids, mixture critical points of, with CO_2, 8t
Orientational correlations, CO_2-expanded acetonitrile, 56–58
Oxidation
  green chemistry, 276
  heterogeneous, catalysis in scCO$_2$ media, 279, 281–282
Oxidations of gas-expanded liquids
  axial ligands (AL) and activity of catalyst, 158–159

catalysis in supercritical carbon dioxide ($scCO_2$), 146
catalytic chemical processes, 147–148
Co(salen) reactivities with different AL, 159f, 160f, 160t
enhanced solubility of O_2 in CO_2-expanded liquids (CXLs), 156, 157f
hydrogen peroxide-based, 20–21
media for catalytic oxidations, 156, 158–159
oxygen, 19–20
relative efficacies of $scCO_2$ and CXL for Co(salen), 156, 157f
selective, 22
structure of Co(salen) complex, 158f
See also Propylene oxide (PO); Terephthalic acid (TPA) process
Oxygen, binding to metal(salen) nanoparticles, 286–287

P

Palladium nanoparticles. *See* Nanoparticles
Parallel gravimetric extraction, solubility measurements, 247–248
Particle formation
ASES (aerosol solvent extraction system), 24–25
DELOS (depressurization of expanded liquid organic solution), 23
gas antisolvent solvent (GAS) technique, 23, 24f
gas-expanded liquids (GXLs), 23–26
particle fractionation and coating, 25–26
particle precipitation from reverse emulsions, 25

PCA (precipitation with compressed antisolvents), 24–25
PGSS (particles from gas-saturated solution), 24f, 25
Particle precipitation
reverse emulsions, 25
See also Atomized rapid injection for solvent extraction (ARISE)
Particles from gas-saturated solution (PGSS)
particle formation, 24f, 25
polymer processing, 26
PCA (precipitation with compressed antisolvents), particle formation, 24–25
Peng–Robinson equation of state (PR–EoS)
CO_2-expanded solvent mixtures, 48, 50–51
modeling vapor-liquid equilibria, 61–63
Peroxidation, comparing methyltrioxorhenium (MTO) process, 182, 183t
Peroxycarbonic acid
formation in gas-expanded liquids, 132
See also Alkylcarbonic acids
PGSS (particles from gas-saturated solution)
particle formation, 24f, 25
polymer processing, 26
Phase behavior
biphasic ionic liquids/CO_2 systems, 222f
high-pressure, and equilibria data, 113–114
ionic liquid and refrigerant system, 117–118
See also Ionic liquids (ILs) and refrigerants
Phase equilibrium
biphasic ionic liquids/CO_2 systems, 224–225, 228–230

effects of, on catalytic reaction rate, 230, 232
experiments, 193–195
gas-expanded liquids, 5, 7
ionic liquid and refrigerant system, 117–118
simulations results, 45, 47–51
single component systems, 47
α-Pinene
chemical structure, 192*f*
hydrogenation, 192–193
hydrogenation experiments, 196, 197*f*
hydrogenation profiles, 197*f*, 198*f*
hydrogenation reaction rates in one and two phase mixtures, 199
liquid-vapor equilibrium behavior in mixtures of CO_2 + hydrogen, 198–199
molar ratio of hydrogen to, in liquid phase, 195, 196*t*
Pt catalysts for hydrogenation, 193, 198*f*
vapor-liquid diagram, 195*f*
volume expansion of liquid, under CO_2 pressure, 194*f*
See also Terpene hydrogenations
Platinum nanoparticles
dodecanethiol stabilized, 293, 295*f*
See also Nanoparticles
Polarity
adjustable, of CO_2-expanded solvent mixture, 269
gas-expanded liquids (GXLs), 11, 13–14
Kamlet–Taft (KT) parameters, 13–14, 15*f*
Onsager's reaction field theory (ORFT), 84
Pollutants, human health, 275
Polyacrylamide, emulsion-templated crosslinked, 249, 251*f*, 252
Poly(acrylate), scCO_2 for porous crosslinked, 245

Poly(ether carbonate) (PEC)
copolymers, solubility in CO_2, 247
Polyethylene glycol) (PEG)
blending with poly(vinyl alcohol) (PVA), 253–254
producing di- and triblocks, 249, 250*f*
Polymerizations, gas-expanded liquids (GXLs), 21
Polymer processing, gas-expanded liquids (GXLs), 26
Polymer-supported catalysts
hydroformylation of 1-octene, 215
See also Soluble-polymer supported ligands
Poly(methacrylates), scCO_2 for porous crosslinked, 245
Poly(vinyl acetate) (PVAc)
hydroxyl-terminated PVAc, 252
solubility in CO_2, 248
Poly(vinyl alcohol) (PVA)
blending with poly(ethylene glycol) (PEG), 253–254
electron micrographs of open-cell porous hydrogel, 254*f*
hydrogels as biomaterials, 253
Porosity, aligned, 245
Porous materials
aligned, 245
CO_2-philic fluoropolymers or silicone-based, 247
CO_2-philic surfactants for CO_2-in-water (C/W) emulsion formation, 250*f*
directional freezing, 245, 246*f*
electron micrographs of open-cell porous PVA hydrogel, 254*f*
emulsion-templated crosslinked polyacrylamide materials, 251*f*
factors influencing polymer solubility, 246–247
high internal phase C/W emulsion (C/W HIPE), 249, 251*f*, 252
high throughput solubility measurements in CO_2, 247–248

hydrogels of poly(vinyl alcohol) (PVA), 253
inexpensive and biodegradable CO_2-philes, 248–249
limitations in initial C/W emulsion templating, 253–254
method for producing surfactants for $scCO_2$, 248–249, 250f
methodology producing, by templating concentrated C/W emulsions, 252–253
OVAc-functionalized dye dissolving in CO_2, 250f
parallel gravimetric extraction, 247
poly(ether carbonate) (PEC) copolymers, 247
poly(vinyl acetate) (PVAc), 248
polymer solubility in CO_2, 246–254
PVA, blended PVA/PEG, and chitosan materials, 253
sugar acetate with aligned structure, 246f
and supercritical fluids, 244–245
surfactants stabilizing highly concentrated C/W emulsions, 252
templating of supercritical fluid (SCF) emulsions, 249, 252–254
Porphryin complex, activated enzyme site, 149, 150f
Post-reaction catalyst separation, gas-expanded liquids (GXLs), 26–27
Powders. *See* Atomized rapid injection for solvent extraction (ARISE)
Precipitation
mixing controlled, kinetics of SCF processes, 312
phase homogeneity during, 311
supercritical fluid (SCF), 310
See also Atomized rapid injection for solvent extraction (ARISE)
Precipitation with compressed antisolvents (PCA)
nanoparticle formation, 282–283
particle formation, 24–25

schematic of instrumentation, 284f
size of nanoparticles by PCA, 283, 286
structures of compounds used in PCA studies, 284f
supercritical fluid antisolvent platform, 310
Pressure, effect on hydrogenation rate in biphasic ionic liquid/CO_2, 227, 228f
Pressure-composition diagrams
CO_2-expanded solvents, 48, 50f, 51f
ionic liquid and refrigerant system, 123–124, 126f
Pressure-density curves, CO_2-expanded solvents, 50–51, 52f, 53f
Process, advantages of CO_2-expanded liquids (CXLs), 4
Propan-2-ol, mixture critical points of, with CO_2, 8t
Propylene glycol, rate of diazodiphenylmethane (DDM) decomposition, 134, 135f
Propylene oxide (PO)
catalyst methyltrioxorhenium (MTO), 179
catalyst stability in successive runs using 2F-pyridine as axial ligand, 184f
catalytic destruction of H_2O_2, 185
C–H stretching vibrations in IR spectrum showing increasing solubility with increasing N_2 pressure, 182f
first monophasic epoxidation reaction using CXL media, 180
influence of pressurized gas on PO yield, 181f
light olefin epoxidation with MTO, 185
MTO durability, 185–186
MTO process advantages, 182, 183t

pressure-intensified MTO process
  for, 180f
pressure-intensified PO process,
  182–183
process and origins, 178–179, 181–
  183
substrate expanded liquid (SXL),
  182
theoretical in-service lifetime of
  MTO, 183–186

Q

Quantum dots
  nanoparticle fractionation, 298–299
  See also Nanoparticles

R

Radial distribution functions (RDFs)
  CO_2-expanded acetonitrile, 57f
  CO_2-expanded solvents, 55, 56f
Radical initiated processes, oxidations,
  162–163, 165–166
Reaction expanded liquid (RXL),
  ozone and CO_2, 148
Reaction rates, hydrogenation of 1-
  octene, 227, 228f
Rearrangement mixture
  processes using Beckmann
    rearrangement, 236
  temperature-dependent vapor
    pressures of, 236, 238f
  vapor pressure measurements, 236
Recyclability
  metal complexes, 281
  polymer-supported catalysts, 210–
    211
  soluble polymer-supported
    rhodium, 213, 214t
Redlich–Kwong type, cubic equations
  of state, 119–120

Refrigerants
  combining ionic liquids and, 114,
    116
  model 1,1,1,2-tetrafluoroethane (R-
    134a), 113–114
  solubility and separation of, 113
  structure of R-134a, 115f
  structure of R-134a, 115f
  See also Ionic liquids (ILs) and
    refrigerants
Respirable powders. See Atomized
  rapid injection for solvent
  extraction (ARISE)
Reverse emulsions, particle
  precipitation from, 25
Rotational correlation times,
  simulation and experimental for
  CO_2 and acetonitrile vs. mole
  fraction CO_2, 53, 54f
Ruhrchemie/Rhone–Poulenc
  (RCH/RP) process,
  hydroformylation of higher olefins,
  203, 219

S

Safety, gas-expanded liquids (GXLs),
  28–30
Selectivity
  catalyst design, 149, 150f
  oxidation of p-xylene, 173, 176
Separations, gas-expanded liquids
  (GXLs), 26–27
Shear viscosity, simulation vs.
  experimental for CO_2/acetonitrile,
  53, 55f
Shell process, terephthalic acid (TPA),
  162, 163f, 164f
Silicone-based materials, CO_2
  solubility, 244, 247
Silver nanoparticles
  alkanethiol-stabilized, in hexane vs.
    pressure, 296f, 297

diameters vs. pressure in CO_2
expanded hexane, 305*f*
dodecanethiol and lauric acid
stabilized, 293, 294*f*
dodecanethiol stabilized, vs. system
pressure, 303, 305*t*
fractionation using spiral tube
assembly, 297–298, 300*f*
See also Nanoparticles
Simulations
Center for Environmentally
Beneficial Catalysis (CEBC),
42
constant volume (NVT) and
constant pressure (NPT)
simulations, 52
dipole-partitioned NN radial
distribution functions (RDFs)
for CXL acetonitrile at various
pressures, 57*f*
force field models, 44–45
hydrogen bonding in CXL
methanol and CXL acetic acid,
58–59
mixtures, 48, 50–51
Monte Carlo and molecular-
dynamic (MD) simulations, 43–
44
O(OH)–H(OH) RDF for CXL
methanol, 59*f*
O–H RDFs for CXL acetic acid at
various pressures, 60*f*
optimization for industrial reactor
use, 43
orientational correlations in CXL
acetonitrile, 56–58
parameters for non-bonded
interactions, 46*t*
Peng–Robinson equation of state
(PR–EOS), 44, 48, 50–51, 61–
63
pressure-composition diagram for
CXLs methanol, acetic acid,
toluene and 1-octene, 51*f*

pressure-composition diagrams for
CXL acetonitrile and ethanol,
50*f*
pressure vs. density curves for CXL
acetonitrile and ethanol, 52*f*
pressure vs. density curves for CXL
methanol, acetic acid, toluene
and 1-octene, 51, 53*f*
promising alternative catalytic
media, 42–43
RDF between N atom sites on
different acetonitriles for CXL
acetonitrile, 56*f*
RDFs, 55, 56*f*
rotational correlation times in
CO_2/acetonitrile, 53, 54*f*
shear viscosity in CO_2/acetonitrile
mixtures, 53, 55*f*
simulation results for phase
equilibrium, 45, 47–51
simulation vs. experimental for
rotational correlation times, 53,
54*f*
simulation vs. experimental for
shear viscosities, 53, 55*f*
simulation vs. experimental for
translational diffusion constants,
52, 54*f*
single component systems, 47
structural properties of, 55–59
translational diffusion constants in
CO_2/acetonitrile, 52, 54*f*
transport properties, 51–53, 55
vapor-liquid coexistence curves for
single component systems, 47*f*
volume expansion for various
CXLs, 49*f*
See also CO_2-expanded liquids
(CXLs)
Single component systems, vapor-
liquid coexistence curves, 47
Sodium dodecyl sulfate (SDS)
polymerization of
tetrafluoroethylene (TFE) in

CO_2-expanded acetic acid using SDS, 264
See also Fluoropolymers
Solid components, multi-phase equilibrium, 14
Solid-liquid-vapor equilibrium (SLV), multi-phase equilibrium, 14
Solubility
  effect of CO_2 expansion on catalyst, 214*f*, 215
  hydrogen, in solvents under CO_2 pressure, 196, 198
  polymer, in CO_2, 246–254
  refrigerant gases, 113
  See also Porous materials
Soluble-polymer supported ligands
  catalysis, 208–209
  effect of CO_2 pressure on hydroformylation of 1-octene, 215*t*
  hydroformylation of 1-octene with, 210*t*
  hydroformylation with, 208–215
  potential benefits, 209
  recyclability, 210–211, 213, 214*t*
  See also Hydroformylations
Solvation dynamics
  dynamics of preferential, 98–99
  enrichment, 105–109
  enrichment factors for acetonitrile component, 104*f*
  first solvation shell coordination numbers, 103*f*
  per-atom contributions, 104*f*
  shell distributions functions, 102*f*
  See also Solvatochromism and solvation dynamics in CO_2-expanded liquids
Solvatochromic shift
  experimental and calculated, for Coumarin 153, 86*f*
  comparing preferential solvation using solvatochromic probe

trans-4-(dimethylamino)-4'-cyanostilbene (DCS), 98, 100
composition-dependent trends in solvatochromic shifts and time-resolved fluorescence, 109
dynamics of preferential solvation in GXLs, 98–99
enrichment dynamics, 105–109
enrichment factors for acetonitrile component, 104*f*
experimental spectral response functions and simulated solvation correlation functions for three compositions, 105–106
experiment-simulation comparisons, 109–110
first solvation shell coordination numbers of DCS, 103*f*
gas-to-solution emission frequency shift data for acetonitrile + CO_2 mixtures, 98*f*
modeling electrostatic contribution, 99
models and methods, 99–101
molecular aspects of solvation, 96
molecular dynamics (MD) simulations, 97–98
observed and simulated gas-to-solution shifts of absorption and emission frequencies, 101*f*
Onsager's reaction field theory (ORFT), 84, 86*f*
Solvatochromism and solvation dynamics in CO_2-expanded liquids
per-atom contributions from two solvent components in solute-solvent electrical interaction energies, 104*f*
region of maximum mass density vs. composition, 105
simulated solvation correlation functions for range of compositions, 106–107
solvation response function, 100

solvation shell distribution functions, 102*f*
solvatochromism and local composition, 101–105
time-dependent changes in numbers of atoms of solvent type in first solvation shell, 108*f*
time regimes for electrostriction and solvent redistribution, 109
Solvent extraction process. *See* Atomized rapid injection for solvent extraction (ARISE)
Solvent polarity, Onsager's reaction field theory (ORFT), 84
Solvents
  classification, 4–5
  comparing classes and expansion behavior, 6*t*
  liquid expansion vs. wt% CO_2 in liquid phase, 6*f*
  See also Modeling dilute solutes in compressible solvents
Solvolysis reactions, alkylcarbonic acid catalysis, 142, 143*f*
Spectroscopy
  Coumarin 153 (C153) absorption/emission maxima vs. CO_2 concentration, 88, 90*f*
  See also Cybotactic structure of gas-expanded liquids (GXLs); Solvatochromism and solvation dynamics in CO_2-expanded liquids
Spiral tube assembly, size fractionation method, 297–298, 300*f*
Stokes–Einstein relationship, diffusivity and viscosity, 11
Structural properties, CO_2-expanded solvents, 55–59
Substrate expanded liquid (SXL), propylene to propylene oxide, 148, 182
Sugar acetate, porous, with aligned structure, 245, 246*f*

Sulfur trioxide, interactions with caprolactamium hydrogen sulfate, 236, 237, 239
Supercritical antisolvent (SAS) precipitation
  nanoparticle formation, 282
  supercritical fluid anti-solvent platform, 310
Supercritical CO_2
  background in catalytic studies in, 151–152, 156
  benefits and limitations, 42
  benign, 3–4
  Co(salen*) studies in, 153*f*
  formation of water-in-CO_2 (W/C) and CO_2-in-water (C/W) emulsions, 248–249
  heterogeneous oxidation catalysis in scCO$_2$ media, 279, 281–282
  mechanism of oxidation of phenol by Co(salen)(axial ligand AL)/O_2 system, 154, 155
  oxidation of di-*t*-butylphenol (DTBP), 279, 281*t*
  oxidative catalysis in, 146
  polymerization of fluoropolymers, 260
  porous crosslinked poly(acrylate) and poly(methacrylates), 245
  porous materials and supercritical fluids, 244–245
  saturation of O_2 conversion dependence in, 152, 153*f*
  solubility and CO_2-soluble materials, 244
  temperature dependence of conversion and selectivity, 153*f*
Supercritical fluids (SCF)
  improvement of existing SCF precipitation, 310–312
  mixing controlled precipitation kinetics of SCF processes, 312
  porous materials and, 244–245
  precipitation, 310

templating SCF emulsions, 249, 252–254
Surfactants
solubilization and emulsification, 248–249
See also Porous materials
Sustainability, catalyst design, 148–149

T

Template polymerizations
metal complexes, 278–279
See also Metal complexes
Terephthalic acid (TPA) process
benchmark MC process, 173, 176
4-carboxybenzaldehyde (4-CBA) contaminant and TPA in Shell process, 162, 163*f*
Co(OAc)$_2$ with *N*-hydroxysuccinimide catalyst, 166, 169, 172–173
contaminant CBA, 173, 176
crystal structure of diaquabis(*N*-hydroxysuccinimidato)cobalt(II), 173*f*
cyclic olefin cleavages, 169, 172
effect of catalyst concentration on Co/Zr(acac)$_4$ catalyzed toluene oxidation, 166, 170*f*, 171*f*
factors affecting purity of TPA, 160
intermediates in oxidation of p-xylene to, 161
isothermal titration calorimetry confirming Co^{II}(NHSI)$^+$ and Co(NHSI)$_2^0$, 175*f*
ketones as promoters for Co(OAc)$_2$/Zr(OAc)$_2$/Zr(OAc)$_4$/ketone catalyst, 165, 168
lessons from Shell process, 162, 163*f*, 164*f*
MC-like processes in dense CO$_2$, 178
MC (MidCentury) process, 160
oxidation of toluene with Co/Zr/ketone catalyst, 167*t*
oxidation using Co(OAc)$_2$/NHSI catalyst in acetic acid CXL, 166, 172*t*
production, 159–160
products from oxidations of cycloalkanes by Co(OAc)$_2$/NHSI, 172, 174
radical initiated processes, 162–163, 165–166
solubilities in HOAc CXL of cobalt catalyst with Zr salts, 162, 164*f*
p-xylene oxidation, 178*t*
See also Oxidations in gas-expanded liquids
Ternary systems
comparing consistent expressions and literature, 73–75
dilute binary systems as special cases of dilute, 73
imperfect gases, 77
truncated composition expansions for fugacity coefficients of dilute, 69–73
See also Modeling dilute solutes in compressible solvents
Terpene hydrogenations
citral in carbon dioxide, 193
experimental, 193–195
hydrogenation experiments, 196, 197*f*
hydrogenation in expanded liquid vs. supercritical phase, 199
kinetics and volume expansion, 196, 198
limonene, 192*f*, 193
α-pinene, 192–193
profiles of α-pinene using Pt catalysts, 197*f*, 198*f*
reaction rates in one and two phase mixtures, 199
schematic of volume contraction of liquid phase, 198, 199*f*

solubility of hydrogen in
acetonitrile, acetone and
methanol under CO_2, 196,
198
vapor-liquid diagram of $H_2 + CO_2$
$+ \alpha$-pinene, 195f
VLE behavior of binary mixtures
of citral with CO_2, 200
volume expansion and phase
equilibrium, 193–195
See also α-Pinene; Limonene
Tert-butanol, rate of
diazodiphenylmethane (DDM)
decomposition, 134, 135f
1,1,1,2-Tetrafluoroethane (R-134a).
See Ionic liquids (ILs) and
refrigerants
Tetrafluoroethylene (TFE)
polymerization, 262–264
polymerization in CO_2-expanded
acetic acid, 263–264
polymerization in $scCO_2$, 263
See also Fluoropolymers
Thermal stability, colloidal
poly(tetrafluoroethylene) (PTFE)
particles, 270–271
Thermodynamic consistency,
truncated expansions, 76–78
Toluene
catalyst solubility in CO_2-
expanded, 214f, 215
parameters for non-bonded
interactions, 46t
pressure-composition diagram for
CO_2-expanded, 48, 51f
pressure-density curves for CO_2-
expanded, 51, 53f
vapor-liquid coexistence curves for
single component system, 47f
volume expansion for CO_2-
expanded, 48, 49f
Toluene oxidation
Co/Zr(acac)$_4$, 170f, 171f
Co/Zr/ketone catalyst, 165–166,
167t

Trans-4-(dimethylamino)-4'-
cyanostilbene (DCS)
solvatochromic probe, 98,
100
See also Solvatochromism and
solvation dynamics in CO_2-
expanded liquids
Transferable potentials for
phase equilibria force field
(TrAPPE), methanol and
acetone pair interactions,
87
Transition metal chemistry.
See Metal complexes
Translational diffusion constants,
simulation and experimental for
CO_2 and acetonitrile vs. mole
fraction CO_2, 52, 54f
Transport properties
CO_2-expanded solvents, 51–53,
55
gas-expanded liquids (GXLs), 10–
11
Truncated composition expansions
binary mixtures, 67–68
fugacity coefficients of dilute
ternary systems, 69–73
solvation behavior at infinite
dilution, 68
thermodynamic consistency, 76–
78
See also Modeling dilute solutes in
compressible solvents
Turnover frequencies (TOFs),
hydroformylation, 28
Turn-over frequency, hydrogenation
reaction, 226–227

U

UCC process, hydroformylation of
higher olefins, 203
Upper critical end-point (UCEP),
global phase behavior, 116–117

V

Vapor-liquid coexistence, single component systems, 47
Vapor-liquid equilibrium (VLE)
  binary mixtures of citral with CO_2, 200
  gas-expanded liquid (GXL) system, 5, 8*f*
  ionic liquid and CO_2, 7, 10, 10*f*
  VLE data of ionic liquid and refrigerant system, 124*t*
Vapor-liquid-liquid equilibrium (VLLE)
  experimental and model predictions for ionic liquids and refrigerant system, 121–123, 125
  global phase behavior, 116–117
  multi-phase equilibrium, 14, 16
Viscosity
  methanol and CO_2 with pressure, 10–11
  Stokes–Einstein relationship, 11
Volume contraction, schematic, 198, 199*f*
Volume expansion
  CO_2-expanded solvents, 48, 49*f*
  experiments, 193–195
  gas-expanded liquids, 5, 7
  ionic liquid, 228, 229*f*, 229*t*
  See also Terpene hydrogenation
Volumetric properties, CO_2-expanded liquids (CXLs), 96, 97*f*

W

Water, rate of diazodiphenylmethane (DDM) decomposition, 134, 135*f*

X

p-Xylene
  oxidation to terephthalic acid, 160, 161
  oxidation using cobalt acetate/*N*-hydroxysuccinimide, 166, 172*t*
  selectivity in oxidation, 173, 176

Z

Zirconium acetate, radical chain reaction, 163, 165